HENRY DARWIN ROGERS, 1808–1866

History of American Science and Technology Series

General Editor, LESTER D. STEPHENS

Henry Darwin Rogers, 1808–1866

American Geologist

PATSY GERSTNER

The University of Alabama Press

Tuscaloosa

The University of Alabama Press
Tuscaloosa, Alabama 35487-0380
uapress.ua.edu

Hardcover edition published 1994.
Paperback edition published 2014.
eBook edition published 2014.

Inquiries about reproducing material from this work should be addressed
to the University of Alabama Press.

Manufactured in the United States of America
Cover photograph: Henry Darwin Rogers, from *The Annual of Scientific
Discovery: or, Year-Book of Facts in Science and Art from 1858*; Boston:
Gould and Lincoln, 1858

∞

The paper on which this book is printed meets the minimum requirements
of American National Standard for Information Science–Permanence of
Paper for Printed Library Materials, ANSI Z39.48-1984.

Paperback ISBN: 978-0-8173-5819-8
eBook ISBN: 978-0-8173-8840-9

A previous edition of this book has been catalogued by the Library of
Congress as follows:
Library of Congress Cataloging-in-Publication Data

Gerstner, Patsy, 1933–
    Henry Darwin Rogers, 1808–1866 : American geologist / Patsy
Gerstner.
        p.   cm.—(History of American science and technology series)
    Includes bibliographical references and index.
    ISBN 0-8173-0735-4 (alk. paper)
    1. Rogers, Henry D. (Henry Darwin), 1808–1866. 2. Geologists–Unit-
ed States–Biography. I. Title. II. Series.
QE22.R64G47   1995
550'.92—dc20
[B]   94-4192

British Library Cataloguing-in-Publication Data available

For Jack and Patty Edmonson

Contents

Figures

Preface

Henry Darwin Rogers was one of the first professional geologists in the United States. He taught the subject, he practiced it, and he earned his living from it. As director, he led two of the earliest state geological surveys, those of New Jersey and Pennsylvania. Pennsylvania was also one of the largest. With a small group of assistants rarely exceeding twelve people, he covered approximately 45,000 square miles of Pennsylvania, much of it rugged and unexplored wilderness, in order to describe and explain the geological structure of the state and its potential for economic development.

The study of the geology of the United States had scarcely reached adolescence when Rogers began the survey of New Jersey in 1835 and of Pennsylvania in 1836. New Jersey held only minor interest for him, but in Pennsylvania, he was absorbed with the study of the Appalachian Mountains, which he saw as great folds of sedimentary rock. Rogers believed that an interpretation of these mountains would lead to an understanding of the dynamic processes that had shaped the earth. From his effort to explain the folds came the first uniquely American theory of mountain elevation.

Rogers was practical, freethinking, individualistic, introspective, and outspoken. A man of slight build and medium height, he had a characteristic dour expression that belied a gentle love of family and friends and a deep compassion for those who were oppressed. But Rogers was a man of many faces and moods, and while some people saw only the gentle side of his nature, others saw something else. Rogers was a perfectionist who demanded much of himself and others, often allowing his demands to damage his relations with his colleagues. Physical illness and mental depression aggravated his relationships with some, and an unfortunate but characteristic inability to finish things destroyed the confidence of still others. Furthermore, Rogers tended to be unyielding in his geological theories and, therefore, unable to incorporate a growing and changing body of information into his thoughts. This rigidity, coupled with the peculiarities of his personality, worked to separate Rogers from the mainstream of American geology. After a meteoric rise to prominence in the mid-1830s, he fell from favor almost as dramatically in the 1840s. Nevertheless, his work and his ideas contributed to the maturing of geology in the United States, and his story reflects much of what affected American geology between 1830 and 1860. His story is also that of the Pennsylvania State Geological Survey and as such echos many of the problems faced by the early surveys and by the men who directed them.

To anyone who has studied the history of geology in the United States in this period, the name Henry Darwin Rogers is a familiar one. So are the names of his brothers, James Blythe, Robert Empie, and William Barton, who all worked at some point in their careers as geologists. James, Robert, and William are perhaps better known as chemists and William as the founder of the Massachusetts Institute of Technology. Of them all, Henry was the only one to spend nearly his entire life working as a geologist. Little, however, has been written about him with a view to understanding the totality of his theories and his place in American science. It is hoped that this biography will provide greater insight into Rogers's career.

The preparation of the present book has been interrupted by many personal and professional responsibilities, and the project has, therefore, taken a long time to complete. I have received so much help over the years that it is now impossible to acknowledge all of my debts. I can single out only a few persons and hope that others will understand that their help at various stages was no less critical or appreciated.

The Cleveland Medical Library Association, for which I have worked for twenty-five years, has always encouraged research and made it possible for me to have an initial leave, during which the early research for this book was done. Part of this early research took me to England and Scotland, where my work was made possible by a grant from the American Philosophical Society. The assistance of these institutions is deeply appreciated.

The Archives at the Massachusetts Institute of Technology is the major repository of Rogers's papers, and Helen Samuels and her staff made each visit a pleasure. Martin Levitt at the American Philosophical Society contributed substantially to my work, as did Clark Elliott at the Harvard University Archives and Gladys I. Breuer at the Franklin Institute.

I want to say a special word of thanks to Susan Hill, Interlibrary Loan Librarian at the Cleveland Health Sciences Library (of which the Cleveland Medical Library Association is a part). She heads one of the busiest medical interlibrary loan departments in the country, but she cheerfully took time to get books, often rare or obscure, for me. I hope that she found it an interesting diversion from her usual searches. Her help was invaluable.

During the course of my research, I have had several meetings and conversations with Donald Hoskins, the current director of the Pennsylvania Geological Survey. His own interest in Rogers has been a source of information, and our discussions have never failed to add a new dimension to my studies and to renew my enthusiasm for Rogers. Darwin Stapleton and James Edmonson read the manuscript in various stages, and their insights were a tremendous help. In addition, both offered suggestions and gave me leads at various times during my research that were of the greatest importance. As friends, they listened patiently to my talk of Rogers over the years, and to them and to Marilyn Wolfe, Joel Orosz, Glen Jenkins, and others who listened as only friends can do, thank you. And finally, to Robert Schofield, teacher and friend, I will always owe any contribution I make to the history of science.

HENRY DARWIN ROGERS, 1808–1866

FIGURE 1. Henry Darwin Rogers (From *The Annual of Scientific Discovery: or, Year-Book of Facts in Science and Art from 1858* [Boston: Gould and Lincoln, 1858], frontispiece)

To Have Some Certain and Definite Object in View, 1808–1829

As the unsuccessful Irish rebellion of 1798 neared its end, those who had supported its goal to end British rule in Ireland found their freedom, and sometimes their lives, in jeopardy. One of them was twenty-two-year-old Patrick Kerr Rogers, who had contributed several antigovernment articles to a Dublin newspaper and who was probably a member of the Society of United Irishmen, whose militant efforts had led to the attempted rebellion.[1] Rogers was able to get to the safety of his family's home in Londonderry in the Presbyterian province of Ulster and from there to make his way to Philadelphia, where many other Irish refugees had come to find freedom and safety.

Rogers was no stranger to controversy and strife. His Presbyterian ancestors had settled in Ireland in the seventeenth century and had known generations of religious, political, and economic troubles in the country. As Patrick was growing up, his father, Robert, and his mother, Sarah Kerr Rogers, were involved in a struggle for liberal doctrinal reform in the conservative Presbyterian church.[2] Although limited information is available about earlier generations, the lives of Robert, Sarah, and Patrick make it evident that the Rogers family was one characterized by determination in the face of adversity, whose members were unafraid to champion causes of liberty and freedom of thought. These same characteristics would be strong attributes of Patrick Rogers's third son, Henry Darwin Rogers.

Patrick Rogers arrived in Philadelphia in August 1798. Once settled, he decided on a career in medicine and entered the University of Pennsylvania, where he supported himself as a tutor.[3] While still a student, he met and married Hannah Blythe on January 2, 1801. She was as determined an individual as Patrick and had a similar background. Her father, a publisher and stationer, was involved in antigovernment activities, especially through the *Londonderry Journal,* which he established in 1772. Blythe never allowed his ownership to be known, and the family remained relatively safe. Nevertheless, after his death in 1787 and the death of his wife in 1794, their children felt compelled to leave the country and emigrated to the United States, where they settled in Philadelphia, welcomed there by a cousin who had fled Ireland before them because of his affiliation with the United Irishmen.[4]

A year after his marriage Rogers received his medical degree. His thesis was on the chemical and therapeutic properties of the tulip poplar, a tree of such interest to many early American naturalists that Thomas Jefferson planted a few at Monticello, two of which still stand.[5] Where or just when his interest in chemistry developed is uncertain, but one of Patrick's teachers at the University of Pennsylvania was James Woodhouse, professor of chemistry, who was known as an inspiring lecturer and who directed the attention of many students to the subject.[6] Patrick Rogers maintained a broad interest in chemistry throughout his life and imparted that interest to all of his sons. On completing his medical education, Rogers established a medical practice in Philadelphia and is said to have lectured on a variety of subjects in the next few years, including botany, the history of medicine, and medical philosophy. It was also reported that he gave public demonstrations of the effects of nitrous oxide, or laughing gas.[7]

Rogers's first son, James Blythe Rogers, was born in February 1802.[8] A new family and a new medical practice put Patrick into serious debt. When his father died in 1803, he anticipated that the settlement of the estate would provide him with enough money to pay his debts and provide a small inheritance. Rogers returned to Ireland, where he spent a year settling his father's affairs, only to find that there was just enough money to cover obligations and no surplus.[9] Not only was he disappointed by the failure to find some financial security for his family through an inheritance but when he returned to Philadelphia, he found it was impossible to reestablish his medical practice at its former level. His income was not sufficient to support his wife and son and a second son, William Barton Rogers, born on December 7, 1804. Although lectures provided supplemental income, over the next several years he tried various other ways to better his financial position. One of his most ambitious undertakings was an attempt to develop a circulating medical library in Philadelphia with volumes acquired from booksellers on credit.[10] The library failed within two years, and although he was able to return many of the books to the sellers, he sank deep into debt.

As his financial insecurity grew, so did his family with the birth of Henry Darwin on August 1, 1808. The name Darwin was chosen because of Patrick's interest in Erasmus Darwin, whose work *The Botanic Garden: A Poem in Two Parts* (1789–1791) Patrick is said to have quoted from on many occasions. More desperate than ever to find a secure financial footing in Philadelphia, Rogers applied for the professorship of chemistry at the University of Pennsylvania, a position made available by the death of Woodhouse in 1809, but failed to secure it. Of necessity, lectures remained a major part of the family's support. A series on chemistry and natural philosophy in 1810 was expanded in 1811 to two courses, one devoted entirely to natural philosophy and the other entirely to chemistry.[11] These lectures, given at the Chestnut Street Lyceum, emphasized the practical application of the subjects. The continuation of the sessions and their expansion in scope suggest that they were well attended. Rogers noted only that they were attended "no doubt for amusement, or from courteous or friendly motives by the director of the mint, Robert Patterson, and several of the professors of the University of Pennsylvania."[12]

Although successful, the lectures did little to improve the family's financial situation. To the contrary, the purchase of apparatus and the time taken away from his medical practice to prepare for them increased the debt. Rogers contemplated asking for help from charitable sources, and only "sensibility to reputation" and "dread . . . of disgrace" kept him from it.[13] His personal effects slowly dwindled as he attempted to remain solvent, but there seemed little hope, and his friends advised him to leave Philadelphia.

Rogers took his young family to Baltimore to join his brother Alexander, who had come to the United States with Patrick after their father's estate was settled. On March 29, 1813, a fourth son, Robert, was born. Patrick's life took on a more settled tone in Baltimore when he opened a successful apothecary shop soon after arriving. A few years later he was recognized as a qualified physician when his right to practice medicine was approved by the Maryland Medical and Chirurgical Faculty in January 1816. A year later, he was elected to membership in the Medical Society of Maryland, and he also served as physician to the Hibernian Society of Baltimore from 1816.[14] His newfound professional peace was interrupted briefly when, in 1816, he engaged in a running argument with Dr. James Smith over the relative merits of inoculation and vaccination in protecting against smallpox, Rogers favoring the older and less well-thought-of inoculation.[15] The skirmish did him no real harm, and his financial condition gradually improved. It was never good, however, because he was still burdened with debts from Philadelphia, about which his creditors constantly reminded him.

The growing children fared well in Baltimore. The three oldest boys were of school age, but Rogers had his own ideas about education and decided against traditional schooling for them. Instead, he limited their attendance at school and provided much of their education himself. As William recalled later, his father's

chief and favorite employment in the intervals of business was the instruction of James, Henry and myself. Henry was then too young to be sent to school, at least so my father thought. On this subject his views were peculiar, and I have ever regarded them not only as benevolent but wise. The same anxiety that led him to postpone mere book instruction to the natural development of the physical and intellectual powers in Henry's case caused him to restrict our attendance on school, at a later period, to half days. So that, with the exception of a short period during which James and myself walked about two miles to Baltimore College to receive instruction in Latin, we never spent any of our afternoon hours in school. Henry, I am sure, was exempt during the whole of his schoolboy life from attendance in the afternoon. It thus happened that our education was conducted in great part at home, and by the daily personal attention of our kind and judicious father; and to this cause I may justly ascribe the thoroughness of our knowledge on all subjects which we studied, though in the apparent extent of our attainments we were by no means in advance of our playmates trained in the ordinary system of school drudgery, and confined to their books for the greater part of the day.[16]

Although they may have been on a par with their peers, William also recalled that, as a child, Henry was known among his friends for mathematical and mechanical skills.[17]

In spite of his success in Baltimore, Rogers wanted an academic position to provide greater security for his family. After an unsuccessful attempt to secure a position at the new University of Virginia, he found a place at the College of William and Mary in Williamsburg, Virginia, where he replaced Robert Hare as professor of natural philosophy and chemistry in 1819.[18] Hare had taken a similar post at the University of Pennsylvania, where many years later he was succeeded by Rogers's oldest son, James Rogers. The family, with the exception of James, who had just entered the University of Maryland to study medicine, moved to Williamsburg in October. Rogers found his work challenging and, in 1822, he published a textbook for the use of his classes in natural philosophy.[19] Although he felt the college would forever be unprosperous and unsuccessful, that it suffered from its location and "other disadvantages," he remained there until his death in 1828.[20]

Very little is known about the family life of the Rogerses in Williamsburg. William entered William and Mary in 1819 and graduated in 1821. In 1820, Hannah Rogers contracted malaria and died. About that time, James left his studies in Baltimore and joined the family in Williamsburg, where he also attended William and Mary for a short time, but he soon returned to Maryland to complete his medical studies.[21] Henry and Robert were still too young to care for themselves when their mother died and were often cared for by family friends while their father was busy with academic work. Robert became so devoted to Reverend and Mrs. Adam Empie who cared for him during this time that he later adopted Empie as his middle name.

In, or just before, 1825, Henry attended William and Mary as his brothers had done. He and William moved back to Baltimore before the year was over, however, in search of suitable careers.[22] While they tried to decide on their future course, Henry went to work for a retail merchant named Leche and lived with his family. It was an unhappy arrangement from the start because Leche blamed poor business on Henry and other young men who worked for him. After a few weeks William reported to his father that Leche had begun to display "a peevish, fault finding and arbitrary temper" toward Henry, and that Henry was "detained in the store every evening until almost eight o'clock at night and sometimes later so that the only time he could call his own was the short interval between that and bedtime and the whole of Sunday." But even these few hours were not Henry's to use as he wished, for Leche expected him to devote them to prayer.[23]

Henry's health had been a long-standing concern for the family. His constitution was not strong, and he suffered frequent problems with his chest and throat. Therefore, before going to Baltimore, his father had counseled him on exercise and rest in accordance with his condition. Henry's health grew worse because of the conditions imposed on him by Leche, and although he was willing to do anything connected directly with his employment, Henry felt that he could not do what he and his father considered necessary for his health and at the same time devote his few Sunday hours to prayer. Leche refused to be swayed by anything Henry said. When he went so far as to suggest that Patrick Rogers had failed to instill the proper values in his young son, William sent an urgent letter to his father, telling him what had happened and asking him what Henry should do.[24] Patrick felt sure that "the constraints which Mr.

Leche would impose on Henry, would, if not fatal to his life, prove destructive of his health forever," and he advised his son to leave the Leches immediately. He expressed complete confidence in Henry's judgment, telling William that Henry "has been at home one of the most obedient of children, and of ready acquiescence to his parents wishes . . . , [with] qualities of unimpeachable veracity, sincerity, and integrity."[25]

After leaving Leche, Henry and William considered a job in Virginia, deciding against it because their father opposed it. Instead, they stayed in the Baltimore area, where, in the fall of 1826, they opened a school at Windsor, Maryland, a few miles from the city. They did not expect much profit from the school, but by early December they had twenty-seven students, including their younger brother Robert, which meant about $600 income for the year. Although far from financial security, William thought it was enough to support them. Furthermore, Henry's health seemed much improved in Windsor.[26]

In January 1827, William also began a course of lectures on natural philosophy at the Maryland Institute in Baltimore. The Maryland Institute had been founded in 1825 and was patterned after Philadelphia's Franklin Institute, which had opened a year earlier. Its purpose was that of "disseminating scientific information connected with the mechanic arts, among the manufacturers, mechanics and artizans [sic] of the city and state."[27] This was to be accomplished with popular lectures, exhibits of "the products of Domestic Industry," scientific apparatus, a collection of minerals, and a library. William's course was well attended, and he held out hope that he would receive a permanent appointment there as lecturer in natural philosophy.[28] Were this to happen, he thought, Henry could then operate the Windsor school alone, an arrangement that would be a distinct financial advantage to each brother. Henry wanted very much to be able to put a couple of hundred dollars away each year toward acquiring a profession, and although he was not yet sure what that profession would be, he knew that he and his brother would "be glad . . . to have some certain and definite object in view." Medicine and law were among the professional choices available to them, but neither appealed to the brothers, especially not medicine. There were, according to Henry, already too many doctors in Baltimore and little chance of success unless one was willing to settle in an unhealthy climate.[29]

The brothers had enough income to provide for their needs and were comfortably settled in the home of the Fitzhugh family. In contrast to the Leches, with whom Rogers's experience had been unfortunate, the Fitzhughs were concerned for his health and well-being and made many efforts to provide a pleasant living environment, including music. Music had been an integral part of life in the Rogers household, and the brothers were delighted whenever Fitzhugh was able to borrow a "fine-toned violin" for Henry's use and William could borrow a cousin's flute.[30] They were also delighted by James's return to the city. James had established a medical practice in Harford County, Maryland (northeast of Baltimore). He had not liked it and was glad to find work as a chemist in the chemical manufacturing business of Isaac Tyson in Baltimore.[31]

William was appointed again to lecture at the Maryland Institute in the fall of 1827, and the following spring he proposed to open a high school at the institute. The school at Windsor was losing students and on the verge of closing. Therefore, William invited Henry to join in his proposed school. Henry felt that such a "connection with William, though it must for the present be in a subordinate capacity, will eventually redound to my advantage."[32] The school, reflecting the purpose of the institute, would, William said, "impart such knowledge and . . . induce such habits of mind as may be most beneficial to youth engaging in mechanical and mercantile employments." Students would thus be prepared for any trade ranging from chemist to engineer. Advanced students were to have the advantage of attending the scientific lectures in the institute as a reward for their diligence.[33]

A high school with similar purposes was already in operation at the Franklin Institute in Philadelphia, and William and Henry visited it in early May of 1828 in preparation for opening the school in Maryland. The Franklin Institute High School, which had opened in September 1826, catered to the members of the middle class who wanted an education comparable to that of the wealthy. The institute school, relatively inexpensive, followed a plan introduced in Philadelphia about 1809 by Joseph Lancaster. Instruction was provided by monitors, tutors, and teachers. The full course was three years in length, during which time the student "took Greek and Latin every year, three years of mathematics and French, two years of Spanish and drawing, plus courses in history, geography, political economy, astronomy, natural philosophy, natural history, chemistry, bookkeeping, and stenography. Additional training in science was to be provided by the Institute's regular evening lectures."[34] The Franklin Institute's high school prepared the student for a trade *or* served as an avenue to college.

William had already drawn up his plans for a high school before visiting the Franklin Institute, and he found no reason after his visit to change those plans. It was not his purpose to include the classics or provide a route to college, and he was not swayed by the more multifaceted approach being attempted in Philadelphia. The school opened in late May or in June in an "airy and tolerably commodious" apartment in the institute.[35] It compared favorably with the Franklin Institute school in regard to cost. At eight dollars per quarter, it charged only one dollar more. In class size, however, it had an advantage over the Philadelphia school, where classes numbered about 300 people. The class size at the Maryland Institute school was limited to 50. When the school opened, William and Henry taught about six hours each day but found these hours an unbearable burden likely to affect their health. Consequently, they worked out a plan to alternate teaching duties, each teaching every other day.[36]

Although William's interest in practical education was demonstrated by his involvement in the institute in general, the school represented the Rogers's brothers first known effort to educate young men for careers in the various trades. Educational reform of this nature became an abiding interest of both Henry and William and led eventually to William's successful efforts to establish the Massachusetts Institute of Technology.

In the summer of 1828, Patrick Rogers set out from Williamsburg to visit his sons in Baltimore. His visits had become eagerly anticipated annual events for the brothers and especially so for Henry, who had an exceptionally strong bond with his father. "I am continually wishing for your enlivening company," he told him shortly before his scheduled visit,

I feel an eager longing for those cheerful moments which an intercourse with you has never failed to bring. I believe I shall never cease to look to you as a guardian spirit. This sense of security which I always have when possessing your advice has afforded me many of my happiest hours; and, now that I am embarking in an arduous business, the value of your counsel will be highly prized.[37]

The anticipated visit never took place. Patrick suffered an attack of malaria on his way to Baltimore and died at Ellicott's Mills, Maryland, on August 1, 1828, Henry's twentieth birthday. Although Patrick's brother, James, who lived in Philadelphia, wrote to assure the brothers to "at all time command my service and my money too," he could not fill the void left by Patrick's death.[38]

Within weeks of the death of Patrick Rogers the lives of William and Henry changed. William left the Maryland Institute to assume the chair of natural philosophy and chemistry at William and Mary that his father had occupied. Robert went with him. Henry and James remained in Baltimore. Henry hoped to remain at the institute, replacing William as a principal speaker in the program of scientific lectures and continuing the school, but the administration was slow to decide. He was not well during this period, and his poor health, his father's death, and an uncertain future weighed heavily on him in the early winter. A deep depression overcame him. He wrote to William:

I have been subject for two weeks past to the most deep despondence. A sense of friendless destitution is ever rising to shadow with its gloom my liveliest aspirings; it requires for its suppression the utmost exertion which my fortitude can sustain. Oh, how I sometimes deplore the necessity of my absence from you. Each succeeding day seems only to heighten my regret. You will not think me unreasonable in my repining when you reflect on my utter loneliness,—on the harassing incertitude of mind arising from the inexplicable delay in the arrival of apparatus, and on the precarious condition of my health.[39]

One of Henry's closest friends in Baltimore was Michael Keyser, whose father owned a china warehouse where William had worked for a while, and he helped Henry through this period. Some months later Henry wrote to Keyser and reminisced about seeing him "in that room . . . where I while a valetudinarian have so frequently met [with you], when eager to dispel in society my listlessness and despondency; whose evenings of elated though unperturbed enjoyment, have eased my days of solitary moodiness and gloom."[40] Rogers's future life would often be interrupted by such periods of despondency, which usually followed rejection, illness, or some change in his life.

Before the year ended, things began to look up for Rogers. The institute decided to continue the school and the lectures, and Henry was joined there by his brother James. Although James had become superintendent of Tyson's business, he gave up the job in 1828 to assume the chair of chemistry at the Washington Medical College in Baltimore, but his new responsibilities allowed him enough time to lecture at the institute.[41] Although Henry worked for a reduced stipend, which was a personal hardship, the winter spent in Baltimore had far-reaching effects on his life, for that winter he had the opportunity to hear Frances Wright speak. Wright was an outspoken advocate of radical social, religious, and educational reform. Her lectures commonly denounced organized religion and advocated woman's rights, abolition, and a variety of other then unpopular reforms. Rogers was drawn to the lectures because he, too, felt the need for any number of changes. He favored abolition, was not put off by the suggestion of religious reform, and found education sadly in need of new directions.

Wright was especially interested in educational reform and in cooperative, or Utopian, communities such as those developed by the noted British social reformer Robert Owen, whom she had met in New York some years earlier. Owen, concerned about the plight of the worker in a rapidly industrializing world, had established a model factory town at New Lanark, Scotland, where the social, health, and educational needs of the worker could be met. In 1825 he established a similar settlement in the United States at New Harmony, Indiana, a place where work was confined to eight hours a day and where the leisure hours were filled with useful instruction and pleasant diversions.

New Harmony got off to a good start, and Owen was able to persuade businessman, geologist, and philanthropist William Maclure to join him in the venture. Maclure was interested in many of the same things that interested Owen and had the wherewithal to give more than verbal support to projects of which he approved. Maclure was a follower of the Swiss educator Johann Heinrich Pestalozzi and had been promoting his ideas in the United States since 1806, when he arranged for one of Pestalozzi's assistants to come to Philadelphia. Pestalozzi advocated radical change in the way children were taught, emphasizing less drudgery, less concern with the classical subjects, and more learning through experiences. He advocated practical education aimed at providing the child with a vocation. Science was generally looked upon as the most practical of all replacements for the classics, since knowledge of one or more of the sciences was fundamental to so many trades. As science was viewed as the path to knowledge, knowledge was seen as the path to equality.

Owen's sons had been educated at Hofwyl near Bern, Switzerland, a school under the direction of another educational reformer, Philipp Emanuel von Fellenberg, who also emphasized vocational education. Together, Maclure and Owen intended to make New Harmony an educational Utopia. But even with Maclure's help, which included bringing several prominent Philadelphia scientists to New Harmony, the community faltered under mismanagement and dissension before the dream of Owen and Maclure had a chance to mature. Nevertheless, the concept of New Harmony and efforts at social and educational reform lived on among the followers of Owen, one of them being Frances Wright.[42]

Wright had attempted to establish a community similar to New Harmony at Nashoba, near Memphis, Tennessee, but specifically for black men. Education, she believed, would prepare them to find jobs and to become part of the mainstream of society when slavery was abolished. Nashoba never really got a good start, and Wright left it in 1828. She went to New Harmony to join Owen's community and worked closely there with Owen's son, Robert Dale Owen. The two of them started a newspaper, the *New Harmony Gazette*, which addressed a variety of social issues. When New Harmony began to falter, Wright, undaunted, went to

New York, where she continued her crusade. She spoke on reform in the lecture halls of the East and Midwest, Baltimore being one of her stops.[43]

Wright gave four lectures in Baltimore in 1828. They were on free inquiry, on the importance and nature of knowledge, and on religion. Rogers was fired with enthusiasm by this woman he described as "nearly six feet high, majestic in her mien, and with a countenance betokening a long indulgence in the most refined and philosophic thought, with her short hair unbound and in ringlets on a head which would have braced Minerva, standing before a multitude in the delivery of strains written in a style of unsurpassed elegance."[44]

Although he was nominally, because of his family, a Presbyterian, Rogers never expressed a deep interest in religion, and he found Wright's derisive comments about organized religion "nothing short of spellbinding." It was Wright's comments on educational reform that got Rogers's closest attention, however. Science was very much a part of her plans for educational reform, and when Rogers heard her speak in Baltimore, she outlined a plan for science education that he found interesting. She sought to see a hall of science established in every major city, places that would be free and that would encompass lecture halls, libraries, and apparatus for scientific education.

As had been apparent in his work with William at the Maryland Institute, Henry was convinced that science was the most practical and useful of all possible studies, far superior in its significance to any kind of classical education that had no obvious or immediate practical value. Science prepared people for vocations, and, like Maclure and Wright, Rogers found this essential to equality. Furthermore, the presence of apparatus in the science halls as outlined by Wright provided the opportunity for each person to learn by experiences, and for Rogers and the other educational reformers, experience was a key to education.

Rogers's ideas on the latter were certainly influenced directly by the way in which his father educated him, but he was also strongly influenced by the philosophy of Thomas Brown. Brown was a philosopher, physician, and poet whose ideas were part of the Scottish Common Sense Philosophy espoused by Thomas Reid and Dugald Stewart. Instead of arguing that things are not real, as some philosophers did, their philosophy held that things are real and that it is possible to perceive them through various sensations. After Brown's death and the publication of his *Lectures on the Philosophy of the Human Mind* and a *Sketch of a System of Philosophy of the Human Mind Comprehending the Physiology of the Mind* in 1820, Brown enjoyed considerable popularity in the United States. Like many people, Rogers was impressed with Brown's emphasis on sensory experience, and he thought that the kind of learning center advocated by Wright, where sensory experience was provided by tools and apparatus, was "a happy extension and application of the sound philosophy of Brown."[45]

Within a few years, Rogers would become one of the Owenites. Although his family did not find the Owenites nearly as interesting as he did, Rogers credited his interest in their ideas to the "precious freedom from the despotic sway of false and perverting doctrines" that he owed to his father and to William.[46] For the moment, however, Rogers's classes at the Maryland Institute remained his principal concern. Enrollment dwindled dramatically early in 1829, and it became clear that the institute affiliation would soon end. A law authorizing the city of Baltimore to establish a public school system was passed by the state legislature in 1826, and the first public schools opened in 1829, a probable factor in the decline of the institute school. When the winter session ended in March, the school closed, and Henry went to Williamsburg to join William, while James, who had his work at the medical school, stayed in Baltimore.[47] Henry was actually happy to leave Baltimore because he felt it was too busy and, in his words, too filled with human folly. He spent the spring and summer in Williamsburg, which he found, in contrast, enlightened and peaceful. It was a place where he felt he could pursue the more important objects of the mind. Before the summer was over, however, Henry left Williamsburg for a trip north, anxious to see, according to James, a little of the world.[48] He went to Philadelphia, his childhood home, where his uncle James lived, and, probably through some of James's connections, became a candidate for a position at Dickinson College in Carlisle, Pennsylvania.

Acquiring an Intimacy with Geology, 1829–1833 $\qquad$ **2**

In 1829 Henry Vethake, popular professor of mathematics and natural philosophy at Dickinson College, announced his intention to leave. With little deliberation, the Dickinson trustees hired Henry Rogers to replace him. Rogers expected this appointment to be the beginning of an academic career that would enable him to develop his skills as a teacher of chemistry and natural philosophy. He no doubt expected to be able to implement some of his ideas on education as well, but not long after his arrival at Dickinson, it became clear that his tenure there would be a difficult and unhappy one destined for failure. Because it failed, he embarked on a career course that would eventually turn his attention to geology.

Rogers learned about the opening at Dickinson in the late summer or early fall, either through John W. Vethake, a friend from the Maryland Institute who had taught at Dickinson in 1826–1827, or through Henry Vethake, the incumbent, who was known to the Rogers family. The opportunity was a golden one for the young man, and his family sought to help him win the position. Rogers's uncle, James, was well established in Philadelphia and had many friends there. Through his friends, he was able to obtain a partial list of the trustees of Dickinson in order to make contacts in Henry's behalf.[1] Rogers's brother, James, spoke to Dr. James Henry Miller, his friend and colleague from the Washington Medical College in Baltimore and an 1808 graduate of Dickinson, in the hope that Miller could exert some influence at Dickinson on Henry's behalf. The first indication that an appointment at Dickinson might be trouble for Rogers came when Miller warned James that the administration at Dickinson was difficult "and that everything in Carlisle goes by favoritism."[2] Furthermore, Miller felt that there was little he could do to help Henry's quest because he had left the college after an angry confrontation with the trustees. In 1827, Miller and several of his colleagues had tried to persuade Dickinson to agree to let the Washington Medical College, which was at that time located in Washington, Pennsylvania, operate under Dickinson's charter. Since similar arrangements between schools geographically separate from each other were not uncommon, Miller was optimistic about the success of the plan.[3] His optimism was misplaced; Dickinson refused. The members of the medical school tried to change the minds of the trustees, but their efforts dissolved in anger. As a consequence, Miller and the medical school moved to Baltimore, where it was located at the time James Rogers joined its faculty.

In the 1820s, expansion, curriculum change, and increased income were on the minds of the Dickinson trustees, but as Miller's comments and experiences suggested, it was a troubled school.[4] When Rogers became a candidate for the Dickinson position in 1829, the college was in the midst of its last years of Presbyterian affiliation, and power rested largely with Reverend George Duffield, who was both pastor of the Carlisle Presbyterian Church and a trustee of the college. He had held the first position since 1816 and the second since 1820. A conservative and evangelical approach to religion led Duffield to use a firm hand in college affairs and to insist on moral and mental discipline. Students rebelled over Duffield's moral discipline, which encompassed the banning of all social and recreational activities. Since faculty members were allowed no control over the students, they felt isolated and powerless. The trustees tried to control the students but were not successful. In general, Duffield's policies created a state of confusion on all levels.

Frustration and anger drove many faculty members from Dickinson, among them Henry Vethake, whose career had begun in 1813 at Queen's College (now Rutgers) and continued at the College of New Jersey (now Princeton) in 1817. As a member of the Dickinson faculty since 1821, Vethake was a distinguished and popular faculty member who taught chemistry in addition to mathematics and natural philosophy.[5] His popularity at the school was such that most of the students departed with him.[6] Nevertheless, the trustees were determined to carry on, and on October 17, 1829, they approved the hiring of Henry Rogers to replace Vethake, offering him the professorship of mathematics and natural philosophy at a salary of $500 a session.[7]

Mathematics, to instill reasoning powers in students, was a standard part of the curriculum at American colleges in the 1820s and was often coupled with natural philosophy, which customarily included physics and astronomy and sometimes chemistry as well.[8] Rogers had been well schooled by his father in these topics. He had also taught and lectured on some of the same subjects at the school in Windsor and at the Maryland Institute. In addition to his education and experience, the family's Presbyterian affiliation worked in Rogers's favor. At this time, no one except William and his friend Michael Keyser was likely to have known about his interest in reform. Had the very conservative Dickinson trustees known, they would not even have considered him for the position.

In spite of Dickinson's problems, Rogers accepted the job and began teaching in the winter session that started in November 1829.[9] The number of classes that Rogers taught is not clear, since no college catalog is extant for the session, but a list he sent to William shows that he was responsible for a wide range of topics within mathematics and natural philosophy, including corpuscular philosophy and dynamics, astronomy, mechanics, electricity, pneumatics, algebra, logarithms, mensuration, trigonometry, plane and spherical surveying, navigation, and chemistry. He was also asked to teach a class on moral philosophy.[10] This class was normally taught by the principal of the school, but with the college so unsettled, no one could be found who was willing to accept this position.

Rogers accepted all his responsibilities willingly and enthusiastically and even found time to do a series of popular lectures for "a little scientific society" in Carlisle. He gave these lectures early in 1830, reporting to William that his first lecture made him nervous but that his performance suggested he could be a very good teacher.[11] His self-appraisal was accurate, for in later years he was often cited for his dynamic speaking skills.

Although things went well for him, his depression returned, caused by a combination of Carlisle's internal confusion and his separation from William. He lamented to William that he had "few social delights, as I am destitute of your presence and feel an extreme reluctance for society."[12] William was not quite four years older than Henry, but Henry relied on him as on a father. Although James was the oldest brother, he was never robust in health, and he lacked the strong personality that characterized William. Therefore, on their father's death, William was accepted by all the brothers as the head of the family, and they continued throughout their lives to look to him for guidance and help. While Henry relied on William in this fashion, he and William, as brothers, shared a bond of similar hopes and dreams for their future.

During his depression, Rogers began to think that he ought to be of more help to his fellow men and that he should be improving his own knowledge and sharing it with others. He wrote to William:

Of late, I have minded not the petty vicissitudes around me, for change is busier within me; in new powers of vision I behold new scenes and new paths in the field of enterprise. . . . Are there not frequent periods of self-upbraiding when your sagacity discloses how profitless to real good are all the fine talents and extensive knowledge you possess? For myself, I feel an exalted incentive to pursue knowledge. A fever has been born in my heart that will never leave it.[13]

Educational and social reform began to emerge as causes for the twenty-two-year-old professor, and concern for the welfare of man, eloquently impressed on him by Wright, forged deep into his consciousness. Ideas on educational reform would soon bring him into conflict with the trustees at Dickinson and would cost him his job, but even before the trustees became aware of his heretical ideas, he was in trouble over their social restrictions.

Social events, considered immoral by Duffield, were forbidden for students, faculty, and townspeople. Nevertheless, such events were held and were attended by some college trustees, but when Rogers attended a party and joined in a dance with "the choicest society of the place," he was severely criticized and told in no uncertain terms that he had acted wrongly and failed to show "the college [as] a school of religious discipline." Outraged, Rogers wrote an angry letter to William decrying the "nefarious priesthood" at Dickinson, their deep hypocrisy, and their oppressive attitudes. He threatened to leave, but once his anger subsided, he thought it best to meet the trustees at least halfway. Although the immediate situation was diffused, Rogers, stung by Duffield's rigidity, remained concerned about his future at Dickinson.[14]

In March 1830, Samuel Blanchard How was appointed principal of the school. Rogers found him to be "a gentleman of dignified and polished urbanity and . . . an accomplished scholar" and concern about his job eased.[15] At the same time, Rogers's teaching responsibilities were altered in a way that pleased him. He was appointed professor of chemistry for the session to begin in May, and his responsibilities for teaching mathematics and some parts of natural philosophy were turned over to a new member of the faculty, Alexander W. McFarlane. With a new principal, Rogers was relieved of the class on moral philosophy, but a class on natural history was added to his teaching.[16] An 1834 Dickinson catalog lists mineralogy, geology, botany, and animal and vegetable physiology as part of the natural history class. It was probably the same, or nearly the same, when Rogers taught the class in 1830.[17] If so, it was his first extensive experience with geology in general, but since mineralogy was closely allied with chemistry, it was probably not his first encounter with that subject.

The winter session ended in March. Rogers had a month before the spring session began during which he visited William in Virginia and his uncle in Philadelphia, but his time was primarily occupied with plans for a popular journal that he had been thinking about during the winter months. It was called the *Messenger of Useful Knowledge* and was financially backed by James Hamilton, Jr., of Carlisle. Rogers published the first issue at Carlisle in August 1830.[18] Hamilton, who was active in Presbyterian church affairs and educational matters in general in Carlisle, occasionally collaborated with Rogers and probably helped with general editing of the journal as well. The journal was an ambitious project "designed to unite the several features of a Journal of Science, a Literary Review, and a Register of News and Events."[19]

Articles and reviews in the *Messenger* covered a wide range of subjects, from astronomy to zoology, and the publication became a forum in which Rogers could present some of his ideas on educational reform. He hoped to develop a market for the journal in Philadelphia and in Baltimore, where his brother James was still living and lecturing at the Maryland Institute. His efforts were, however, disappointing.[20]

Under the pressure of his teaching and efforts to market his journal, Rogers began to complain of dyspepsia, a general name applied to various gastric disturbances. In an effort to find relief, as soon as his teaching ended in September, he went to Rochester, New York, for "Halsted's" medical treatment. Since medicine at the time could do little for this complaint, which was poorly understood, people often sought relief from alternative healers like Halsted. Halsted became well known in later years for relieving dyspepsia with massage of the abdominal area, but when Henry visited him his treatment was a special dietary regimen consisting of Halsted's own dyspepsia crackers, buttermilk, an occasional boiled egg, and roasted apples. Like many who undertook such plans, Rogers soon gave it up, claiming that it did not work.[21]

Almost as soon as the winter session began at Dickinson in November, Rogers began to have problems with the faculty and administration. He was asked to take on more classes, including one devoted entirely to geology, which he was not willing to do. In addition, one of the professors, probably McFarlane, wanted his lecture room and his apparatus. Since McFarlane was teaching some of the things Rogers had taught the previous year, he no doubt felt it was his right to take over the equipment, but Rogers was miffed because he was not consulted in the matter.[22] His troubles began in earnest, however, when he wrote an article on educational reform for the December issue of the *Messenger* extolling the ideas of Johann Heinrich Pestalozzi. The article decried the traditional and narrow form of education that required the learning of languages, alive and dead, to the detriment of the child's natural curiosity. Following Pestalozzi's principles, Rogers said that children "must be taught to employ their senses for the *accurate observation of things,* which forms the basis of all knowledge; . . . to express with correctness the result of their observations, and . . . to reason justly on the various objects of perception and of thought." Rogers told his readers that everyone was born with a desire for education but that the desire was effectively killed by the schools with their emphasis on rules and regulations, the classical but impractical languages, and the weary and dreary hours of concentration. He argued that there was no place within the traditional framework of education for the practical and useful.[23]

Rogers recommended Pestalozzi's doctrines as "simple and beautiful," but Principal How and the Dickinson trustees failed to see either the simplicity or the beauty. Instead, they saw the only result of such an education as the decline of mental and moral discipline. Their view was supported by a study by the president of Yale University, Jeremiah Day, published in 1828, that defended the traditional education. This *Yale Report on the Classics* was an influential tool in demeaning the ideas of educational reformers, arguing that the mental discipline of a classical education, though perhaps not always practical, was essential.

A less fertile ground than Dickinson for educational reform could not have been found. Rogers must have known the impact that his article would have, but like his father, he was never afraid to say what he thought. Rogers expected some safety from the fact that there were others at Dickinson attracted to Pestalozzian ideas, particularly the professor of languages, Charles D. Cleveland. Additional publicity for Rogers's view, however, made him a necessary focus of the trustees' attention. First, *The Chronicle of the Times and Disseminator of Useful and Entertaining Knowledge* in Baltimore reprinted his essay. The *Chronicle* was published by Julius Ducatel, professor of chemistry, and George H. Calvert, professor of moral philosophy at the University of Maryland. Ducatel, who later headed the first geological survey of Maryland, had taught at the Maryland Institute, where Henry may have met him. Certainly, James Rogers knew Ducatel, and it is likely that their acquaintance led to the republication of Henry's article.[24] Second, Rogers's view received publicity in Philadelphia and Carlisle newspapers in the form of an exchange between Rogers and an unknown person.[25] Although details of the encounter are missing, this further public demonstration of Rogers's position was an unwelcome one.

Rogers's support of Pestalozzi brought his career at Dickinson to an end on February 10, 1831, when the trustees decided to terminate Rogers's service because

it is essential to the interests of the College that the members of the Faculty would act in perfect accordance in relation to all matters of discipline and internal regulation and whereas it is known to the Board of Trustees that this state of things does not at present exist, but that Professor Rodgers [*sic*] has differed from the other members of the Faculty about the concerns of the College and its management.[26]

Henry left Dickinson as soon as the session ended in March and spent the next few months trying to find a new direction for his life. His first stop was Philadelphia to visit his uncle. He chanced to arrive just as a series of lectures on geology by George Featherstonhaugh was about to begin.[27] Featherstonhaugh had published several articles on the subject and was about to inaugurate one of the country's first journals devoted mainly to geology, the *Monthly American Journal of Geology and Natural Science*. Rogers found the lectures "quite pleasing," and he was so eager to hear them all that he stayed in Philadelphia longer than planned.

After his stay in Philadelphia, he joined his family in Williamsburg. He thought he might be able to support himself by giving lectures there, but the population of Williamsburg always declined as the residents fled from the approaching summer's heat, and he soon discovered that there were not enough people left in town to support lectures.[28] Nevertheless, Williamsburg presented some opportunities for his new interest in geology, and he spent part of his stay studying the geology and mineralogy of the area.

He returned to Carlisle in the early summer in order to prepare the July, and what proved to be the final, issue of the *Messenger*, but the need to find a new job was uppermost in his mind. He thought briefly about moving the *Messenger* to Philadelphia and publishing it there with one of his brothers, but he soon became interested in a career as a railroad surveying engineer. He gave up his idea about continuing the *Messenger* and tried unsuccessfully to sell his interest in the paper to George Fleming.[29]

Rogers joined his brothers William and Robert, who were traveling in search of a job for Robert on a railroad survey.[30] In early August, they found their way to Hoboken, where they met with William Gibbs McNeill, whom William had known since 1828 and who was one of

America's leading railroad engineers. He was in Hoboken, working on the Paterson and Hudson Railroad. McNeill had been hired to survey proposed routes for the Boston and Providence Rail-Road Company, which had just been established by an act of the state of Massachusetts and was looking for people to hire.[31] Robert was signed on as a member of McNeill's crew to join them in Boston, whereupon the brothers set out on a leisurely trip to Boston.

Their route took them from Hoboken to the Catskills, where they "visited the falls, slept under three blankets, and surveyed in silent astonishment, the illimitable map which nature there unfolds to the gaze of the elated traveler."[32] From the Catskills they visited West Point and New York City. By now Henry Rogers was giving serious thought to a career for himself as a railroad engineer and while in New York discussed this with a friend who was an engineer. Whatever transpired between Henry and his unidentified friend must have convinced him, because from New York the brothers went to Boston, where Henry was hired by McNeill. It seemed as though this might be the beginning of a new and lasting career for the two brothers, although they both recognized that they would be starting at the bottom, carrying compasses, levels, and surveyor's chains.[33]

Henry and Robert began work in September, expecting to be in the field for about a month. They were assigned to work on the route to Providence and also on one to Taunton. They fell quickly into a routine, using country inns as the base of their operations and moving on every two or three days to a new location.[34] Often they amused themselves by sketching local topographical features and making observations on the natural history of the area, "especially the nature of the trees and rocks." The amiable Robert looked at things with a sense of humor, and when struck by some perceived incompetences in the railroad surveys, he wrote to William that "among the numerous contemplated railroads from this place, one is to Providence, the next I suppose will be one to Heaven in addition to the many already instituted by the clergy, yet I hope they will not employ the same Engineers."[35]

McNeill asked Henry to record a series of geological observations on the route, which drew from him the comment "so much for a little science." From this labor came his first geological publication, a "Report Exhibiting the Geological Features of the Country Between Massachusetts and the Narragansett Waters."[36] The report was general, emphasizing, as one would expect, the locations of good building stone for the roadbed and the general geological conditions of the areas for the proposed routes.

As winter approached, work for the railroad came to an end, and the two brothers went to New York City. Frances Wright was in New York, and this must have been a powerful attraction to Henry. Wright, who was idolized by the liberal members of society (and was equally regarded as suspect by the conservative majority), was involved in a crusade for the cause of the workingman in New York City, in which she had been joined by Robert Dale Owen. When Wright was in Baltimore, Rogers met and talked with her on several occasions, so she was aware of his interests and his ideas. Rogers was also acquainted with Owen, and although the circumstances of their first meeting are unknown, Rogers's interests in educational reform and in practical education provided a common ground.[37] After arriving in New York, Rogers was often in the company of Owen, Wright, and their followers, and he soon emerged as their spokesman on the significance of science in everyday life. Nothing could have pleased Rogers more, for he now felt he could truly be of help to his fellow men.

Although Henry spent part of the winter in Williamsburg with his brother William, many weeks were spent in New York. By the time spring came, he was deep into the preparation of articles on education for the *Free Enquirer,* which was the outgrowth of an earlier attempt by Wright to publish a journal similar to the *New Harmony Gazette* and under the editorial supervision of Owen and Wright. The first of Rogers's eight articles appeared on April 28 and the last on June 30. He now had the perfect platform from which to expound on educational reform. Instead of mentioning Pestalozzi, however, he called Von Fellenberg the great reformer, an indication that the Owenites had strongly influenced his thinking. Still, the general idea that had been apparent in the *Messenger* was the same. His articles argued that educators must be content to lead children in discovery, rather than to force learning, and that children should be allowed to proceed at their own rate, always beginning with the observations of familiar things and building on those.[38] The sciences, the heart of practical education, should be taught in the same manner, through the use of large numbers of specimens, mechanical contrivances, and models for the child to explore and observe.[39] Rogers's message was enhanced by a clear sense that education was fundamental to the moral fiber and freedom of mankind that was typical of the Owenites. "The true struggle for human liberty," Rogers said, "is in the field of education, by the pen and through the press, it is in the hall of knowledge and on the leaf of science."[40]

On May 13 at New York's Concert Hall, Rogers delivered a lecture on the importance of science. The lecture was described by Robert Dale

Owen as a "Preliminary" address to a series on the subject of science that Rogers would begin in October. The fall lectures were to be given for three hours each Sunday, which Owen thought was preferable to devoting the entire day to unproductive religious observance.[41] Owen expected that Rogers would devote his full time to these lectures in the fall and that he would spend the summer in Europe to prepare himself to lecture on the latest scientific discoveries.

The idea that Rogers would go to Europe for the Owenites had developed sometime during the winter. The plan was for Rogers to teach at the Association of the Industrious Classes, a mechanics' institute run by Robert Owen on Gray's End Road in London.[42] There he would lecture to workingmen on Sundays throughout the summer on various scientific subjects. The rest of the time he would study science in the learned institutions of London. This opportunity to gain more knowledge and to have a chance to pursue his educational convictions by teaching practical science to workingmen was very important to Rogers. He sensed, however, that neither William nor his uncle, from whom he would need money for the trip, would favor the idea. His uncle did not share Rogers's enthusiasm for reform, and while William approved of certain things on the reform agenda, especially with regard to education and abolition, he, too, was unlikely to condone Henry's plans to associate himself so completely with the radical Owenites. Although the family held liberal religious views, the idea that Henry would lecture on the Sabbath would not be well received.

Henry won William's support for his plan, but it took a concentrated effort to get his uncle's approval.[43] Even though James acquiesced to the plan, just days before he left, Henry still felt compelled to assure him that it was the right decision:

My own feelings assure me that I, not for one moment, have been careless as to how you and William would look upon my schemes. These schemes I have long had in contemplation; but in coming to my decisions about the career I was selecting, I fully appreciated the distress which I saw I would occasion you. My decision was by no means rashly made; I may say that for the last two years I have almost incessantly deliberated upon the matter. . . . I have well studied the state of opinion among that part of society who favour my plans, and feel convinced that they will not fail. I cannot see that I have much to fear from popular odium. . . . My main object is to be useful. Again, were my schemes to fail and all the world to scout, my true happiness would still be greater than any I could have by taking a course contrary to my convictions.[44]

Rogers sailed for London on May 16, 1832, with a small party including Robert Dale Owen and his bride and Robert's brother David Dale.[45]

David Dale Owen had been at New Harmony from 1828 until 1830, when he joined his brother in New York. He had prepared himself in chemistry, anticipating practicing chemistry as part of the scientific community at New Harmony. He also learned geology. Although the decline of New Harmony caused him to turn to a career in lithography and printing, common interests in chemistry and geology led to a friendship with Rogers, and both came back from England to lead state geological surveys.[46] During a long career, Owen headed the surveys of Indiana, Kentucky, and Arkansas.

A trip to England in 1832 by sailing vessel was one of several weeks' duration but not necessarily an unpleasant one for passengers, like Rogers's party, who could travel in the cabin class rather than in steerage. Food was abundant, and passengers found various ways to amuse themselves.[47] Unfortunately for Rogers, he was seasick for most of trip and was unable to do much more than lie in his bed. Rogers's ship sailed around the southern coast of England, through the straits of Dover, and up the Thames River directly to London, and by midsummer, the party was settled in the city.

Confronted for the first time with the sight of the city and the conditions of the laborers he wanted to help, Rogers was shocked by the terrible contrast between their squalor and the splendor of the British upper classes. He also discovered that few of either class seemed to accept the idea that social change required moral and intellectual education. On a more immediate level, he was concerned about labor exchanges established by Robert Owen as places where people could trade products without money. Rogers thought they skewed the real values of the items and encouraged people to try to get more than they were willing to give. Thus, he argued, they did not promote any moral sense among those exchanging products.[48] Whether this represented a serious rift with the Owenites or not, Rogers's ardor for their cause cooled over the summer as more and more of his attention was taken up with scientific pursuits.

Rogers had intended to stay in England only until late September, but he stayed a full year, not leaving for the United States until June 1833. He did attempt to sail home in late September, as scheduled, to undertake his lectures, but the ship was repeatedly driven back by fierce storms, finally forcing him to give up entirely.[49] He expected again to leave in November, but events were changing his life too rapidly. He wrote to William that he wanted to spend the winter in London but would need money to do so. William wrote to their uncle on Henry's behalf, requesting any assistance that he could provide and agreeing to be accountable for any monetary help extended to Henry.[50] James sent the money to his nephew, but he had seen enough of what he considered unorthodox behavior. Believing that Henry wanted to stay for the sole purpose of his socialistic concerns, he wrote to William: "I am now pretty sure this is the last time in my life, I ever shall, directly or indirectly, have any agency in the promotion of a scheme or prospect of any kind of which I disapprove. How grievously do I repent having anything to do with this matter. . . . Henry has permitted himself, by dwelling much on a certain subject, as has been the call of many excellent men—to let it get almost exclusively possession of his whole soul." Nevertheless, James promised to write a friendly letter to Henry. "I shall not scold him. Anything I may be prompted to say in the honesty of my feeling I will reserve until I see him. I think I can still fully appreciate his great worth and purity of purpose."[51]

What was happening to Rogers in London would eventually assuage the anger that James felt, for the extended sojourn directed Rogers to a new career. "Comfortable at Mr. Owen's, 4 Crescent Place, Burton Crescent, London," Rogers continued to lecture for Owen, but he spent more and more time through the winter attending scientific lectures and rapidly establishing friendships that led to many introductions. In short order, he came to know many of the leading scientific men of England. He attended lectures on electricity by Michael Faraday, whose style of lecturing reminded him of "Paganini's playing: so easy, so adroit, so much execution."[52] He went to lectures by Edward Turner, professor of chemistry and lecturer in geology at the University of London. Geology was still not uppermost in his mind, but before the year ended, Turner had introduced him to the Geological Society of London, of which he was a secretary.

Rogers readily found a place among the members of the Geological Society. He was well spoken and had a solid background in chemistry, which was recognized as basic to mineralogical work. Rogers's background in mathematics may also have recommended him to the attention of some members of the society who were educated in a mathematical tradition at Cambridge University and applied mathematical reasoning to geology in a variety of ways.[53] Rogers's ease in conversing on these subjects and a broad range of other scientific, social, and philosophical topics assured the members of the Geological Society that they had an American in their midst who was worthy of note. They freely bestowed their attention and tutelage on Rogers in the coming months.

The first meeting of the society that Rogers attended was on December 13, 1832. He was convinced immediately that geology was a career that he wanted to pursue. He told his uncle that he would "have fine chances for making myself a geologist by the free access I may have to the

Society's superb museum."[54] His interest in geology burgeoned as he met the established and emerging British leaders of geology, including Henry de la Beche, Charles Lyell, Charles Babbage, Adam Sedgwick, Roderick Murchison, and John Phillips among others. Rogers was completely taken with the pursuit of geology, and soon began to study it at the British Museum Library.[55] He suddenly found himself in agreement with his uncle's concerns over his work with the social reformers. He wrote to him that because of his acquaintance with so many eminent men of science and the opportunities open to him, "it has become my consuming ambition to retrieve my mistakes by devoting myself to those studies which will please my friends and procure me an honourable name."[56] Although remorseful at having disregarded the concerns of his family about his trip to England, he consoled himself, his brother, and his uncle by noting that "I find that I am acquiring an intimacy in geology [and] fossils especially much beyond what I could ever have gained at home and likewise in chemistry."[57]

The paramount problem for Rogers was the impending thought of returning home. He felt that he had "glorious means before me for studying geology" and was annoyed at the thought of leaving.[58] He hesitated to stay longer because William and his uncle continued to provide a great deal of his support, but he decided that he could manage without their help. Nevertheless, he received money from William and his uncle in December, and the Christmas holidays found him enamored of English traditions and off on an excursion south of London to study geology.[59]

Rogers's brothers felt that his advances in England were remarkable, and they were obviously proud of him. They made every effort to convince their uncle that Henry's continued stay in London was important. While he conceded the point and continued to finance the stay, James still felt that Henry's actions were too erratic, and Henry's continuing affiliation with Owen almost certainly disturbed him. James had little to worry about. Although Rogers lectured at one of Owen's institutes throughout the spring, his classes were small, and he noted that there was "a great want of taste with the working population here."[60] His lectures now included geology, and this, plus the fact that he thought the teaching experience was helpful, were his primary reasons for continuing with Owen, since nearly every indication of interest in social reform vanished from his correspondence at this time.

As the winter and spring passed, Rogers became a favorite of the English scientific community. He attended gatherings at the home of George Greenough, president of the Geological Society of London, further developing his associations with British geologists.[61] Although David Dale Owen was also in London, and Richard Harlan, an American known for his paleontological investigations, would soon arrive, the versatility and ease that were part of Rogers's personality allowed him to fit easily into the society in a way that other American visitors were unable to do. Henry told William that his

intercourse with the men of science is every day becoming more easy and valuable to me. I go, free of ceremony, to almost any of the societies, once every week to the Royal, and now that Faraday and I are familiar, without even a member's ticket, to the Royal Institution. . . . I went to Oxford under excellent auspices . . . [where] they entertained me for two days most hospitably. . . . I was present at their society, where I met the whole body, went to several lectures, saw all the colleges, museums, Bodleian Library, etc. . . . [I] am invited to dinner oftener sometimes than I wish.[62]

After he left England, one of his new friends wrote that "Rogers was a very modest unassuming man [who] . . . left a very favorable impression on all who became acquainted with him."[63]

On March 27, the members of the Geological Society showed their respect and affection for the young American when they nominated him for fellowship in the society. Since formal election would not occur for another month, he now had an excuse to stay even longer in England. This gave him a chance to join de la Beche in Devonshire, where the latter was making a geological survey. De la Beche had convinced the British government only recently of the importance of a geological survey as a necessary complement to the topographical mapping then under way by the Ordnance Survey. His survey led to the establishment of the Geological Survey of Great Britain. Thus, Rogers had a unique opportunity to learn about geological surveying, and forecasting the future, he told William that in addition to being a great privilege to accompany de la Beche, "it will fit me . . . to do the like at home."[64] De la Beche was so impressed with Rogers that he gave him the proofs of the new, third edition of his *Manual of Geology* so that Rogers could republish it with his own notes on returning home. Rogers was, however, never able to find a publisher in America willing to take on this work.[65]

On May 1, 1833, Rogers became the first American to be elected a fellow of the Geological Society of London. He was urged to attend the Cambridge meeting of the British Association for the Advancement of Science (BAAS) in June, and William encouraged him to stay.[66] This was only the association's third meeting, and many of the leading members of the Geological Society were working hard to make sure that it was a sound body, representing the best of science. Geology was one of six sections into which the association was divided, and the invitation to Rogers represented an acknowledgment of his stature. He continued to impress those he met, and before he left England, he was asked to prepare reports on the geology of North America for the Geological Society and for the next meeting of the BAAS, to be held in Edinburgh in 1834.

Rogers left England in June of 1833. He was no longer the dedicated social reformer who had arrived in England a year earlier. He now considered himself a geologist. He had made many friends, and his future was bright. He was sure he would find a place in geology at home. After arriving in the United States, Rogers stopped briefly in New Haven to give a paper written by Michael Faraday for the *American Journal of Science* to the journal's editor, Benjamin Silliman.[67] From there, he went immediately to Philadelphia to join Robert and William. He settled into a residence on Chestnut Street and would call the city home for the next twelve years.

3 Promoting an Interesting Branch of Science, 1833–1835

Rogers returned to the United States with very definite ideas about his future. He wanted to study the geology of the country, especially through the agency of state geological surveys. He also returned with a strong sense that science in America lagged far behind that in Europe, and he wanted to work toward an improvement in American science so that it could enjoy a stature similar to that which he had observed in England.

On the latter front, Rogers was convinced that science could grow and prosper only if it was supported by strong organizations and journals. He thought an organization like the BAAS would bring all the sciences together for mutual benefit, and since a journal was essential for the dissemination of information, he wrote to Benjamin Silliman early in 1834 about both an organization and a journal. Silliman had started a scientific journal in 1818, called the *American Journal of Science and Arts* and popularly known as "Silliman's Journal." It was later called the *American Journal of Science,* the title that I will use hereafter. Rogers encouraged Silliman to persevere with the journal even though it had met with "slender encouragement." "We must believe," he told Silliman, "that the United States can soon sustain a prosperous periodical on the edifice of our domestic science." Rogers, who soon after his return to the United States had written an article on a chemical subject, offered to contribute to the improvement of chemistry with more articles that he would send to the journal. He also told Silliman that because he was "much interested in the cultivation of our native geology," he would "in this also be happy to make use of your journal."[1]

He encouraged Silliman to think about the need for a national scientific organization, but Silliman thought that the idea was premature, and Rogers contented himself with participation in the existing societies of Philadelphia. There were several of them, and within a year of his return from England, Rogers was a member of the Academy of Natural Sciences, the Geological Society of Pennsylvania, and the Franklin Institute.[2] Shortly thereafter he also became a member of the American Philosophical Society.

Of the organizations Rogers joined, the Franklin Institute proved to be the most important to his future. His uncle James was one of the earliest members of the institute, and Rogers was directly familiar with it from his and William's visit there in 1828. Although he and William had decided that the goal of education at the Franklin Institute was not quite what they wanted to establish in their high school at the Maryland Institute, Rogers felt a kinship with the institute on the basis of its interest in practical education. It was, however, not the tradition that caught his attention at this time. He was intrigued by a particularly energetic group of men within the institute. Like Rogers, they were concerned that American science had not achieved as high a level as that in Europe, and they intended to do something about the disparity. This group of "kindred spirits ready to discuss the principles or the applications of science, and prepared to extend their views over the whole horizon of physical and mechanical research" was led by Alexander Dallas Bache.[3] Bache and Rogers were nearly the same age, and they quickly became friends.

Bache was the great-grandson of Benjamin Franklin. After graduating from West Point, he served with the Army Topographical Engineers before settling in Philadelphia to become professor of natural philosophy and chemistry at the University of Pennsylvania. In later years, Bache organized the United States Coast Survey and was the first president of the National Academy of Sciences. His scientific and technical pursuits were already recognized when he joined the institute in 1829, five years after its founding, and he rose quickly to a position of authority within the institute. Among others, Bache's "kindred spirits" at the institute included Franklin Peale (professor of mechanics), Charles B. Trego (geologist and legislator), James P. Espy (meteorologist), Stephen Alexander (mathematician), Sears C. Walker (astronomer), Nicholas Biddle (financier), E. H. Courtenay (mathematician), and Robert Patterson, Jr. (whose father was a founder of the institute and director of the United States Mint, a position that Robert, Jr., subsequently held). Rogers had met Courtenay when he and his brothers visited West Point during their summer travels in 1831, but there is no evidence that he knew the others before returning from England. Nonetheless, Rogers's belief in the need for a strong scientific journal to support science, his interest in scientific organizations, and his success within the British scientific community attracted their attention, and they proposed his name for membership on January 18, 1834.[4]

Bache and his followers intended to use their respective positions and the Franklin Institute to raise the level of science in the United States. Until Bache met Rogers, he opposed the inclusion of geology in the institute's program, ostensibly because it would diffuse the level of interest in technology but in reality because he thought American geologists were amateurs and "quacks," ill equipped and ill trained to understand or develop the theoretical basis of the science.[5]

Geologists in Philadelphia typified the kind of work being done at the time. Philadelphia had emerged as a major center of geological study under the leadership of Samuel George Morton and his protégé Timothy Conrad, George Featherstonhaugh, Peter A. Browne, Richard C. Taylor, and Richard Harlan. Morton, a physician, had turned his attention to geology in the 1820s and was especially interested in the use of invertebrate fossils as a guide to identifying strata. Building initially on studies by fellow Philadelphian Lardner Vanuxem, Morton published several articles on the geology and fossils of New Jersey and Delaware. Conrad, whose specialty was fossil conchology, was strongly influenced by Morton and his work, as were many others.[6] Browne, a local businessman and member of the Franklin Institute, had been urging a geological survey of Pennsylvania since 1820 and had published short papers on various aspects of Pennsylvania and Virginia geology. In 1832, he organized the Geological Society of Pennsylvania to help promote a survey of the state.

Featherstonhaugh, originally from England, had turned his attention to geology in the late 1820s after a successful career in farming. He established himself in Philadelphia as a geologist, and it was a series of lectures by him that Rogers heard on his visit to Philadelphia in 1831. A prolific writer on a wide range of geological subjects, Featherstonhaugh was also the publisher, in 1831 and 1832, of the *Monthly American Journal of Geology and Natural Science*. The journal was, among other things, a vehicle for papers by members of Browne's Geological Society and a source of the society's transactions until the society began to publish them separately in 1834. While Rogers was in Europe, Featherstonhaugh had promoted the idea of a national geological survey, and in 1834, the federal government appointed him to survey the area between the Missouri and Red rivers.[7] Richard Harlan, a physician, like Morton, was interested in vertebrate paleontology and had visited England while Rogers was there. Although an argumentative man by nature, Harlan did much for the development of vertebrate paleontology in the United States and for the promotion of an exchange of ideas on the subject with European scientists.[8] Richard C. Taylor published several papers on Pennsylvania geology, particularly the coal regions, and was considered the leading expert on the state's geology.

In light of the above, Bache's appraisal seems unnecessarily harsh, but his point was that geologists in the United States had failed to rise to the level of making what he would later call "bold generalizations." Rogers, on the other hand, was fully capable of doing so and proved his ability by his relationship with British geologists and by their invitation to him to prepare papers for their learned societies. Rogers was, in Bache's eyes, a professional geologist.

An early sign of Bache's interest in Rogers came when he arranged for Rogers to use the institute's lecture hall for a series of lectures on geology "with a view to the promotion of an interesting branch of science."[9] These illustrated lectures began in mid-December 1833 and continued for three weeks. Shortly afterward, Bache and Rogers jointly published the article "Analysis of Some Coals of Pennsylvania" in an effort to aid the rapidly developing coal-mining industry of the state through the scientific analysis of the coal.[10]

Rogers's alliance with Bache and his friends at the Franklin Institute had immediate impact on geology outside the institute when, in the spring of 1834, the University of Pennsylvania added a course on geology to the curriculum and asked Rogers to teach it. Education in geology focusing on mineralogy had sporadically been a part of the curriculum since the late eighteenth century, and some geology may have been included in Bache's classes on natural philosophy and chemistry. The last course at the university limited to the subject of geology, however, had been one called "Geology," taught by Robert Patterson, Sr., in 1824.[11] On March 4, 1834, the university trustees appointed a committee to consider the feasibility of a new course on geology and, at a special meeting on March 10, less than a week later, resolved that an arrangement should be made with Henry Rogers for such a course to begin almost immediately.[12]

The sudden decision to offer a class on geology took the established geological community by surprise. Featherstonhaugh, Conrad, Harlan, and the others knew nothing about any plans to add geology to the university curriculum.[13] They resented the appointment of a newcomer and wondered how it had happened. The only plausible explanation is that Rogers's new friends at the Franklin Institute arranged it. Since Rogers shared their ideas on science, they felt certain that he could do for geology what they wanted done for science in general in the United States. They would, therefore, do all they could to help him, and Bache and Courtenay, who were members of the university faculty, and Nicholas Biddle, who was a trustee, had enough influence to arrange Rogers's appointment. The university course began immediately and was completed by the end of July. As was customary for new faculty members, Rogers was awarded a master of arts degree by the university at the commencement on July 31, 1834.

Evidence of the growing relationship between Rogers and Bache's circle lies in Rogers's membership in "the club." Sometime in 1834 Bache, James P. Espy, Stephen Alexander, Edward Courtenay, and Sears C. Walker of the institute, with Bache's friends Joseph Henry and John Torrey, who taught at Princeton, organized this informal and secret group known by no other name than "the club." Rogers was either a member from the beginning or joined the group soon after its initial meeting.[14] Back from Europe for only eight months, Rogers had been accepted by a rising group of new professionals in science. A year later Rogers, along with Espy, was heralded by the Franklin Institute as "enlarging the boundaries of knowledge and contributing to the best interests of the members of this institution."[15]

Soon after his return from England, Rogers had started to work on his papers for the Geological Society and the BAAS. He finished the former in the spring of 1834. Entitled "Some Facts in the Geology of the Central and Western Portions of North America, Collected Principally from Statements and Unpublished Notices of Recent Travelers," it was a sweeping review of the area based on information gleaned from a fur trader, from the Lewis and Clark expedition and from the explorers Thomas Nuttall and Stephen Long.[16] De la Beche had apparently impressed the young, aspiring geologist in England with one rule, and that was to be wary of reports that were not personally verified. Rogers thus felt compelled to explain to de la Beche his reasons for using just such sources. In a long letter, he justified his action by arguing that the accounts came from experienced and acute observers and filled a serious gap in knowledge about North American geology.[17]

His paper for the BAAS, which he had promised to send for the meeting at Edinburgh in 1834, proved to be more difficult than the one for the Geological Society. Because it was to deal with the eastern part of the United States, the paper would have to discuss a more extensive range of formations that would include the coal and rocks below it.

Although the coal had been studied by several persons, Rogers found the studies inadequate. The rocks below the coal had received little attention and were a poorly understood jumble of formations that together were called "Transition rocks," after a European classification. It would obviously take time to prepare a suitable discussion of these formations, and Rogers would have to do a considerable amount of fieldwork himself. Casual mention of field trips to New York in 1834 suggest that he wasted no time in getting started. Nevertheless, he knew that he could not complete these studies in time to present them to the Edinburgh assembly. Therefore, he decided to prepare two papers for the BAAS, saving the study of the older formations for the following year and sending a study of the rocks above the coal to Edinburgh.

The rocks above the coal had been more carefully studied, and for this part of his paper, Rogers could draw on the work of ten or more persons already recognized for their knowledge of the deposits. These included the Massachusetts geologist Edward Hitchcock, Timothy Conrad of Philadelphia, and Conrad's mentor, Samuel George Morton. Morton was especially helpful because he allowed Rogers to read his as yet unpublished *Synopsis of the Organic Remains of the Cretaceous Group of the United States,* which was a summary of his own extensive work along the East Coast.[18] Rogers was also able to use information from the state geological surveys of Maryland and Tennessee, the former under the direction of Ducatel, who had reprinted Rogers's essay on Pestalozzi. By relying on information from the pens of respected and well-known geologists, Rogers completed his report on the rocks above the coal with relative ease and sent it to John Phillips, secretary of the British Association's Geological Section.[19] He was determined to see that the paper was not overlooked when the members of the BAAS met, and he also sent a copy to Roderick Murchison in the event that for some reason Phillips could not present it. Worried that his paper might not receive much attention, he told Murchison that he was "particularly desirous that you should peruse it at an early day, in order that light may be cast by you upon some of the questions described—moreover that it may have a friend on the spot who will save it from being forgotten."[20] His paper was read by both Murchison *and* Phillips, the former reporting that he "read the most striking parts commenting upon them and endeavoring to impress the Section . . . [with] the importance of the subject." Furthermore, at a general session in the evening Adam Sedgwick "explained its nature and content."[21] Thus it received all the attention that the young American wanted or for which he might have hoped.

The report began with an apology for America's slowness, when compared with Europe, in investigating its geology, but noted that the interest expressed by British geologists was surely bound "to cheer and quicken our progress." The main part of the paper was devoted to formations called Tertiary. In England, Charles Lyell had divided these formations into the Eocene (the oldest), the Meiocene, the Older Pleiocene, and the Newer Pleiocene, using the number of fossils in each that were like modern animals as a guide to the relative position of each in the stratigraphic column (figure 2). The Eocene had the fewest fossils like living species, the Pleiocene the most. Rogers found Lyell's method useful and applied it to the American Tertiary, but he hesitated to equate the American Tertiary formations with those in Europe as many of his contemporaries were doing. Rogers believed that there was not yet enough information available to make such a decision with certainty.

Rogers was particularly concerned about the use of fossils in correlating formations between areas. Since the English surveyor, William Smith, had shown in the early nineteenth century that specific fossils were characteristic of each rock formation, it had become customary in stratigraphic studies to use fossils rather than the physical characteristics and mineral content of rock to identify formations. The rock types, such as chalk or limestone, were often repeated within a sequence in one area, and they did not always occupy the same position in every sequence. Consequently, they appeared to be unreliable stratigraphic indicators. Most of Rogers's contemporaries were thus using fossils to make correlations between rocks in widely separated areas, and they equated the Tertiary formations of both continents on that basis.

Some years earlier, Rogers's English friend John Phillips had questioned whether animals had, in fact, appeared everywhere at the same time.[22] Rogers was strongly influenced by this suggestion, and he believed that his own investigations produced no convincing a priori argument that life had appeared everywhere at the same time. Rogers therefore adopted the principle that until such time as the local sequences of rocks everywhere had been thoroughly studied and the progress of life in every quarter determined, it would remain dangerous and possibly misleading to make correlations on the basis of fossils.[23] Rogers knew that identifying rocks by their physical characteristics was also dangerous, but he believed that, with careful observation, rock type was the best first line to be taken in identification. Fossils could be used as auxiliary evidence, but they were always, for Rogers, of secondary importance. This conviction became a major impediment to the acceptance of Rogers's work in the years to come.

As a consequence of his concern about the development of similar

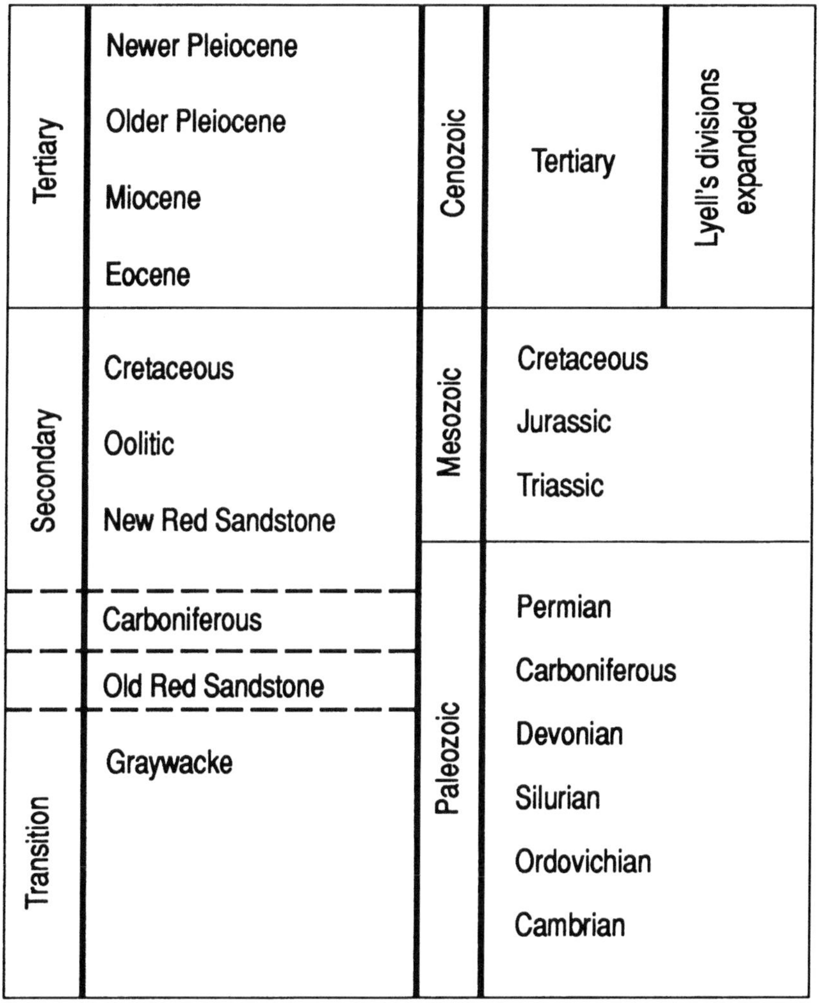

Tertiary	Newer Pleiocene	Cenozoic	Tertiary	Lyell's divisions expanded
	Older Pleiocene			
	Miocene			
	Eocene			
Secondary	Cretaceous	Mesozoic	Cretaceous	
	Oolitic		Jurassic	
	New Red Sandstone		Triassic	
	Carboniferous	Paleozoic	Permian	
	Old Red Sandstone		Carboniferous	
Transition	Graywacke		Devonian	
			Silurian	
			Ordovichian	
			Cambrian	

FIGURE 2. The first two columns represent significant divisions of fossil-bearing formations as understood in Great Britain about the time Rogers began his work. The second two columns represent global sequences established after 1835. The more recent geological periods, including the Pleistocene, or ice age, are not included, but are above the Tertiary.

animals everywhere at the same time, Rogers did not want to use the same names used in Europe for American formations because such usage would imply correlation whether it was there or not. He suggested using "Newer Tertiary" for the New Pleiocene of Lyell, "Middle Tertiary" for the Older Pleiocene *and* the Meiocene (because he thought there was not enough information about these formations in the United States to distinguish one from the other), and "Older Tertiary" for the Eocene. These names, he argued, reflected chronological relationships between formations without implying correlation between the formations in different areas.

This same concern is apparent in another part of his paper dealing with so-called Secondary formations below the Tertiary and particularly with the Cretaceous, which was the youngest of the Secondary formations. This part of Rogers's paper was based on the work of Morton. On the basis of fossils, Morton had determined that a formation he had traced along the Atlantic coast of the United States was the equivalent of one called the Cretaceous in Europe. Rogers was especially concerned about Morton's assignment of a New Jersey formation called the Greensand to the lower part of the Cretaceous. The lack of what, in his view, was significant agreement between the mineral character, or for that matter the fossils, of the Greensand and the European Cretaceous, led him to suggest that the deposits might belong to a formation distinct from the Cretaceous. To avoid confusion and the premature suggestion of relationship between the American and European formations, he again suggested the use of a nomenclature that would only reflect the relative chronological positions of the rocks.[24] Unfortunately, his paper, as published, ended abruptly and without any further elaboration about what that nomenclature should be.

Edward Turner, John Phillips, and Roderick Murchison each wrote to tell Rogers that his paper had been well received.[25] Phillips called it an "exceedingly valuable contribution to our knowledge of American geology," and Lyell agreed with Rogers's caution in making any attempt at correlation between areas, emphasizing that Rogers was right in pointing out that there was no way to know that species characteristic of an area had changed in the same way and at the same rate in widely separated areas.[26] Murchison told Rogers that the reading of his paper was well attended and that the size of the meeting indicated the importance attached to the subject of American geology by geologists in England. He encouraged him to send the second part. "I can assure you," he said, "that the Section pronounced *so favourable* an opinion upon your work that there can be no doubt of its being printed in our volume and the general meeting (in consequence of our request) have called upon you to continue your labours to send us for a subsequent year the report upon the older rocks of your country." Phillips also urged Rogers to prepare a continuation.[27] The reception undoubtedly pleased Rogers, and his friends at the Franklin Institute were reassured that Rogers was everything they thought he was.

Because Rogers's university course had been judged a success by the University trustees, a trustee committee was appointed in November to study the possibility of establishing a professorship of geology and mineralogy. The committee reported favorably on this plan in December, and Rogers was elected on January 6, 1835, to fill the professorship. What financial arrangement Rogers had with the university concerning his first class the year before is not known, but after the professorship was established, he was permitted to sell tickets to his course, a common means by which university professors were paid for their work.[28] His classes, which were given in the "lecture room of Professor Bache," began on the third Monday in January.

Rogers prepared a small book of forty-three pages for use by his students entitled *A Guide to a Course of Lectures*. It was little more than an outline based heavily on works by Charles Lyell and de la Beche. According to the *Guide,* Rogers intended to cover the materials that compose the earth, the causes of change, and formations that make up the earth's crust. The first ten pages offered descriptions of the most common minerals found on the earth, the study of which, Rogers said, would lead to a study of the mixture of the minerals that make up the rocks. He noted two classes of rocks, the stratified and the unstratified, the former layered in structure and deposited in water (sedimentary rocks), the latter crystalline and formed by heat (igneous rocks). He also called attention to a class of crystalline, stratified rocks, which he said most geologists thought were formed by heat but which Lyell believed were formed in water and modified by heat and which Lyell had called metamorphic. Rogers included a classification of the older fossiliferous stratified rocks taken from de la Beche, and of the younger rocks from Lyell, as well as a glossary of geological terms from Lyell.[29] Few details of the course's content are given beyond the above. No clue suggests what Rogers planned to cover under the heading of causes of change other than the note that there was geological evidence of both alluvial and diluvial deposition and that the former was the result of stream action, while the latter resulted from a great sweep of water across the land. More insight on Rogers's thinking about diluvial action, however, is afforded in a paper he published about the same time.

The paper, "On the Falls of the Niagara and the Reasonings of Some Authors Respecting Them," was intended as a response to a paper by George Fairholme, a British visitor to the United States.[30] Fairholme had published a paper in 1834, suggesting that the trough through which the Niagara River travels after dropping over the falls had been created as the water slowly cut back the rock over which it fell until the falls reached their present location.[31] Rogers argued that the trough was the result of a sudden and violent forward rush of waters across the land rather than slow backward cutting. At the time, the idea of catastrophic change such as Rogers described was more common among geologists than the notion of a slow process such as that described by Fairholme. Although geological opinion would gradually shift away from catastrophic diluvial action, it remained a central element in Rogers's understanding of the geological history of the globe.

About the time his course started in 1835, Bache asked Rogers to become the editor of the Franklin Institute's journal, an invitation that Bache did not extend casually. Bache had been working toward the revision of the institute's journal for more than a year. He wanted it to be a true international journal expressing the highest standards of American science. In order to make it so, Bache felt a need to find an editor who understood and shared his ideals. Rogers was an obvious choice to him, but Rogers turned down the offer because he had too many other commitments.[32]

At this point, few Americans knew anything about the content of the papers Rogers had sent to England, although it was general knowledge in Philadelphia that he was writing them. The first opportunity for Americans to learn anything about the paper he wrote for the British Association came when a summary of it appeared in the *Edinburgh New Philosophical Journal* early in 1835, followed shortly thereafter by a nearly identical summary in the *American Journal of Science*.[33] The summary, however, was limited to a critique of Rogers's comments on the Cretaceous, and it was, as Phillips said later, a very "imperfect abstract" put in the journal by an unknown person.[34] This "imperfect abstract" became the focus of a bitter attack on Rogers by an American who called himself "Amphibole," an attack ostensibly prompted by Rogers's appointment at the University of Pennsylvania.

The anonymous Amphibole was the Philadelphia paleontologist Richard Harlan.[35] He was well known in Philadelphia through his publications and his activities at the Academy of Natural Sciences and the American Philosophical Society, and he was a founding member of the Geological Society of Pennsylvania. Harlan perceived himself to be the leader of the geological community in Philadelphia and the logical heir to any teaching position in geology in that city. He was bitter and angry about Rogers's appointment at the university. Furthermore, he was working on what he considered to be a summation of what had been done in geology in the United States, and he resented Rogers's attempt to summarize American work. As a matter of fact, Harlan's work, the first installment of which was published in the *Transactions of the Geological Society of Pennsylvania* in 1835, was quite different from Rogers's work in that it dealt almost exclusively with paleontology, and to at least one Philadelphian, Samuel George Morton, it was a seriously flawed work. Morton called Harlan's report "garbled" and decidedly unfair in its failure to mention his [Morton's] work "owing to a personal dislike of many years standing." Although no information on Rogers's paper was available to Americans at the time Harlan's paper appeared in the *Transactions*, Morton felt sure it would "shew the scientific world *who* has done most actual labor in this vineyard."[36]

In January 1835, Harlan, as Amphibole, published his attack.[37] He opened with a long statement on the lack of ethics in secret appointments and went on to proclaim the Britons "Good natured souls" for thinking Rogers's report elaborate and valuable, because it was filled with errors and information stolen from others. Although Harlan was clearly angry, he cannot be faulted entirely for suggesting the worst in regard to the origin of Rogers's information, because the summary in the *Edinburgh journal* omitted references to other persons' works that Rogers had carefully included in his full paper. Others were misled by this omission, too. When Morton saw the abstract, he wrote in utter disgust to his friend Gideon Mantell, the British paleontologist, that Rogers had gotten all of his facts from him and from Conrad.[38] Morton quickly retracted his statement when he saw the paper as published by the BAAS, telling Mantell that Rogers "has done all I could desire" and that Rogers's paper was "a judicious and candid document."[39]

The editors of the official published report of the BAAS meeting at Edinburgh felt it necessary to make some reference to the attack but did so in restrained terms, ignoring the charge of plagiarism and trying to excuse any errors in Rogers's work by explaining that Rogers had no opportunity to make revisions before publication and that the association never expected the author to verify each comment by personal observation and study.[40] Gideon Mantell spoke more directly for Rogers's friends in England when he wrote to Benjamin Silliman to say that "a scandalous . . . attack on Rogers with the signature of amphibole, printed at Philadelphia, has to the great annoyance of many of us been sent by post from Liverpool, costing for shipping and postage 3/10 cent: it is really scandalous: the writer will get no good by it."[41]

Rogers privately dismissed Harlan's attack as trifling.[42] Nevertheless, the attack was more important to Rogers than this reaction indicates, and a short time later, Bache, who was visiting England, told Phillips that Rogers hesitated to send another paper because of the Amphibole incident. Phillips, certain that the publication of the full paper had erased any doubts about Rogers, wrote to encourage him to send the paper. He told Rogers that he had thrown the Amphibole paper in the fire, as others had done, and that "it would be a fatal error if you should suppose your reputation unsafe in the hands of the English geologists." Rogers's reputation was so unmarred by Harlan's attack that the Geological Society of London planned to acknowledge his work by awarding a prize to him, although there is no record of such a prize being given.[43]

While Harlan's attack appears on the surface to be little more than the anger of one man who felt that he had been unjustly overlooked for an important appointment, the reasons for the attack were deeper. American science, including geology, changed dramatically in the 1830s as the pursuit of science became a professional occupation. Instead of the work of men who devoted part of their time to scientific studies while earning their livelihood elsewhere, science became the full-time occupation and sole support of an increasing number of men.[44] Although some of Rogers's predecessors in the geological community of Philadelphia did eventually make their living as geologists (Conrad, Featherstonhaugh, and Taylor, for example), Rogers would be the first to do so in Philadelphia. Furthermore, he had been trained in geology in England, and although the education was not the formal classroom learning that would characterize later geologists, to Americans it represented an advanced education that could not be readily duplicated in the United States. He was a professor of geology at the University of Pennsylvania and the author, by request, of articles for the British audience. All in all, Rogers represented an unfamiliar kind of geologist to those who had worked at geology in Philadelphia for so long, and he represented a threat to this establishment, which had found an institutional base in the Geological Society of Pennsylvania.

Harlan's attack was neither the first nor the last of the problems between Rogers and the veterans. The first arena of conflict had been in Virginia, where the Geological Society members made studies ranging from Harlan's visit to the caves of that state in 1831, to Richard Taylor's study of a coalfield near Richmond, to Conrad's studies of Virginia fossils, to studies of the gold regions by Featherstonhaugh, including one study sponsored by the Geological Society of Pennsylvania. Feather-stonhaugh, in particular, had spent a considerable amount of time explor-ing the geology of Virginia, beginning in 1832, and he had published several articles on the state's mineral wealth and especially on its marls. Marls are natural fertilizers that occur in many areas along the East Coast, and their potential importance to farming was just being recog-nized in the early 1830s.[45]

Throughout much of 1834, Henry and William Rogers worked inde-pendently on the geology of Virginia. In 1833, William Rogers, already established as a chemist and natural philosopher at the College of William and Mary, was approached by Edmund Ruffin, an authority on Ameri-can farming practices and editor of the *Farmers Register*. Ruffin asked him to study the marls of Virginia with a view to testing the growing belief that they would improve the soil. The result was a series of articles in the *Farmers Register* that confirmed their importance.[46] At about the same time, Henry Rogers was studying Virginia's mineral springs, and he and William together were pursuing other aspects of Virginia geology, in-cluding stratigraphy and fossils. Their combined activities were an inva-sion of the work the Geological Society members saw as rightfully theirs. Conflict began to show in several ways.

Henry Rogers planned to publish a paper on the chemical composition of the Virginia mineral springs. Instead, in mid-October, he read a paper to a meeting of the Academy of Natural Sciences of Philadelphia on his investigations, claiming that he found it necessary to present his findings in this manner because the result of his labor was about to be published under someone else's name.[47] The "someone else" was Featherston-haugh. An extract of a letter from him on the mineral content of warm springs in Bath County, Virginia, had just been published in the *Transac-tions of the Geological Society of Pennsylvania* as a preliminary statement to what Featherstonhaugh intended to be a longer article. Rogers had earlier shared some thoughts of his own on the warm springs with Feather-stonhaugh, and he was convinced that these formed part of the basis of Featherstonhaugh's letter and proposed paper. Although Featherston-haugh was not at the meeting, Harlan came to his defense, claiming that the letter was actually printed without Featherstonhaugh's approval, so that if it contained Rogers's ideas, it did so accidentally.[48]

Rogers let the issue drop, but further evidence of a growing rift is found in William and Henry's joint effort to elucidate the Tertiary forma-tions of the United States. Like Conrad, they were studying the Tertiary fossils of Virginia in 1834 and 1835. By the time they began, Conrad had already established himself as an expert on these fossils and was actively working in the area. Late in 1834, Henry Rogers thought it would be useful if Conrad would accompany him on a study trip to Virginia. Conrad agreed to make the trip in the summer of 1835, but before the trip took place, Harlan's attack on Rogers was published. Rogers immediately became suspicious that Conrad had been involved in the attack with Harlan and that Conrad might not want to help him in Virginia. His suspicions were confirmed when Conrad set off on a southern trip without telling him.[49] Henry and William completed their studies of the Virginia Tertiary without Conrad and presented them in a paper read to the American Philosophical Society on May 5, 1835, in which, among other things, they described the Eocene in Virginia for the first time.[50] They criticized Conrad's almost total reliance on specific fossils as indicators of stratigraphic position, arguing, as Henry had done before, that animals may not have appeared at the same time everywhere.

A great blow to any hope of a useful relationship between Rogers and the others came with the establishment of the Virginia Geological Survey in the spring of 1835 under the direction of William Rogers. Members of the Geological Society of Pennsylvania had wanted to see a survey of Virginia undertaken for a long time. In 1833, Peter Browne wrote to the governor of Virginia urging the state to appropriate funds for a survey, and the following year, Featherstonhaugh did the same. The work the members were doing in Virginia on the marls, the gold regions, and the coal deposits would contribute to a survey, and Featherstonhaugh coveted its leadership.[51] It was William Rogers who became the head of the Virginia survey, however, and his efforts rather than theirs had gotten the legislature to establish the survey. The geological study of Virginia was an important goal of the members of the Geological Society, and it had been taken out of their hands at a crucial moment.

In 1835, Featherstonhaugh published a report on a government-sponsored survey of the area between the Missouri and Red rivers that he had recently completed. Reviews of this work in the *Transactions of the Geological Society of Pennsylvania* and in the *Journal of the Franklin Institute* show the gulf that separated Rogers and his friends at the Franklin Institute from the members of the Geological Society of Pennsylvania.[52] The Geological Society's review was a short and glowing tribute to a fine study. The anonymous reviewer for the institute, perhaps Bache himself, had quite a different reaction. He first pointed to numerous errors and generally unintelligible language that, in his estimate, rendered the publication unsuitable. He placed the blame for the inadequate report squarely on Featherstonhaugh's shoulders, saying that good information can be expected only when those "selected to make these surveys are eminently qualified." The reviewer's final remarks spoke to the great fear of the Franklin Institute men that science in America would never be improved as long as people could accept the work of men like Featherstonhaugh as serious science:

> Mr. Featherstonhaugh may, therefore, think I have treated his bantling with undue severity; but as every citizen of this country is interested in vindicating its literary and scientific, as well as its political character, I cannot refrain from the remark, that if a work, such as this, can be presented to the National Legislature of the American people, and by it be ordered to be printed, and thus circulated with the stamp of its approbation through the world, if it is extolled with the most adulatory encomiums by a great portion of the newspaper press, and receives no castigation in the pages of those journals, whose peculiar province it is to be leaders and guides in matters of science, it is time for America to abandon all pretensions to rivalry with Europe on such subjects, and sink at once into that station of mental inferiority, which has been sometimes so contemptuously and so unjustly assigned to her, by a certain class of transatlantic writers.[53]

Henry Rogers's triumph over the Geological Society and its members began with his appointment as director of the New Jersey survey in 1835. Although the society's members had not exerted any particular effort toward establishing the New Jersey survey, surveys were a part of what they had worked for, and they were again denied a place in them. The grand disappointment, the loss of the Pennsylvania survey to Rogers, was to come in 1836, but by the end of 1835, Rogers was able to dismiss Featherstonhaugh, Harlan, and the others without a hint of hesitation, saying that they were "impotent against the weight of reputation."[54]

Rogers's position with the surveys and, in the case of Pennsylvania, the survey itself were achieved with the help of the members of the Franklin Institute. When pitted against the new professional, who was supported and encouraged by Bache and his friends, the members of the Geological Society were found wanting.

4 Field Research of a Scientific Kind, 1835–1836

While Rogers was still in England, his work with de la Beche had persuaded him of the importance of geological surveys and had, he told William, prepared him to do the same kind of work at home.[1] Although he did not lack for things to do after his return, his thoughts were never far from the possibility of conducting a state geological survey. State surveys of North and South Carolina, Massachusetts, and Tennessee had been completed by 1833, and other states were considering surveys. Among them were Virginia, New Jersey, and Pennsylvania, and Rogers became involved with all of them. Of the three states, Virginia had the largest area, since modern-day West Virginia had not yet been separated from the whole, and it was the first to attract Rogers's attention.

In the late spring of 1834, Henry went to Virginia where he, William, and possibly Robert as well spent the summer in geological fieldwork, part of which resulted in Henry and William's publication on the Virginia Tertiary.[2] The purpose of the trip, however, went beyond their immediate studies of the Virginia Tertiary, for they wanted to see a survey established there as much as the Geological Society members did. Before leaving for Virginia, Rogers wrote a letter to John F. Frazer, a friend and member of the Franklin Institute and a Bache protégé, that indicated he was planning a geological reconnaissance in Virginia. Furthermore, he cautioned Frazer to "say as little as possible about my absence," which suggests that he and William planned to work for a survey independently of Browne.[3] In midsummer, Henry told his uncle that he was contemplating a move to Virginia and that he and William were trying to raise enough interest in Virginia to establish a geological survey.[4] Subsequently, William prepared a petition to the Virginia legislature for several counties in the marl district, urging a survey of the state.[5]

Since he had published his articles on Virginia marl, William's reputation in geology had grown in Virginia, and this plus his residence there and his academic position at William and Mary made him a more logical spokesperson for a survey than any member of the Geological Society. When appointed to a select committee of the legislature to investigate a survey, William was able to persuade the members that a survey was needed, and he was put in charge of the preliminary geological reconnaissance recommended in the committee's report when the survey was established on March 6, 1835.[6] He was then appointed state geologist in 1836 when the systematic survey began. William had good qualifications for the Virginia position, and since Bache, who had met William shortly after Henry joined the Franklin Institute, considered him one of his "kindred spirits" in science, his appointment was, no doubt, welcome news at the institute.[7]

In the meantime, having left the plans for the Virginia survey in his brother's hands, Henry was pressing both the New Jersey and Pennsylvania surveys. The survey of New Jersey was established at nearly the same moment as the one in Virginia, and as with that one, a few years had passed since a survey had been first suggested in 1832 by the governor of New Jersey, Peter D. Vroom. Although no action was taken at that time, Vroom renewed his plea for a survey at a meeting of the New Jersey legislature in October 1834, and less than three months later Rogers submitted a plan for a geological survey to the legislative bodies of both New Jersey and Pennsylvania.[8] As will be seen in the discussion of the Pennsylvania survey to follow, the submission of the plan in that state was part of an effort led by Bache to launch a Pennsylvania survey. Whether Rogers's submission of his plan to the New Jersey legislature was part of a similar effort on Bache's part is unknown, but Bache was involved in Rogers's appointment to head the survey.

A committee appointed to study the governor's message to the legislature reported favorably on the survey because it "would tend to develop still further the wealth and resources of the state."[9] On February 19, 1835, the New Jersey legislature passed an act establishing the survey and allocating $1,000 for a geological reconnaissance to be completed within that calendar year. Less than a week later, the bill was approved by the New Jersey Legislative Council representing both houses. At this point, it is clear that Bache and his friends were ready to expend as much energy as necessary to make sure that Rogers was appointed director of the survey. Bache wrote to Vroom, pointing out that Rogers's qualifications had already been recognized by the trustees of the University of Pennsylvania, as witnessed by his appointment as professor of geology. Making full use of Rogers's European experience, he emphasized that "Mr. Rogers has had the advantage, enjoyed by no other American, of making his studies in geology under the auspices of some of the leading men of the English school. Fully imbued with the science and familiar with its practical parts, he has labored constantly since his return to this country to illustrate its Geology."[10] But if Bache wanted to promote Rogers in New Jersey, obviously the best way to do so was through someone well known in appropriate circles in New Jersey. Joseph Henry, who had been at Princeton since 1832, fit the bill nicely, and since he believed that Rogers and his good friend, the chemist and naturalist Jacob Green, were the only two people qualified to lead such an undertaking, he needed no urging. As part of Bache's inner circle of friends, Henry knew Rogers and supported him because he was well qualified and because "for the good of science" he was anxious that "a proper person be appointed."[11] He wrote a letter of recommendation for Rogers that he asked his personal friend Richard Stockton Field, a member of the New Jersey Assembly and the legislative committee responsible for the survey bill, to give to the governor along with Rogers's proposal for the survey. Because Field was not in the same political camp as the governor and, therefore, might not carry much weight, Henry urged Rogers to be sure the governor received recommendations from others.[12]

After recommending Rogers, Joseph Henry heard that Jacob Green was a possible candidate as well. Henry was concerned that Green might feel that he had failed him by recommending Rogers for the job. He wrote to Green, saying that had he known about his interest in the position, his support would unquestionably have been his rather than Rogers's. Intent on seeing a "proper person" hold the position, however, and also unsure whether or not Green really was a candidate, he asked that if Green were not a candidate, he "do so much for the good of the cause of science as to mention to your Brother your opinion of the qualifications of Rogers."[13] Green's brother was James Sproat Green, district attorney of New Jersey and trustee of the College of New Jersey (Princeton), and a person whom Joseph Henry believed had considerable influence with the governor.

Others were being considered by Vroom, including Thomas Green Clemson (a Philadelphian with interests in mineralogy and a member of the Geological Society), James Pierce, and Timothy Conrad.[14] The governor was slow to make a decision, and Rogers grew uneasy about his chances. He asked Joseph Henry if he would meet with the governor and point out to him the importance of the state geologist's being a qualified chemist and topographer as he was. Rogers wanted to be sure the governor understood that he was capable of doing "field research of a scientific kind such as I have witnessed with De la Beche." He also wanted the governor to know that his brother had been appointed to head the Virginia survey, which "is certainly a reason why I could perform that of Jersey better [than] any other person." The efforts on Rogers's behalf were rewarded when he was appointed on April 24, 1835.[15]

Rogers began work almost immediately. Although the appropriation for the survey was small, meaning a small salary for Rogers, he was willing to accept it because he was anxious to have the job and, as he had told William earlier, because New Jersey was a poor state.[16] Rogers had no independent income, but he was frugal and had a modest income from his lectures and teaching with which he could support himself. Furthermore, he fully expected that Pennsylvania would establish a survey and that he would receive a substantial income as its director.

A geological survey of Pennsylvania had been proposed by Peter Browne in 1820 and again in 1824, but in neither case did his proposal generate much interest. At the time of his earliest suggestion, the idea of a geological survey of a state was very new. Although North and South Carolina undertook surveys between 1820 and 1824, the expense of such a survey was difficult to support, and there was not another state-sponsored survey until 1830, when one was undertaken by Massachusetts. Nevertheless, Browne's efforts to organize a state survey of Pennsylvania escalated in 1826, when he addressed a meeting at the Franklin Institute on the importance of knowing about the economic resources of Pennsylvania.[17] He followed his lecture at the institute with many lectures to county natural history societies in an effort to promote interest, even suggesting at one point that, if the state could not finance a survey, it could be financed by individual subscriptions.[18] Once again his dream failed, a failure due in large part to the lack of interest and lack of funds on the part of the state of Pennsylvania.

Browne, no doubt, expected more encouragement for the survey to come from the Franklin Institute than was actually the case. Although he was initially an influential member of the institute, his importance declined precipitously in 1826 when his efforts to establish a school of practical arts at the institute failed, and the high school William and Henry had visited was established instead. Browne did not want a school that insisted on courses in the classics, but he was opposed by Robert M. Patterson, institute member and trustee of the University of Pennsylvania, who would not support a curriculum lacking such courses. Rival factions in the institute developed around this issue, but Patterson's role in the controversy, which demonstrated a powerful affiliation between the institute and the university, won out, and Browne's idea was defeated.[19] This was the beginning of the end of Browne's influence with members of the Franklin Institute and his hope of their support on the matter of a survey, although he was probably involved in an unsuccessful 1831 initiative at the institute to conduct a manufacturing census. This census was intended to reveal, among other things, the resources of the state as a first step toward a geological survey.[20]

Browne had a strong sense of the need for support from some organized body in order to get attention for a survey, and when it became clear that the Franklin Institute would not be of help, he organized the Geological Society of Pennsylvania in 1832 to promote a survey.[21] When the Geological Society was formed, a survey was conducted by questionnaire aimed at gathering information on the mineral resources of the state.[22] It was probably no more successful than the Franklin Institute's survey, but the society's overall efforts succeeded in getting the state legislature interested in a survey. The state had been receiving pleas for such a survey from some counties, almost surely the result of Browne's efforts, even before the Geological Society began its campaign.[23] It was, however, only after the Geological Society was formed that momentum developed.

On December 7, 1832, Browne presented a memorial to the house of representatives proposing state support for a geological, mineralogical, and topographical survey of the state.[24] It was presented by Benjamin Say, member of the house representing the county and city of Philadelphia and younger brother of Thomas Say, one of Philadelphia's most prominent naturalists. Although Thomas Say was not a member of the Geological Society, he was a longtime member of Philadelphia's scientific community and, through the Academy of Natural Sciences, a close associate of many of the Geological Society's members.

This memorial was followed in close succession in 1833 by similar pleas from Montgomery, Lancaster, Crawford, and Schuylkill counties, and the Chester and Montgomery county cabinets of natural sciences became members of the Geological Society, probably to further the idea of a survey.[25] Benjamin Say was made chairman of a legislative committee to study the various memorials, and on March 23, 1833, he introduced a report to the house prepared by John B. Gibson, Richard Harlan, and Henry Tanner on behalf of the Geological Society. Gibson was president of the Geological Society, and Tanner was its treasurer. The report proposed a topographical, geological, and mineralogical survey that would result in a report, the publication of a complete series of maps, profiles, and sections, and the creation of complete cabinets of specimens to be placed in various institutions around the state. An act providing for the survey, making the Geological Society responsible for it, was introduced with the report. It called for a total expenditure of no more than $15,000, not to exceed $5,000 in any of three years.[26]

This was not the best time to ask the state for money. The financial condition of the Commonwealth of Pennsylvania was shaky in 1833. George Wolf had been elected governor in 1829. A staunch advocate of internal improvements, he had encouraged the financing of massive canal and road building by borrowing money. To pay interest on the money borrowed, Wolf had to implement several new taxes, but with income thus assured, he was able to continue borrowing as needed. Much of the borrowed money came from the Second Bank of the United States, directed by Nicholas Biddle. The bank was chartered by the federal government, and federal funds deposited there represented a significant portion of moneys available for loan. In 1833, President Andrew Jackson, fearful that the bank was in a position to wield too much power, was locked in battle with Biddle, whom he disliked intensely. In 1832, Biddle had petitioned to have the bank's charter renewed, although it was not due to expire until 1836. The president refused and threatened to withdraw federal funds from the bank. To prepare for this eventuality, Biddle began to call in loans, a process that caused a public financial panic and depression in 1833.[27] A strong movement to recharter the bank took shape in 1834, and sensing success, Biddle offered to loan the state more money, but Wolf declared that he would borrow it from other banks. To prevent him from doing so, Biddle called in loans from these banks so that they could not finance a loan to the state. This effort backfired, however, and heightened support for Jackson's repudiation of the bank, and the bank's charter was not renewed. Wolf was able to borrow money from banks outside the state, and reasonable financial stability returned to Pennsylvania in 1834.

As the financial condition improved, the legislature continued to receive pleas for a survey. In February 1834, the Chester County cabinet memorialized the house, the matter being referred again to a committee that made a report a little more than a month later.[28] This committee was made up entirely of legislators who stressed the potential economic significance of the survey more forcefully than the Geological Society had done. If the survey had economic ramifications for the rapidly growing state, it might be fundable from moneys set aside for internal improvements. The report was accompanied by a commentary from Benjamin

Silliman that also stressed the economic merits of a survey along with its scientific significance for a state known to be rich in resources. Since recess was scheduled for June, the committee felt that it was too late for the current legislature to consider such a survey but offered a resolution calling on the next legislature to consider the question as soon as possible. More memorials from Pennsylvania counties were received, and the Geological Society continued its efforts, but the next legislative session did not pursue the issue with much enthusiasm.

In January 1835, Rogers submitted his proposal for a survey to the state of Pennsylvania. The proposal is not extant, but a comprehensive letter outlining his ideas does exist. The letter was written by Rogers to J. Vanderkamp of Philadelphia.[29] Since Rogers seldom expressed ideas until they were well formed and was not given to changing his mind, this letter must be nearly identical to the plan submitted to the state legislators. It was a broad plan stating the purposes of the survey, how it should be conducted, and its staffing. The survey would, Rogers said, investigate the extent and limits of various rocks, point out those with particular significance economically, create maps, and carry out chemical analyses of the important materials. Although Rogers knew there were great deficiencies in the only topographical map of the state in existence, he hesitated to recommend a topographical survey as part of the geological survey because such a survey would increase the cost. Rather, he said, one or more of his assistants on the survey should be responsible for any topographical work that needed to be done. Rogers believed that a survey of Pennsylvania's nearly 45,000 square miles would take about ten years to complete and that an appropriation of $5,000 per year would be necessary for it to function. Of this, $2,000 would be the salary of the chief geologist, and $1,500 would be paid to each of two assistants. He suggested that a survey should begin with a rapid reconnaissance of the state in order to create a systematic plan for further seasons' work.

When Rogers discussed the staffing, the first point he made was that the governor should appoint a *scientific* geologist to direct the survey. This same argument had been used by Joseph Henry in New Jersey, and in addition to its general significance, in the case of Pennsylvania, it was an argument directed against the Geological Society and ran contrary to the suggestion that the survey ought to be under the direction of the society rather than of an individual. Of the two assistants he thought necessary, one, he said, should be a competent chemical analyst and the other, a good mapmaker and topographer.

A state survey of New York was being discussed at this time. There the size of the state had led to the suggestion that it should be divided into districts for study, a practice that was followed when that survey began. Fearful that partitioning might be considered for Pennsylvania, Rogers argued against it. For one thing, he said, partitioning would bring disharmony and confuse the issue of stratigraphy. It was bad enough, he told Vanderkamp, that "the geologists of the country, guided by their predilections for a peculiar classification of rocks adapted to the formations *of Europe,* are daily misinterpreting the age and relations of our strata to the obvious detriment of the science."[30] With the state partitioned, the confusion would be magnified.

After receiving Rogers's plan for a survey in January, Governor Wolf became more interested in a survey, and in his annual and last message to the legislature at the end of 1835, he made a stronger plea for a survey than had ever been made before, stressing its economic importance to the future of the state. "For a comparatively trifling expenditure," he said, "we would secure a denser population and add incalculably to our individual and general prosperity."[31] He noted that other states, like Massachusetts, were reaping benefits from their surveys and for "an annual sum, such as the treasury could spare without injury to other interests," Pennsylvania could do the same.

A new governor, Joseph Ritner, assumed office at the beginning of 1836. Rogers began to lobby actively for the survey by meeting with him and with several senators.[32] Persons, whom Rogers described simply as friends of the survey, arranged for him to deliver a lecture on the advantages of the survey and how it should be conducted.[33] The lecture was delivered in the Hall of Delegates at the state capitol on January 25, 1835. One of the friends who arranged the lecture was Charles Trego, a representative from Philadelphia, who had introduced the resolution to allow Rogers to use the hall for his lecture.[34] Trego was a relative newcomer to the legislature, having been elected to the Pennsylvania house in 1835. He was a member of the Geological Society, the Franklin Institute, and Bache's exclusive "club." Trego took a deep interest in the promotion of a survey and had been instrumental in transmitting Rogers's plan to the governor and in arranging the meeting between Rogers, the governor, and legislators that had preceded the capitol lectures. The part of Governor Wolf's year-end message that dealt with the survey was referred to a committee chaired by Trego. Trego's report for the committee was read before the house on February 3, 1836.[35]

The points made in Trego's report are similar to those made by Rogers in his letter to Vanderkamp. While this similarity suggests that Rogers influenced Trego's report, the arguments are not altogether unexpected for a report whose aim was to encourage support for a survey. Rogers's influence is supported, however, by a compelling single item in Trego's report concerning the specific plan for the survey and its prosecution. Trego's report recommended that, in the survey's early stages, a reconnaissance of the state should be made following five great lines across the state.[36] The first was to begin at the Delaware River opposite Bordentown and follow a course paralleling the Susquehanna valley to the New York line; the second was to start at Philadelphia and reach the New York border via Northumberland and Potter County; the third would begin at the mouth of the Susquehanna and end at Fort Erie; the fourth would cross from the Maryland line in Adams County to the Mahoning valley and the Ohio line; the fifth would extend from the Virginia line via Fayette and Washington counties to the border. Where Pennsylvania and New Jersey meet, these reconnaissance lines are precise continuations of those established by Rogers for New Jersey.[37] Since the New Jersey report had not yet been published, the only way Trego could have arrived at such perfect agreement between the lines was if Rogers had plotted them himself. Therefore, it is reasonable to assume that Rogers worked closely with Trego in preparing the report.

Trego's report got strong backing in the house from the outspoken Thaddeus Stevens among others. Stevens, described as "one of the most fiery, most aggressive, and most uncompromising leaders in Pennsylvania affairs" was known for the strong support he gave to internal improvements. The governor, Stevens, and other supporters of the survey looked at it as a form of internal improvement, since the economic benefits to be gained from a survey would improve the economic condition of the state and its citizens. Frederick Fraley, also an advocate of internal improvements presented the bill in the senate as Stevens did in the house.[38] At the moment, the attention of the legislators was focused on the bank charter controversy and concerns about Masonry, the secret lodges and rites of which frightened many people and led them to fear that the Masons might have infiltrated the government. Thus with virtually no opposition and the strong and outspoken support of Trego, Fraley, and Stevens, and the general support of Governor Ritner, the survey was established on March 29, 1836:

Be it enacted by the Senate and House of Representatives of the Commonwealth of Pennsylvania in General Assembly met, and it is hereby enacted by the authority of the same, That the Governor is hereby authorized and required, within thirty days after the passage of this act, to appoint a state geologist of talents, integrity, and suitable scientific, and practical knowledge of his profession, who shall appoint as his assistants two geologists, also of integrity and competent skill, one of whom shall also be a scientific and practical mineralogist, and the said state geologist shall also appoint a competent, practical, analytical and experimental chemist to assist him in his duties.[39]

Peter Browne and members of the Geological Society of Pennsylvania had worked very hard to promote a survey. Through their ceaseless efforts to inform the public about the advantages of a survey and a steady stream of memorials to the legislature, one as late as 1834, they had kept the idea alive in the legislator's minds.[40] The society is, therefore, usually given the primary credit for the final establishment of the survey, with the election of Joseph Ritner a strong contributing factor.[41] Trego's membership in the Geological Society is offered as his reason for promoting the survey. There is a problem with this scenario, however, because Trego would never have encouraged Rogers's participation in establishing the survey if he were representing the members of the Geological Society, with whom Rogers was at odds. Still, Trego, as a member of the Franklin Institute and supporter of Bache's goals for science had good reason to support Rogers. There is little question that the success of the Pennsylvania survey in the legislature, as well as its character and its structure, bore the clear imprint of Henry Rogers and his friends at the Franklin Institute. Pennsylvania, along with New Jersey and Virginia, represented the opportunity to create a new level of geological study in the United States. It was important to Bache that a scientifically competent person head the survey. When one was found in the person of Henry Rogers, it became important to Bache and others at the institute to see that the survey was organized. Trego's help was essential, and Fraley's assistance in the senate was not an accident. He and Bache had been friends since childhood, and he, too, was a member of the Franklin Institute and of Bache's inner circle. Indeed, he had helped Bache to a position of power within the institute.[42] Had it not been for the scientific goals set by Bache and his followers, the Pennsylvania survey might well have been delayed many more years.

Unlike the situation in New Jersey, there seems to have been little competition for the job of state geologist in Pennsylvania. There were many possible contenders, but the attraction of the imminent New York survey may have kept some, like Timothy Conrad who joined the New York survey, from entering their names in competition. The biographer of Trego suggested that the governor offered the job to Trego, but he gave no evidence to support this assertion, and in light of Trego's support of Rogers, it is unlikely.[43] Featherstonhaugh certainly wanted the job, but Rogers expected little real competition from him or any of the other possible contenders, telling William that "when the time comes to pre-sent myself as conductor of the survey I shall have every chance of being appointed."[44] He was, indeed, officially appointed soon after the survey was established, and the governor described him as a person "whose high literary and scientific attainments, and character for industry and intel-ligence, give the best pledge of his capability."[45]

Rogers's official duties as state geologist closely matched the duties as outlined in Trego's report, and although Trego's report had not suggested a specific allocation for the survey, a reasonable one was made.[46] The enabling legislation called for an annual budget of $6,400, $2,000 of which was for the state geologist's salary, $1,200 for each of two as-sistants, $1,000 for the expenses of a chemist, and $1,000 for incidental expenses. Although it is difficult to make a perfect comparison between the initial funding of the Pennsylvania survey and other contemporary surveys because the ways in which states handled allocations varied and details are not always available, Pennsylvania compares favorably or bet-ter with other surveys that began about the same time. The annual ap-propriation for New Jersey was, as we have seen, much smaller. In Ohio, where a survey began in 1837, a total of $12,000 was allocated for an unspecified period of time. Of this total, a little more than $2,000 was spent the first year and over $9,500 the second and final year. In Virginia, William's survey was given $5,000 per year. In New York, $20,000 was allocated for each of four years. New York, however, employed eight geologists and a draftsman, compared with Rogers's two field assistants and one chemist. Only one assistant was authorized for William, while Ohio's allocation was to cover the costs of two or more assistants. In both Ohio and New York, it was expected that some of the funds would be spent for a natural history survey as well as a geological one. The salaries paid to the geologists varied as well. Two thousand dollars were allocated for Henry Rogers's salary the first year in Pennsylvania, while in New Jersey he had been given only $1,000 for all operations, including his salary and that of his assistant; William Rogers was given $3,000 per year to be apportioned between himself and his assistant. Each of the eight geologists in New York was paid $1,500.[47]

All the while the negotiations for the Pennsylvania survey had been going on in 1835, Rogers had been contending with his geological recon-naissance of New Jersey. In spite of the meager finances of the survey, he was able to hire Maxwell Walker as his assistant and his young brother, Robert, to do chemical analyses of marls found in New Jersey.[48] Robert was doing similar work for William in Virginia. Robert had begun work toward a degree in medicine in 1833 and was still studying under Robert

Hare at the University of Pennsylvania. He received his medical degree in 1836 but had long before decided that his first interest was chemistry. The opportunity to work with his brothers was a welcome one.

Henry Rogers and Walker carried out the reconnaissance of New Jersey along five lines crossing the state's geological formations, and Rogers prepared cross sections of the formations along each line for his first report. He was able to divide the state into two general chronological areas along a line from the Delaware River just below Trenton to the Raritan River near the mouth of Lawrence's Brook. The rocks of the Northwest section were older than the coal formations, thus making it clear that a popular search for coal in these areas was futile. Those southeast of the line were more recent and encompassed the marls.

On February 1, 1836, Rogers wrote to Governor Vroom of New Jersey to let him know that the report of the reconnaissance was nearly complete and that he would come to Trenton to present it to the governor whenever Vroom wanted him to do so, although he did ask for a period of two weeks from that date in order to complete some of the chemical analyses for the report. The report was presented on February 16 and published the same year.[49] Rogers had been instructed by the New Jersey governor to direct his efforts to the study of economic deposits in the state "rather than to the promotion of higher branches of science naturally connected with a work of this kind."[50] As a result, there was a heavy concentration of work on the marls of the state because of their potential importance to agriculture. Rogers's first report was a substantial one, with discussion of the chemical composition and extent of the marls accounting for slightly more than one-half of it. Unlike some other marls characterized by the presence of calcium, those in New Jersey are characterized by the presence of glauconite, a silicate of iron and potassium, which gives them a green color, hence the formation from which they came was called the Greensand. In the course of his work on the survey this season, Robert Rogers may have been the first to note the characteristic potassium in the New Jersey marls. In a letter to William late in 1835 before others had determined its presence, Robert expressed some surprise that a recent experiment had failed to yield expected information on the potassium content of the marl, suggesting that he had already come to expect its presence.[51]

Henry Rogers felt that he could not ignore Morton's attempt to correlate the Greensand of New Jersey with formations in Europe, even though he had addressed the issue in his earlier paper for the British Association for the Advancement of Science. Therefore, in his report on the survey, he reiterated his opposition to the general idea of correlation using fossils as primary guides. Although Rogers remained steadfast in his concern that similar animals might not represent the same age in different localities, he expanded slightly on his argument in the BAAS report by arguing that if fossils were to be used as any sort of support for position or correlation, the relationship had to be based on the presence of identical species. Identity on the generic level was insufficient, and he believed that Morton had been led astray by using generic similarity rather than specific similarity.[52] To avoid the problems engendered by nomenclature, Rogers suggested, as he had done in his BAAS report, a descriptive terminology for the New Jersey formations, and this time he gave the Greensand a name reflecting its chronological position and its mineral content: the "secondary Green sand marl stratum." In later years, Rogers developed the idea of a combination of period name (in this case "secondary") and specific identifiers (Greensand marl) as a basis for nomenclature.

Rogers had tired of the New Jersey survey by the end of its first season and told William that he "should wish to cut it altogether but for one consideration, that in all likelihood some one would succeed me who would be but too glad of the opportunity to pick my work to pieces," a probable reference to Harlan or Featherstonhaugh.[53] He would have been happy if New Jersey had decided to end the survey with the initial reconnaissance, but he was sure the government would continue it, and it did. The state made a further appropriation of $2,000 for the survey, doubling its financial commitment to the project for the next year.[54]

When the fieldwork in New Jersey ended in the fall of 1835, Rogers not only began writing his report of the season's work but also spent part of the early winter teaching, planning a textbook on geology (possibly an extension of the *Guide* he had prepared for his classes), and helping William, who moved to the University of Virginia late in 1835, with his survey. William had developed a competence in the geologically more recent formations of Virginia where he had studied the marls. Because of the studies he was making in the preparation of his paper on the older formations for the BAAS, however, Henry had more experience with the older formations than William did. William relied on him for advice on these formations and frequently asked him to "help me all you can."[55] At the outset of his work in Virginia, Rogers wanted his involvement with William to be unofficial because of the New Jersey survey and the pending Pennsylvania survey. Nevertheless, before the Pennsylvania survey began, he assured William that even if he were to have the responsibilities of both New Jersey and Pennsylvania, it would not prevent him from doing "half the work of the Virginia survey."[56] Sometime after this bold statement, he became an official assistant to William only to resign within a year when he discovered that he had overestimated his ability to do so much work.[57]

Rogers expected to have Walker accompany him on a brief trip to Virginia in the spring of 1836 and then work with him on the Pennsylvania survey.[58] Therefore, once the New Jersey survey was approved for a second year, he hired Samuel S. Haldeman as his assistant in New Jersey. Haldeman and Rogers had known each other for several years. Haldeman was a student at Dickinson when Rogers taught there, and they had shared a common interest in the social reform of the Owenites. Indeed, Haldeman was an agent for the *Free Enquirer* in Bainbridge, Pennsylvania, in 1832. Haldeman had developed an interest in geology and talked with Rogers about making a geological map of his home area in 1835. With Haldeman's help, Rogers did not expect to spend much of his time on the New Jersey survey, time that he would need for the Pennsylvania survey.[59]

5 Questions of the Highest Importance, 1836–1837

Prompted by the general desire to study the geology of the United States and the specific need to prepare his papers for the BAAS, Rogers had begun to study the geology of Pennsylvania soon after his return from England. Now, at twenty-eight years of age, he was director of Pennsylvania's first state survey, and as director of this survey and the one in New Jersey, he was positioned to accomplish as much as, or more than, any other geologist in America. The two surveys were necessarily directed toward a better understanding of economically significant deposits in the states, but for Rogers, the Pennsylvania survey held special meaning. The mountainous regions of Pennsylvania were both a greater challenge to him as a geologist than the less disturbed area of New Jersey and an opportunity to pursue broader geological issues concerning the processes that had shaped the earth. Rogers was confident, enthusiastic, and eager to begin a systematic exploration.

His first order of business was, however, the organization and launching of the survey. As soon as the legislature approved the survey, Rogers began to plan, hoping to have a crew in the field within a few weeks. Having pursued the New Jersey survey for a year with little help, Rogers's first priority was to find good assistants. The enabling legislation for the survey allocated funds for two assistants plus a chemist. Rogers asked Maxwell Walker, his assistant in 1835 in New Jersey, to come to the Pennsylvania survey, and he found an additional field assistant in the person of his friend John Frazer.[1] James C. Booth, another member of the Franklin Institute circle, was hired to do the chemical work. Booth was ideally suited to it. He had been trained in chemistry in Germany and while in Europe had spent as much time as he could in geological excursions.[2] Although plans were in place by mid-April, Walker never joined the survey. As a result, Rogers had to assign Booth to do general fieldwork with Frazer, and most of the chemical analyses that might have accompanied their work were sacrificed for the first year.[3] Believing, however, that few persons in the United States were satisfactorily trained for geological surveying, Rogers considered himself fortunate to have two competent people. He probably wished that his younger brother, Robert, was available to work for him as chemist in place of Booth. Robert, however, who had assisted William in Virginia in 1835 and Henry in New Jersey, had already agreed to join William again in Virginia by the time Henry knew about Walker and could no longer manage dual jobs.[4]

In addition to his own assistants, Rogers tried to find other assistants for the Virginia survey. Rogers thought that James Espy, the meteorologist and Franklin Institute member, would make a fine assistant for William, but Espy was not willing to work for the amount of money that William could pay. He also thought Richard Taylor would make a good assistant because he was adept at practical work, although in Henry Rogers's opinion "not fit to generalize," but if William offered him the job, Taylor declined.[5]

Rogers, Booth, and Frazer were ready to begin the fieldwork by May 1 and planned an immediate reconnaissance trek across the state. In spite of their carefully laid plans, continuous rains prevented them from beginning their travels until about June 1.[6] The Pennsylvania legislature wanted a report on their progress before adjournment in late June. They had expected to survey a line from near Philadelphia to Harrisburg for this report, but their late start made that impossible if they traveled on foot or by horseback as they would normally do. Unwilling to admit any alteration of their plans so early in the survey but, nevertheless, determined to submit a report on time, the three geologists decided to take the railroad from West Chester to Columbia and the canal from there to Harrisburg. From the general view of the formations thus gained, Booth said that they could talk with the legislators and "maintain the ground that we had been at work since the 1st May."[7] Rogers was able to report successfully on their progress. The reconnaissance then proceeded at a more normal pace on foot and horseback, following a line almost directly northwest through Lewistown, Phillipsburg, Clearfield, Brookville, and Meadville, ending at Lake Erie (figure 3). On their return, they followed a path to the southeast of this line.[8]

Although they fancied themselves adventurers and pioneers, the trip

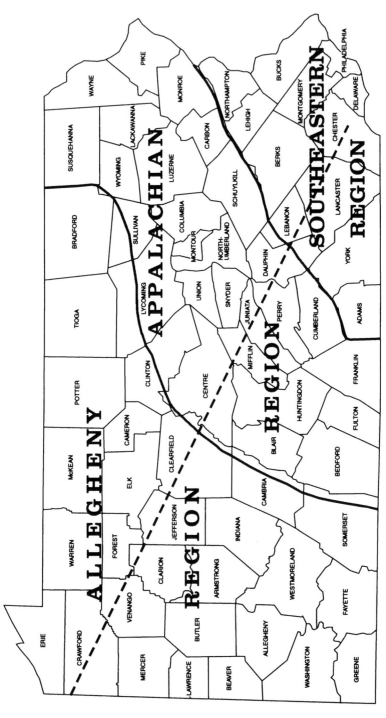

FIGURE 3. The solid lines show the division of the state into three areas for study; the broken line shows the approximate path of the reconnaissance. Lines have been superimposed on a modern map with current county boundaries, which differ from those of the 1830s. (Cartography by The University of Alabama Cartographic Research Laboratory.)

across the state showed the young geologists that their work and surroundings were both lonely and difficult. Booth's letters home give a fine
sense of their experiences. In one letter, he told his mother:

It [the forest of north-central Pennsylvania] consists of a dense assemblage of tall,
straight pines, growing to the height of 100 feet or more, with a slight foliage
near the top,—here & there relieved by an oak, a maple or other forest tree,—an
entangled undergrowth of Kalmias & Rhododendrons, on the outskirts,—large
trees which have fallen ages ago & are now covered with green moss or are
crumbling to earth,—the solitary, echoing sound of the woodcutter's axe,—such
a combination of circumstances tends to raise a chain of sad feelings & thoughts
in the mind of no ordinary marcher;—add to these the wild scream of the bird of
prey, the howling of a wolf or panther & the startling rattle of the snake, & the
picture is complete.[9]

Although Booth and Frazer had some experience with geology, they
were faced with establishing procedures and routines in a challenging
environment, and they relied on Rogers to provide the necessary leadership. Unfortunately, Rogers was often absent from the field on official
survey business in Harrisburg, on trips to Virginia to help William, or on
trips to New Jersey, where the survey of that state was in its second year
under Haldeman's guidance.[10] Rogers expected a great deal from his
assistants, and his absences created hardships. In the initial months of the
survey, however, the excitement and adventure carried them through
difficult times.

After the reconnaissance had been completed, Rogers divided the state
into three regions for further study: the southeastern region; the middle,
or Appalachian region; and the northwestern, or Allegheny, region (figure 3). The southeastern region was divided from the Appalachian region
roughly along the Blue Ridge Mountains; the Allegheny region was
separated from the Appalachian region by the Allegheny Mountains.
While most people called all the mountains of Pennsylvania the Appalachians, Rogers was emphatic in his division of the Appalachians from
the Alleghenies, declaring without elaboration in the *First Annual Report*
on the survey that the two groups were so obviously diverse that they
could not be considered the same.[11]

Rogers chose to concentrate the remainder of the first year's work in
the most geologically disturbed area of the state, his Appalachian region.
He planned to proceed through the area from southwest to northeast
following the general orientation of the mountain ridges in the area.
Starting in Bedford County in the southwestern part of the area, he
would use every possible means of learning the structure of the area.

"Wherever natural or artificial exposures of the strata were accessible, they were sought, whether laid bare on the steep flanks of the mountains, in crags, cliffs or coves, in the banks, or dried channels of the streams, or in excavations on the public works, in mines or in the common pits and wells of the country."[12] In this way, Rogers and the others began to bring some order to their understanding of Pennsylvania's geological structure.

Rogers had several reasons for focusing on the Appalachian region at the outset of the survey. In view of the economic motivation for the survey, it made sense to start in this area because of its heavy concentration of anthracite coal. By 1830, it had become obvious that coal would be the principal fuel for industrial and private use in Pennsylvania. Although there had been an increasing number of coal studies in the early 1830s, such as those done by Richard Taylor, Benjamin Silliman, and Bache and Rogers, and including a study of the coal commissioned by the state in 1833, the state survey was expected to make a major addition to existing information about the coal deposits.[13]

Iron was also a major economic resource for the state because furnaces for smelting the ore were readily fueled by charcoal produced from Pennsylvania's extensive forests and, in a growing number of cases, by bituminous coal. Rogers, therefore, expected to concentrate some attention on iron deposits in the area and "to submit to the legislature . . . probably before the final complete report on the survey, a separate view of the present state and future prospects of the iron manufacture, and iron trade of Pennsylvania."[14]

Another reason for beginning in the Appalachian region was Rogers's belief that this geologically disturbed area was an area where "the leading problems in our geology [are] . . . presented for solution," an area that would furnish "a clue to phenomena more faintly developed elsewhere," and an area the exploration of which would "settle many questions of the highest interest and importance." It was obvious to him that "the forces which upheaved and deranged the strata, had acted with their maximum energy in this quarter."[15] Because the upheaval of the mountains had brought the lowest rocks within the range of study, Rogers believed the area would yield the clue to the dynamic processes that had characterized the history of the earth. Pennsylvania was, he thought, a great natural laboratory for such studies.

Rogers completed his *First Annual Report* before Christmas and submitted it to the state on December 22 in order to meet a deadline of January 1 established by the state. Of the first report, 3,000 copies were printed in English, and because of the extensive Pennsylvania German population, 1,000 were printed in German. Rogers intended to make the report a short one, giving a few general results and even fewer details. He succeeded in just twenty-two pages.[16] About one-half of the report was given over to a discussion of how the survey operated during the year, with only a few pages devoted to actual findings, the most important of which was the discovery of the sequence of formations among the older rocks (those below the coal). As noted earlier, these were poorly understood and often appeared to be no more than a confusing mixture that defied any attempt at organization. Indeed, Rogers, Frazer, and Booth were struck by this same sense of confusion in the early weeks of the season. It was not until Frazer plotted the formations exposed along Yellow Creek in Huntingdon County that a real sequence of formations appeared. On the basis of Frazer's observations, subsequently verified by Rogers and Booth, Rogers identified twelve formations in the *First Annual Report*.[17] Such information was fundamental to all efforts to understand the geological history of the area, for without a clear understanding of the formations and their relationship one to another, the geologists would be unable to develop a realistic concept of the area's structure.

On other matters, Rogers tentatively made the suggestion that the bituminous and anthracite deposits were the same age. He also laid to rest rampant rumors about deposits of silver and other valuable metals. He was pleased to be able to report that virtually all such stories were false, pointing out that the rare finds of precious metals should not obscure the real mineral wealth of the state, the iron ore, coal, limestone, and other "kinds of sand stone and clay, applicable in various branches of the manufacturing arts."[18]

Rogers included some requests in his report that he believed would make his future work in the state easier and more productive. He asked that the state's mandated January 1 deadline for his annual reports be extended, arguing as he had done in New Jersey that he would need additional time in the future to complete the chemical analyses. His request was granted. In the conclusion of the report, Rogers asked for funds for a new topographical map of the state.[19] Although he had decided before the survey began that his assistants would make any corrections necessary on an older topographical map, he had become aware before very far into the survey that this would be an extraordinarily difficult task. The only map that existed was one done by John Melish in 1818. Rogers found that it showed level surfaces where there were mountains, ridges where there were plains, ridges drawn on the wrong side of mountains, incorrect junctions of mountains and of valleys, and incorrect measurements, among other inaccuracies. Rogers had made Frazer responsible for new topographical mapping, but there were too many things to be corrected for one person to do while also being responsible for part of the geological survey.

Rogers was particularly concerned about the inaccurate measurements on the old map and wanted a map based on triangulation in order to determine accurate horizontal and vertical elements. As he would later say, "no map can be said to meet the wants, either scientific or practical, of a geological survey, which does not picture, approximately at least, the *vertical* element as well as the horizontal."[20] Vertical measurements made on the same scale as horizontal measurements were, he thought, essential to an accurate picture of the area. His work as a railroad surveyor may have called the need for good vertical scaling to his attention, but his recognition of its importance was also a reflection of his experiences with de la Beche, who had stressed the importance of it.[21] Unfortunately, the legislature turned a deaf ear to Rogers's request to fund a new topographical map, and it remained the job of one or more assistants to do the best possible job in revising the old map.

In addition to the preparation of the Pennsylvania report, several other things occupied Rogers after the fieldwork for the first year ended in the fall of 1836. For one thing, he had to prepare some kind of report on the New Jersey work for the year. He had paid so little attention to New Jersey that year that his report was a scant two pages filed with the governor on January 3, 1837. In it, he described efforts to define more clearly the general boundaries of the various formations in New Jersey and noted that a cabinet of specimens had been created.[22]

Just after the first of the year, Rogers began his course at the University of Pennsylvania and also gave a series of popular lectures on geology in Philadelphia. His most pressing work throughout the late winter and early spring, however, centered on the preparation of two papers that he finished in the spring of 1837. The first was a summary of the geology of the United States for the American edition of Hugh Murray's *Encyclopedia of Geography*. It was a descriptive account of the geological features of the United States based heavily on his first report to the BAAS, his report to the Geological Society of London, the first New Jersey report, William's early work in Virginia, and his work in the Appalachian area.[23] The second, and to Rogers the more important of the two, was the continuation of his paper on American geology for the British Association for the Advancement of Science, a paper that dealt with the coal and rocks below it in the eastern United States and with the geological history of the Appalachian Mountains.[24] His work before and during the first year of the survey formed the basis of this paper.

His manuscript for the British Association clearly demonstrated that

Rogers's principal goal was to present as complete a statement as possible on the geological structure and origin of the Appalachians. He painted a sweeping picture of the stratigraphy and physical geography of the area but devoted a significant part of the manuscript to a theory explaining the origin of the area and the dynamics of mountain elevation. This manuscript is the most comprehensive overview of the subject that Rogers would ever give. Although he did not send it to the BAAS or publish any other extensive statement on the subject until 1842, the manuscript shows that his ideas about the area were well in hand in 1837.

In his manuscript, Rogers identified the Appalachian area as occupying what had once been a great ocean basin, the perimeters of which were defined principally by gneisses and granites that were generally considered to be among the oldest rocks of the earth's crust and were often described as primary rocks. At sometime in the past, these rocks had been folded, and these folds formed the basin. He believed that the sediment that subsequently filled this basin had come primarily from the northeastern part of the area. He outlined the process thus:

On the *commencement* of the train of sedimentary actions in the northeastern quarter of the basin and perhaps elsewhere, a large mass of coarse *debris* from the adjacent and agitated primary region seems to have been accumulated as the lowermost stratum around the margin of the basin upon the upturned beds of the gneiss and other crystalline [nonsedimentary] rocks. Immediately over this to a depth of probably not less than fifteen thousand feet and possibly much more, the remainder of the series to the coal inclusive were deposited, consisting of materials in part derived from the land, in part precipitated or elaborated from the sea.

Because he could find no unconformity (evidence of some interruption in the deposition), he felt certain the deposition had been continuous from its beginning through the end of the Carboniferous, or coal-forming, period. In some areas, he noted that there was a greater thickness of rocks than in other areas. In many cases, he concluded that this simply meant that the basin of the ocean was very deep. In others, however, the rocks and fossils did not support the idea of a deep sea in spite of the thickness. Rogers inferred that the greater thickness in these areas was the result of the gradual depression of the seabed, allowing for a greater accumulation of material without a deepening of the water.

His calculations led him to conclude that the total thickness of the Appalachian deposits was at least 15,000 feet. In seeking an explanation for how such a thickness of rock was elevated, Rogers took into account several things. The area presented the observer with what appeared to be a series of long folds (or anticlines), running in a northeast/southwest direction, an orientation that Rogers thought reflected the primary anticlines that ringed the basin. The folds exhibited a near-parallelism to each other and appeared to him like rolling waves on a body of water. Rogers determined that there were five great groups of folds, some of which curved, while others were straight. He found that in the southeastern-most part of the area, the folding was sometimes so severe that the northwestern side of the fold was nearly vertical or the whole fold over-turned, that is, the southeastern side had been carried over the north-western one to such an extreme that the position of the strata was reversed, the oldest layer of rock appearing on top of the youngest. Moving away from this area, there was far less evidence of disturbance, and no overturned folds, but he discovered that folds with a southeastern dip of the axial plane were often encountered. Visually, the southeastern arm of such folds is longer and more gently inclined than the north-western one. Moving farther toward the northwest, the plane was more nearly vertical and the arms of the folds more nearly equal, the folds grew farther apart, and the folds subsided in size, gradually becoming flat. In short, Rogers detected a gradation in the folds from the southeast to the northwest.

To explain elevation, it was necessary to account for the folded structure of the Appalachians and the various kinds of folds found there. Although he had gingerly touched on some of his observations in the *First Annual Report,* Rogers had made no effort to explain them, but in this manuscript he suggested that the anticlines and their opposite valleys or troughs, called synclines, "may have arisen from a series of actual *subterranean waves or pulses* [emphasis his], propagated laterally from the zone of chief igneous actions situated in and east of the Blue Ridge," which was the area that showed the greatest disturbance. This area was characterized by the intrusion of igneous (molten) rock, torn and crushed rock, and overturned folds that were, he argued, the result of an explosive release of vapors that had accumulated on top of the molten interior of the earth. The explosion set the waves in motion, and they then traveled in sets across the surface of the molten interior from the southeast, thrusting an elastic crust forward in one direction (much like ocean swells pushed by the wind). This push resulted in the common southeastern dip of the folds. Although he felt that others would argue that the trough that followed the crest of every wave would invariably cancel out the elevation of the rock caused by the wave, he said it seemed likely that the waves would "retain in a greater or less degree their inclination in consequence of the lateral support they would have from pushing against themselves along the anticlinal axes." Because he was able to divide the area into five groups of folds, he believed the waves were propagated in at least five sets, all occurring within a limited time period. The force of the waves grew less as they traveled away from their point of origin and the original force diminished, thus explaining the gradation in folds toward the northwest.

Rogers related the types of coal found in the area to this process as well. He was convinced that all the coal was of the same age. As the result of extensive analyses of the coals of Pennsylvania, Maryland, and Virginia made by him and by William, he was also convinced that there was a regular gradation in them from anthracite in the southeast to bituminous in the northwest. Since anthracite contains less volatile matter than bituminous coal, it was said to be "debituminized," and it was already suspected by many that heat was the cause of debituminization. It therefore seemed logical to Rogers that anthracite should occur in the area where the most violent disruption occurred and where igneous intrusion gave ample evidence of extreme heat.

According to Rogers, the upheaval of the great folds, or anticlinal axes, occurred after the coal was formed. The upheaval caused the water that had filled the basin to drain from it in a cataclysmic rush in all directions. This was the same rush of water that he had earlier argued was responsible for the creation of Niagara Falls. The excavating power of the water changed the configuration of some of the anticlines and stripped away some of the rock, leaving once continuous formations, like the coal, divided.

This aquatic catastrophe was also responsible for the configuration of the Allegheny region. Rogers elaborated on his statement in the *First Annual Report* that the Appalachian Mountains were geologically distinct from the Alleghenies by noting that the Allegheny area was an elevated basin or plateau that rose above the original Appalachian basin. Its basic features were not the result of the upheaval of anticlinal axes, as was the case in the Appalachian region, but were, rather, solely the result of violent denudation by the water that drained away from the area at the time of the rising of the Appalachian anticlinal axes.

When Rogers was in England, mountain elevation was a major topic among geologists, and his transformation from social reformer to geologist could hardly have been complete without a thorough indoctrination into the various theories then being discussed. Until the mid-1830s, most European geologists had considered it impossible to look at the massive upheaval of the earth represented by European mountains and think that anything other than a tremendous cataclysmic, or as some like Rogers preferred to call it, paroxysmal, force could have heaved up the strata in their present configuration. Vertical uplift, frequently regarded as occurring through an agency of volcanism that drove granitic wedges into the strata to lift them from below, appealed to many as an explanation. This provided a good explanation for the fact that mountains often seemed to follow straight lines. Such views were supported by the clearly volcanic origin of some mountainous areas and the fact that volcanic activity frequently occurred along the lines followed by mountain ranges. The theory of the French geologist Léonce Elie de Beaumont, however, generated the greatest interest in Europe and the United States in the early 1830s. Believing that the interior of the earth was gradually cooling from an original molten state, Elie de Beaumont argued in 1829 that, as the interior cooled, the crust accommodated itself to the shrinking interior by crumpling. The result of this contraction was often the elevation (or collapse) of large segments of the crust. In 1831, he proposed that mountain chains were elevated along several axes that formed arcs or circles around the globe.[25] The direction of the strike (the direction in which the axis pointed) of a given chain indicated its period of elevation, all those with similar strike being of the same age. His theory was widely accepted, although his idea that elevation occurred in great circles was one of its least appealing aspects and was discounted by many members of the geological community.

Adherents to views like that outlined above were called catastrophists, and they believed the events described were unlike modern events in the intensity of their force. Assuming an original molten state for the interior of the earth, however, catastrophists often recognized a change in the strength and magnitude of forces affecting the earth as the interior cooled. Catastrophic events might occur over a very long period of time, but the strength and violence of the events diminished.[26] Opposing the catastrophists were the "uniformitarians," led by Charles Lyell. Lyell argued that the forces shaping the earth today were always the same and that there had not been intervals of widespread cataclysmic activity. Insofar as mountains were concerned, Lyell stressed gradual continental uplift, aided by igneous forces in combination with slow denudation by water, although he was willing to argue in his *Principles of Geology* that volcanoes and earthquakes, of the kind and strength known to occur periodically today, were possibly contributing factors in local upheaval.[27]

Although Rogers was impressed by Lyell and adopted his division of the Tertiary rocks, he found no merit in the uniformitarian approach. The violent disruption of the strata that he observed in the southeast convinced him that catastrophic events of a kind unknown to the modern world, had shaped the earth. The process that Rogers envisioned was very different from any other that had been suggested by catastrophists, however. Unlike Elie de Beaumont, who saw elevation as the result of contraction, with force exerted from two sides, Rogers saw evidence in the southeastern dip of the folds that elevation resulted from a push in one direction only. This phenomenon inspired his concept of waves traveling over the surface of the molten interior, propelled in one direction by the initial explosive force and lifting the crust into a series of long folds. Rogers's own analogy of the process to a succession of waves on water pushed forward by the wind creates a powerful visual image of what he saw happening beneath the earth's surface.

Rogers completed his manuscript before the second season of the survey got under way. He never sent it to the BAAS, however, as he had intended and as his friends in England had encouraged him to do. Instead, he put the manuscript into the hands of two friends, William McIlvaine and C. C. Biddle, and took extreme care to ensure that he had proof that he was its author.[28] McIlvaine and Biddle had met Rogers for the first time in 1834 when they attended one of his lectures on geology at the Franklin Institute. Impressed by his skill and knowledge, the two older men, both lawyers, befriended him and remained his friends in the coming years.[29] Rogers read his manuscript to McIlvaine at the beginning of April and subsequently gave a written copy to him and to Biddle. When Biddle and McIlvaine received the manuscript, each one initialed every page of it and wrote brief testimonials stating that it had been committed to their care in April and May 1837.[30] This action indicates that, while Rogers wrote the manuscript for publication, something made the time inappropriate and he wanted to protect his authorship for future publication.

Many factors might account for Rogers's action. Perhaps Harlan's attack made him more fearful than he cared to admit. Perhaps Bache, ever concerned that American science present itself on the highest level, had advised him to be cautious and to wait until he had more data to support his theory before offering it to the scientific community of Great Britain. Furthermore, Rogers was a perfectionist in the truest sense of the word and found it difficult to bring any work to a conclusion because there was always something more to be done, another piece of information to gather, another observation to make. Evidence that Rogers may have hesitated on these grounds is a letter from Rogers to Biddle in which Rogers says that he needed more time to perfect the manuscript.[31] An equally important reason to withhold the manuscript at this time, however, may have been Rogers's own decision, as presented in instructions to his assistants, that no information gained from the survey could be used or made public in any way before it first appeared in the official reports of the state geologist.[32] Some of Rogers's work in the Appalachian area that was the basis for his paper and his theory occurred before the survey began, but his work during the survey was critical. If the assistants could not use information gathered during the survey, he surely could not do it.

By the time April arrived, Rogers was ready to return to the Pennsylvania survey. He had asked for, and had received with Trego's help, an increase in his appropriation that allowed him to hire two more assistants.[33] On April 12, Rogers formally rehired Frazer and Booth, and Haldeman left the New Jersey survey to join Rogers in Pennsylvania. The fourth assistant was Trego, whose long interest in geology and in the survey finally persuaded him to leave the legislature and become an active member of the survey. Haldeman stayed the year, and Trego remained until 1841, but the survey careers of Frazer and Booth were nearly at an end. Both resigned shortly after beginning fieldwork in April.

In an effort to structure the duties and responsibilities of his assistants, particularly now that there were two more, Rogers wrote to them about the general responsibilities of the assistants. In addition to his warning about using any of the information gained from the survey before the official reports were published, Rogers told them that they were expected to spend six months in the field and the rest of the year preparing detailed reports, complete with sections and drawings. They were also to assist in the preparation of the cabinets of specimens and were admonished that if they expected to quit the survey, a year's notice was required.[34]

Booth and Frazer were outraged by the letter and threatened to go to Harrisburg to register a complaint against Rogers.[35] While there is no evidence that they carried through with the threat, both resigned. They might well have been upset by any one or all of the statements in Rogers's letter or their anger may have stemmed from Rogers's manuscript for the BAAS. Because both were Rogers's friends as well as his assistants, Rogers probably showed them the manuscript sometime in late March after its completion and before he consigned it to the care of McIlvaine and Biddle. By the time Booth and Frazer received Rogers's letter, they would have been aware that Rogers had used information from the survey in the manuscript. It is difficult neither to imagine an angry reaction nor to think that such a reaction contributed to Rogers's decision not to publish his manuscript immediately.

That Frazer at least was angry before the letter was received is also a possibility. In later years, much was said about Rogers's assistants' being upset because they did not receive enough credit in Rogers's official reports for their specific contributions to the survey. Frazer had made a particularly significant contribution to the survey with his initial discovery of the stratigraphic sequence. He might have been peeved that Rogers had not made special note of his discovery in the *First Annual Report.*

Booth's reaction to the letter was surely tempered by the fact that he had another job opportunity immediately at hand. Sometime in early March, he had learned through relatives that a position as state geologist of Delaware might be in the offing and decided immediately that he would like the job because "it is better to be a king in a village than a common subject of an empire."[36] When offered the job in April, he accepted without hesitation and left the survey before Frazer resigned. Not only would the job suit him, his friends at the Franklin Institute must have been pleased to see another one of their group take over a state survey.

Frazer's formal resignation in May, to which Rogers replied only that "the course you have thought proper to adopt in relation to my instructions renders the acceptance of your resignation a matter of course," signaled the end of the friendship between the men, and it was the first serious breach of the relationship between Rogers and his close associates at the Franklin Institute.[37] It was not, however, the last. Almost immediately, Rogers found himself locked in dispute with Frazer, Booth, and several other members of the institute over an entirely different matter.

At issue was the kind of rock to be used in the Delaware Breakwater. The breakwater project, the first of its kind in the United States, was an attempt by the federal government to create a protected harbor that would be free from the constant forces of water erosion and would be a safe haven in stormy times.[38] Construction began in 1828. Stone hauled from Pennsylvania quarries was dumped along a base about 160 feet wide and was expected to reach 3,600 feet in length. By the end of 1834, some 640,520 tons of rock had been deposited for the breakwater and nearby icebreaker, but shoaling nevertheless accelerated over the following months. Construction was stopped pending new studies of the tides and currents, and a question was raised about the kind of rock used in the breakwater. The rock in use was a gneiss quarried at the Ridley and Crum Creek quarries in Pennsylvania, owned by Leiper and Company (George G. Leiper, William J. Leiper, and Sam M. Leiper) and J. F. Hill, respectively.

Early in 1837, the Leipers had asked the institute's Committee on Science and the Arts to appraise the gneiss rock they quarried. This committee, which Bache had played a major role in establishing in 1834, was autonomous within the institute and was the institute's scientific arm. The committee report was filed on March 9 and was subsequently published in the institute's journal.[39] An unidentified subcommittee of seven had examined the rock and had pronounced it more than adequate for all purposes for which it was used. Although it was not stated in the report that the Leipers' concern responded to questions about the kind of rock in use for the breakwater, it used a great deal of the quarries' product, and it behooved the owners to do what they could to ensure its acceptance. According to Rogers, William Strickland, the government engineer of the breakwater and a member of the institute, had stated a preference for unstratified trappean rocks quarried near Wilmington, Delaware. As a result, a short time after the Franklin Institute investigation initiated by the Leipers, the committee was asked by the secretary of war to evaluate the merits of the gneiss versus the trappean rock.

Until he left for a European trip in 1836, Bache chaired the committee, but at the time the breakwater question came to the institute, the committee was chaired by Roswell Park, a former member of the United States Corps of Engineers who had been in charge of the breakwater just before coming to Philadelphia in 1836 as professor of chemistry and natural history at the University of Pennsylvania. Park, like Strickland, favored the trappean rocks and appointed Rogers, who also favored them, to a subcommittee of ten to study the issue. The members of the committee included, in addition to Rogers, Booth, Frazer, Strickland, John Wetherill, Sears Walker, Isaiah Lukens, Gouverneur Emerson, and Thomas McEuen. Wetherill was chairman of the subcommittee.

According to Rogers, they were almost all friends of Robert M. Patterson, Jr., who favored the gneiss rock because he had some connection, presumably a business one, with the Leipers.[40] Patterson was a longtime member of the institute but had left Philadelphia in 1828 to become professor of natural philosophy at the University of Virginia. He returned in 1835 when he became director of the United States Mint, the position held by his father in earlier years. A solid member of the Bache circle, he resumed his active participation in the institute immediately. Patterson had offered Rogers a job as melter and refiner at the New Orleans mint earlier in the year, but Rogers had refused the offer.[41] This refusal may have antagonized Patterson, but in any event the two became bitter opponents over the breakwater rock.

Eight of the members of the committee visited the quarries, but Rogers was in the field, and before he joined the committee back in Philadelphia, a report had been framed supporting the gneiss rock. There was considerable irritation on the part of the committee members that Rogers had not been there sooner to give his advice as a geologist. He was told that he "had neglected a public duty."[42] Rogers then visited the quarries with Wetherill, and although Wetherill had apparently gone along with the earlier report, Rogers convinced him that the trappean rocks were a superior choice. Wetherill opted, nevertheless, for personal neutrality on the issue. Since the committee did not find unanimously for one or the other of the rocks, three separate reports were taken to the general committee for action. According to Rogers:

> After numerous long sessions (some of them five hours!) in sub-committee, we took into general committee three several reports, one being mine, the official one being only a set of meager resolutions, signed by five out of ten, and not by W[etherill], who was for pure neutrality. For three sessions of the General committee I had to defend my report and views against all P[atterson]'s influence which is mighty, but I foiled him by a vote of eighteen to fourteen.[43]

Although Rogers was successful in the contest, Patterson ultimately won because the outcome was settled by economic necessity; the gneiss rock was cheaper and continued to be used for that reason.

Rogers was irritated by the battle over the breakwater rock because it had kept him from survey work, but at the same time he felt satisfied that he had stood his "ground against every sort of unfairness, attack and ridicule" and "succeeded in carrying with me nearly all true friends of the Institute."[44] Rogers had won a battle, but the victory cost in terms of his position at the Franklin Institute. He was certain that he had gained the permanent enmity of Patterson and his friends, among whom were Frazer and Booth, Trego, and Franklin Peale, who was another one of Bache's group and a member of "the club." The controversy hastened the end of Rogers's tenure as a member of Bache's inner circle in Philadelphia.

The survey was well under way by the time the quarry conflict was over. Trego and Haldeman were at work and Rogers had hired four subassistants, Alfred F. Darby, Edwin Haldeman, Horace Moses, and Peter W. Sheafer (variously spelled Scheafer or Shaeffer by Rogers). Rogers was able to establish a chemical department this year and hired his brother, Robert, as chemical assistant. In spite of frantic attempts to find replacements for Booth and Frazer, it was July before Rogers replaced them with Alexander McKinley of Pittsburgh and James D. Whelpley of Philadelphia. Whelpley had some geological experience under Silliman, but McKinley apparently had none.[45]

Increased attention for some areas was possible during this season because of the additional assistants, but Rogers continued to give the most attention to the Appalachian region, where he and two of the assistants spent most of the season in Schuylkill County. Greater attention was also given to the southeastern region.

Rogers adjusted his classification of the formations this year because two new formations were recognized and two from the previous year were combined. The result of the changes was a list of thirteen formations, a formation defined for the first time by Rogers as a group of rocks referred to a common origin or period.[46] He had given each of the original twelve formations a descriptive name, for example, white fucoidal limestone, but this year he substituted numbers for names because he believed there was less chance of confusion or erroneous correlation with rocks in other areas if numbers were used.

Rogers's most significant discovery during the second season concerned the thickness of the Appalachian strata. In his manuscript, prepared before the season began, he had suggested that the thickness was at least 15,000 feet. By the time the second season was over, he said the thickness was in excess of 40,000 feet.[47] He was the first to recognize the true immensity of the deposits that made up the Appalachians, and henceforward geologists knew that in order to account satisfactorily for the dynamics of mountain elevation, it would be necessary to explain both how the sediments had accumulated to such an extent and how they had been elevated. Rogers believed that his theory did both successfully.

Rogers's work this year also provided a modification of his opinion, given in the manuscript, that the Appalachian formations had been deposited without interruption. He had found no unconformity before 1837, but during that year, he discovered that there had been an interruption in deposition after the third formation from the bottom[48] (figure 4). Several years later, in a joint paper on the White Mountains, Henry and William announced their opinion that this disturbance, unlike the one that had originated in the southeast and had closed the Appalachian sequence after the coal was formed, had originated in the northeast in the area of the White Mountains.[49] It had lifted the White Mountains and large areas of the northeast above the waters, while deposition continued in other areas of the Appalachian sea with the unconformity he had discovered marking the disturbance in those areas that were not lifted above the sea as the White Mountains had been. These deductions were based on the Rogers brothers' recognition that the White Mountains were composed of highly metamorphosed sedimentary rocks characterized by extensive igneous intrusion rather than of primary rocks predating the sedimentary deposition, as was popularly thought. By recognizing that the mountains were sedimentary, the brothers were able to make the

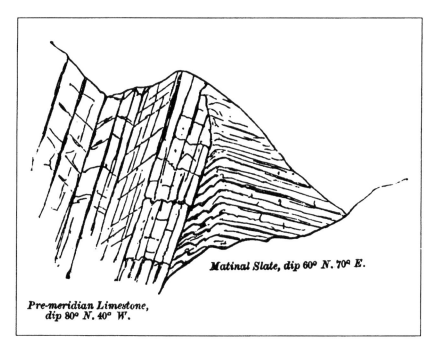

FIGURE 4. An illustration of an unconformity. (From Henry Darwin Rogers, *The Geology of Pennsylvania: A Government Survey* [Philadelphia: J. B. Lippincott, 1858], vol. 1, fig. 586.)

argument that they had formed in the great Appalachian sea. Unlike many of their contemporaries, both brothers believed that metamorphism was the result of heat and therefore a natural by-product of the dynamic process detailed by Henry Rogers in his 1837 manuscript.[50]

Rogers completed the *Second Annual Report* and submitted it early in 1838. Many statements in the annual report of the second season reflect Rogers's concept of the geological processes at work in the area, but because he held a strong belief that the annual reports were inappropriate places in which to discuss theory, his comments appear as isolated statements rather than as part of a whole concept. He was determined also to exclude "from my annual narrative of the operations of the survey, any detailed account of the facts ascertained during each season."[51] He wanted to be able to put together the entire picture of Pennsylvania's geology at one time, thus limiting his commentary in the annual reports in favor of the detailed account that he was even then planning for a final report at the conclusion of the survey. The *Second Annual Report* was, nevertheless, four times longer than the report of 1836.

After completing his responsibility for the season, he settled back in Philadelphia, and for nearly a month afterward, he watched, with a sort of amused detachment, as the legislators debated how many copies of the report to print, a debate that became a regular feature of discussions concerning the survey in future years. This year the decision was to print 10,000 copies in English and 3,000 in German.[52] This quantity was more than three times as many copies as had been printed the first year, but it was not unusual for states to print large numbers of geological reports. Since the surveys were intended for the economic benefit of the states and their residents, it was important that any information derived from the survey be made available immediately to all individuals who might be interested.

Without extensive responsibilities to the survey until later in the spring, Rogers turned his attention to improving his income. While his salary from the Pennsylvania survey met his immediate needs, Rogers sought the security of a larger income, one that would permit him to put some money aside for the future. His teaching and lectures continued to provide supplemental income, but he was determined to find a business that he alone, or with one or more of his brothers, could develop into a source of solid, long-term support.

Rogers cherished the idea of opening a school in Philadelphia, preferably a scientific school emphasizing practical education, that he, Robert, and William would operate, but there seemed to be no real prospect of doing so in the immediate future, and he consequently turned his attention to other possibilities.[53] At the time, a demand for silk had been created by successful New England silk mills, and raising silkworms had became a popular undertaking in the East. The silkworms fed on the leaves of mulberry trees, and in return for a modest investment in things necessary for the cultivation of mulberry cuttings, the entrepreneur could expect a tidy income from the sale of the cuttings to those who raised the silkworms.

Late in 1837, or early in 1838, Henry and Robert went into the mulberry propagation business in Philadelphia in a substantial way. They built two greenhouses, one described as 100 feet long and 16 feet wide, holding 150,000 cuttings; the other as large enough to hold 50,000 cuttings. They employed a gardener to superintend the operation and built a special box to propagate the cuttings.[54] The brothers expected eventually to sell as many as 350,000 cuttings at twenty cents each, but the silk industry collapsed before the end of the decade, leaving them with no market and nothing to show for their efforts.

Where the two brothers got the money for this business is not clear. William Rogers had started to raise silkworms in Virginia at about the same time, and his widow later noted that he, too, was a heavy financial loser in the venture.[55] As head of the family, he may have financed his brothers' business in Philadelphia. There is also a possibility that Rogers got some of the needed capital from Nicholas Biddle, who invested heavily in a variety of businesses in the area and whom he had known for several years through the Franklin Institute. On learning of the potential profit in raising silkworms, Biddle had eagerly entered the business himself with a "large purchase of mulberry trees" and, as a business investor, would have considered a loan of investment money to his friend a good opportunity.[56] Biddle's possible help with this venture is supported by his investment in another of Rogers's enterprises at about the same time. With Biddle's financial commitment, Rogers opened an iron furnace for the manufacture of pig iron in 1838.[57] The furnace was located in Armstrong County, Pennsylvania, on Buffalo Creek, north of Pittsburgh. The furnace was a cold blast charcoal furnace powered by water from Big Buffalo Creek. Although iron manufacturing methods had begun to change by this time, Rogers's furnace was a very traditional one in that it used wood (charcoal) for fuel instead of coal. According to a history of Armstrong County, written in 1883, the furnace had a stack thirty-five feet high and was eight feet wide at the bosh, with a weekly product of about thirty tons. Approximately 100 people were employed, and the operation covered nearly 500 acres.[58]

Biddle's investment in the furnace was made through his close friend Roswell Colt, a speculator and stockbroker. Rogers understood that the investment established a partnership between him and Biddle, but as would become evident later, Biddle and/or Colt understood it as a loan. Over the following years, Biddle and Colt borrowed money on their personal notes to finance the continued operation of the furnace, and Rogers was able to borrow some money on his own signature. He was listed officially as the owner until 1843, when the ownership was changed to Rogers and Colt as part of an effort to continue to finance operations through Colt's ability to borrow money.[59] By that time, the furnace had become a financial disaster for Rogers.

Without any particular effort or foresight on his part, Rogers stumbled onto the one activity that would eventually provide the support he wanted when, in 1838, Biddle asked him to investigate the soundness of an investment he was considering.[60] Over the next few years, Rogers periodically worked as a geological consultant for Biddle, and by the mid-1840s, geological consulting would be his principal source of income.

Although Rogers was always interested in business opportunities, the Pennsylvania survey continued to demand almost all of his time as each new year brought new problems and challenges.

Cautious and Laborious Research, 1838–1840

6

On March 2, 1838, Rogers delivered a public lecture at the capitol in Harrisburg on the progress and needs of the survey. Members of the house and senate were particularly invited and were impressed enough by his comments to approve an increase in the survey's budget by $6,000. This allowed Rogers to hire additional assistants and a draftsman at salaries of $800 each.[1] Samuel Haldeman, Moses, and Darby left the survey at the end of the 1837 season. McKinley, Whelply, Trego, Sheafer, and Robert Rogers returned, and they were joined by John McKinney, Townsend Ward, Harvey B. Holl, James T. Hodge, and Robert Montgomery Smith Jackson. Martin Boye, whom Booth had instructed in chemistry, joined the survey as a second chemical assistant, and he and Robert Rogers set up the survey's chemical laboratory in a building on Grape Street in Philadelphia.[2] Little is known about the background of Ward or McKinney. Holl was visiting the United States from England. Hodge had been an assistant on a survey conducted by C. T. Jackson in Maine and Massachusetts, and R. M. S. Jackson had just graduated from the Jefferson Medical College.

With the additional help now available, Rogers divided the state into six districts, assigning one or two geologists to each one (figure 5). District 1 closely paralleled the original southeastern region and was surveyed this year by Holl and Rogers. District 2, the northeastern part of the Appalachian area where the major anthracite deposits were located, was surveyed by Whelpley and Sheafer. District 3 was the southwestern half of the Appalachian region and was entrusted to McKinley and Jackson. Since so much time had been spent in the Appalachian region during

79

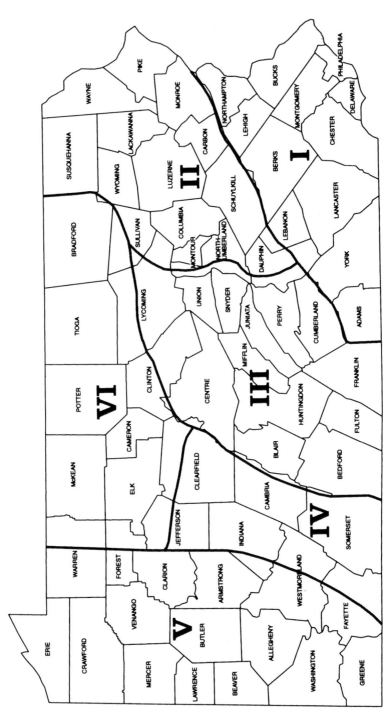

FIGURE 5. The six districts of the state of Pennsylvania. (Cartography by The University of Alabama Cartographic Research Laboratory.)

the first two years, complaints about the lack of attention to other areas had reached the legislature.[3] Under pressure, particularly to make information available about his original Allegheny division, Rogers divided it into three districts and planned to study two of them during the 1838 season. The first of the Allegheny divisions (district 4) encompassed the area between the base of the Allegheny Mountains on the southeast and Chestnut Ridge on the northwest and was surveyed by Trego and Ward. District 5 included counties lying west of Chestnut Ridge and extending north to the state line of New York. Rogers assigned Hodge to the northern half of the area and decided that he and McKinney would do the southern half. District 6, which was largely a wilderness area, encompassed an area north and northwest of the Alleghenies. Rogers did not assign anyone to this district but spent time there himself just "ascertaining the best mode of overcoming the peculiar difficulties which the wilderness character of the region presents to accurate Geological researches."[4]

With the added appropriation he had been given, Rogers was enjoined by the state "to make such inquiries and examinations into the present methods of mining coal and manufacturing iron as the Governor shall deem expedient and proper, to increase the products of the mineral resources of the state."[5] He was asked to present his findings in the form of a written report. These demands reflect the growing importance of anthracite coal to the iron industry in Pennsylvania. Before the mid-1830s, this industry was supported almost entirely by the state's extensive forests that could be converted into charcoal fuel. The great anthracite deposits of Pennsylvania, though plentiful, were virtually unused for iron manufacture because anthracite was difficult to ignite and its use introduced impurities into the iron that made it brittle and difficult to work. The situation began to change in 1828, however, when James Neilson, in Scotland, patented a process for heating the furnace's blast, solving the problem of ignition. His method was employed almost immediately in Great Britain, where it dramatically increased the production of iron coked with bituminous coal. By the early 1830s, experiments to use the hot blast for smelting ore with raw anthracite, rather than coke, were under way, and in 1837, a process using hard coal and the heated blast was introduced in Wales and was reported that year to the British Association for the Advancement of Science. The Franklin Institute reported the new process in its journal almost immediately.

Excitement over the potential use of the vast anthracite deposits of Pennsylvania for the benefit of the iron industry ran high. Pennsylvania recognized the potential by passing a law in 1838 authorizing the incorporation of iron and steel companies, and the state offered a prize to the first ironmaster who kept an anthracite furnace in blast for three months, a prize funded by businessmen and bankers, including Nicholas Biddle.[6] In light of this interest, it is not surprising that the state geologist was expected to concentrate on locating and analyzing iron deposits and to indicate where those deposits were in close proximity to anthracite deposits. Although he was not able to produce the requested report this year, Rogers did assure the secretary of the commonwealth, Thomas H. Burrows, that Pennsylvania was ready to reap the benefits of using anthracite for iron production.[7]

In spite of the need to give more time to other areas, the anthracite regions of the second district remained critical to Rogers because of their importance to the state's economy and also because of his research into the geological dynamics of the area. He promised to let Whelpley and Sheafer hire as much additional help (probably coal surveyors working in the area) as possible, and Rogers spent as much of his time as he could manage in this area as well.[8] He reaffirmed his belief that all the anthracite deposits had once been joined and that the upheaval and subsequent denudation of rock by catastrophic inundation had destroyed the once continuous mass, leaving the now isolated coal basins.[9] The upheaval of the area, Rogers reasoned, had caused a natural division among the anthracite coal deposits, and he divided them into four distinct groups: the great northern basin, an eastern and middle anthracite field, the field of Mahanoy and Shamokin, and the Broad Mountain, Mine Hill, and Pottsville coal basins. The eastern, middle, Mahanoy and Shamokin fields were separated by great anticlines from the fields to the north and to the south, while the eastern and middle fields were separated from the Mahanoy and Shamokin by extensive valleys. In subsequent studies of the area, Rogers used this structural separation in appraising the potential of each area for coal production.

Rogers asked his assistants to do an ever-increasing amount of topographical mapping. Whelpley spent most of his time mapping the coalfields, and was, according to comments made later by J. Peter Lesley who joined the survey in 1839, "the first perfect topographical geologist." Lesley's admiration of Whelpley's work would later lead him to go so far as to say that topographical geology was born in 1838 with Whelpley's work, and that his map of the southern and middle anthracite coalfields, completed in a later season, was "one of the most important contributions to physical science ever made, in any country."[10]

During the early winter, Rogers wrote his annual report for 1838, but before he had a chance to present it to the legislature, his report of the preceding year (1837) was attacked by Timothy Conrad. Writing in the *American Journal of Science* in January 1839, Conrad, who was as bold in his use of fossils in stratigraphy as Rogers was cautious, accused Rogers of making two significant errors in his *Second Annual Report* because he failed to use fossils to identify formations. First, Conrad accused him of making a grave error in confusing certain rocks of the Hudson River area in New York with those found along the Salmon River in Oswego County, New York. "Not a single species," Conrad said, "of shells or plants is common to both . . . , [and Rogers's mistake] shows the great danger of error in endeavoring to identify strata over large areas, if we neglect to appeal to the evidence afforded by paleontology and rely too exclusively upon the ever-varying mineral composition of rocks."[11]

Conrad's second challenge was an indictment on an even broader scale. Rogers had used the term "Lower Secondary" in the *Second Annual Report* to refer to the Appalachian formations as a group. Conrad took exception and argued that fossils proved they were Transition rocks, saying that the Appalachian formations had "scarcely a single feature in common with what has hitherto been termed secondary by all other geologists."[12] The terms "secondary" and "transition" were part of a classification, developed in the late eighteenth and early nineteenth centuries in Europe, in which the rocks of the earth were divided into four major groups, from the bottom to the top: the Primary, Transition, Secondary, and Tertiary. These had been identified on general bases, including the type of rock, whether it was crystalline, unlayered, layered, inclined, and with or without fossils. William Maclure had applied this general scheme to the United States early in the nineteenth century, thereby establishing a terminology and framework in which to place American rocks. The scheme was subsequently strengthened by Amos Eaton, who published the *Index to the Geology of the Northern States* in 1818, which fully established the idea that American formations should, and in his view and that of many others did, fit into this European scheme.[13]

The majority of geologists believed that the Secondary rocks extended downward from the Tertiary (Lyell's divisions of which were discussed earlier), and ended below the Carboniferous group of formations that included the coal proper and usually a formation called the Mountain Limestone and sometimes one called the Old Red Sandstone. Below this were the Transition rocks. Although local sequences within the Transition had been identified, they were a confusing mass of formations that were generally called the Graywacke. A few geologists, including Lyell, did not use the term "Transition," preferring instead to extend the term "Secondary" to encompass the Graywacke. Rogers chose to do the same, calling the Transition the Lower Secondary and the Secondary of other geologists the Middle and Upper Secondary. In a glossary borrowed in principal part from Lyell and accompanying the report that Conrad attacked, Rogers defined the Secondary exactly as Lyell had defined it in his *Principles of Geology* (1833): "An extensive series of the stratified rocks which compose the crust of the globe, with certain characters in common, which distinguish them from another series below them, called *primary,* and from a third series above them called *tertiary.*"[14] Conrad's reaction was based on a common understanding of the Transition, but he had failed to take note of Rogers's definition.

This use of "Secondary" made more sense to Rogers because of his understanding of the physical history of the area. He said in his BAAS manuscript, and again in his *Second Annual Report,* that the coal and formations below it were "a vast group of strata, the successive sediments of one immense ocean, the creations of but one prolonged Geological epoch, commencing almost in the dawn of marine animal and vegetable existence, and terminating with the latest produced deposits of coal."[15] Rogers could find no justification for separating the coal from the formations below it, and as he would make clear in later publications, he believed that in spite of the upheaval that occurred after the coal, the formations immediately above it were still the product of the same ocean in attenuated form. For example, one of the deposits that Rogers was correctly convinced lay above the coal (although some of his contemporaries were not so sure) was called the New Red Sandstone. In two major outcrops, one that includes an area of Pennsylvania and New Jersey, the other in Connecticut, Rogers observed that the formation was gently inclined and relatively thin, with folds characterized by a northern dip, unlike the other formations of the area that were characterized by a southern dip. Rogers argued that the formation was the result of oblique deposition in an estuary that was a remnant of the original ocean.[16] The link between this formation and the ones below it was, Rogers believed, made clear by the term "Secondary" as he used it.

Conrad's attack on Rogers for failure to use fossils more extensively was an argument over methodology and was seen as such by their contemporaries. Although fossils were already recognized for their importance in stratigraphy, and Conrad and other state geologists would further demonstrate their importance, it was still not a completely comfortable approach for all geologists in the late 1830s. Rogers therefore had sympathetic listeners and defenders. In an editorial comment following Conrad's article in the *American Journal of Science,* Benjamin Silliman came to Rogers's defense, emphasizing the opposing methodologies and pointing out that Rogers "considers the argument based on the want of identity in the fossils as inconclusive, until it shall appear that a large number of species from each formation have been compared, and this because he places more confidence in conclusions drawn from following the rocks themselves over wide areas."[17]

Amos Eaton, who was one of America's pioneers in stratigraphy, felt compelled to contribute an article to the *American Journal of Science* early in 1839 also referring to the different methodologies. Eaton had surveyed large areas of the East in the 1820s, primarily under the auspices of Stephen Van Rensselaer, and had developed a stratigraphy by physically following the rocks and placing them in relationship to one another on the basis of their lithological characteristics. Eaton's article was a rambling discourse on his own interpretation of the stratigraphy of New York and Pennsylvania presented in the hope that his findings might help to settle the dispute between Rogers and Conrad.[18] In spite of the fact that he acknowledged the role that fossils could and would play in future stratigraphic studies, by emphasizing the method he had used, Eaton pointed out that a great deal had been accomplished by physically tracing rocks rather than by the use of their fossils. To be sure, some of the things he and others had determined on this basis had been proven wrong, but at the same time, he thought, too much had been accomplished by pioneer stratigraphers to dismiss the method entirely.

Similar thoughts must have gone through the minds of many of the older geologists as they read Conrad's article, and in another issue of the *American Journal of Science* later in the year, in what was almost certainly a reference to this issue and a defense of Rogers, the editors referred to the work of both William and Henry Rogers in flattering terms. They noted that "their cautious and laborious mode of research is likely to insure the approbation of all sound and judicious geologists [and when their work is finished] it does not admit of a doubt that the course of strict induction which they are pursuing will be fully approved."[19]

In one sense, both Conrad and Rogers were right in their approaches. As later geologists would realize, fossils were not always the best guide in establishing a stratigraphic framework in every area, especially in the severely disturbed rocks of Pennsylvania, where strata were contorted and squashed along with the fossils themselves.[20] In a flat, comparatively undisturbed area like New York, fossils could be used to the very best advantage in stratigraphy. In spite of the virtue of each approach for a given area, a secure framework that could be applied to other areas needed to be based in large part on fossils. Work in New York, where there were vast areas of undisturbed rock and a competent paleontologist in the person of Conrad, and later James Hall, provided the opportunity to develop the stratigraphic column for the older rocks of the United States in a way that could not be done easily in Pennsylvania.[21]

The *Third Annual Report* on the survey was presented to the house and senate in February 1839. Discussion seemed routine, and the annual discussion over how many copies to print in English and German took place. Then on March 20, without warning, Richard Brodhead, representative of Northampton County, moved to instruct the house committee on the judiciary system to study the possible repeal of the act providing for the survey.[22] Brodhead, a graduate of Lafayette College and a lawyer, had settled in Easton, Pennsylvania, in 1830. Elected to the Pennsylvania legislature in 1837, he later served as a United States representative and senator. As a conservative Democrat, he felt a legitimate concern over the expense of the survey, but his action suggests he was motivated by questions raised in Rogers's *Second Annual Report* about the potential for coal production in the southern coalfield around Pottsville, Broad Mountain, and Mine Hill, questions that were not put to rest in the report just submitted.

Using his knowledge of the structure of the area and the forces that had shaped it, Rogers had concluded that the geological processes of upheaval that had affected this coalfield rendered "the business of mining these nearly vertical and overtilted seams, precarious in a high degree."[23] This was not welcome news to anyone with mining interests in the area, and it was not a statement likely to win friends for the survey. Anthracite mining was a rapidly developing business in Pennsylvania, particularly with the increased importance of coal in the production of iron and steel. It was the kind of enterprise that got the attention of investors, and they were scrambling to get control of mining lands. Several members of the Henry Charles Carey and Samuel Wetherill families of Philadelphia had invested heavily in mining lands in and around Pottsville. The last thing such investors wanted to hear was that their investments might not be good ones, and they were very willing to complain to sympathetic representatives in the state legislature.[24]

Brodhead and other legislators may also have heard general complaints about the presence of the geologists in the mining districts. Many later writers, including Lesley, noted tension between the geologists and the miners and owners because the latter could not always understand what the geologists were doing or why they were doing it. According to Lesley, the early miners were "ignorant, undisciplined, obstinate, narrow minded and superstitious . . . and hated professional geologists because these had never lived in childhood, pick in hand, under ground,—

because they taught new things hard to comprehend."[25] The difference between the miners and the geologists was stated a little differently a hundred years later, but the point is similar: "geology became a science, while anthracite mining barely succeeded in passing from a gamble or a speculation to a rude art which both above and below the water level was continually hampered by ignorance of geological formations, by technological backwardness, and by inefficiency, carelessness, and colossal waste."[26] It is not surprising that the miners' distrust, economic threats to the investors, and economic threats to the very existence of the mining communities would lead to opposition in the legislature. While Brodhead's attack reflected the concern of the miners and investors, it took an interesting direction in that it focused on the cabinets of specimens that were to be placed in each county in accordance with provisions made in the original survey legislation. For the average miner, these cabinets, or collections of specimens, had the potential to be among the most important and practical results of the survey. The written reports, however thoroughly they might eventually present the geology of the mining area, were not as illustrative of the geology as carefully identified and arranged specimens were likely to be. To be able to see familiar specimens of rock and coal would make the "new things hard to comprehend" more intelligible. They would help in bringing the science of the geologists more within the range of the miners' perspective and aid in their understanding of what the geologists meant when they talked about different formations.

Brodhead's motion was adopted, but no action was taken. Although he was quieted for the moment, he had not said his last word on the subject. Nor was it the last blow the survey would receive this year, for it was followed immediately by the legislature's continuing refusal to fund a topographical survey. Having had no success with his first plea for more topographical work, Rogers had reopened the question of funds for a topographical map of the entire state in 1838. A commission, including Alexander Bache, was appointed by the house of representatives in April 1838 to study the matter. The commission presented its report in June 1839. It was signed by Bache and William P. Alrich, and although it acknowledged the need for such a map, it advocated only enough work to correct the existing map.[27] This recommendation brought to an abrupt end any hope that Rogers had for a topographical survey. Two years earlier Bache would almost certainly have supported Rogers in his request but not now. Bache and Rogers had drifted apart in 1838 for a number of reasons. Bache may have been influenced by the attitude he found among some members of the institute toward Rogers on his return from Europe, but the onetime close friends had come into direct conflict over a proposal for a "general college" at the Franklin Institute.

Late in 1836, Henry and William had worked on a plan for a school of arts on behalf of the Franklin Institute. The institute submitted it to the legislature in 1837. Although the institute's memorial to the legislature was signed only by the collective Board of Managers of the institute, Rogers referred to it privately as his and William's plan.[28] The school's future seemed promising, as city and state authorities looked on the plan with favor, but it was ultimately defeated in the state legislature. Since neither state nor city funds were in the offing, many members of the Franklin Institute favored financing the school through the institute. Rogers agreed, but Bache did not, arguing that the school would succeed only with public support.[29] Although Rogers and Bache remained on cordial terms in the coming years, their relationship was never the same after this disagreement. Bache continued to admire Rogers's abilities as a geologist, but he never again exerted any effort to help the survey.

Fieldwork for the fourth year of the survey began in mid-April 1839. The corps of assistants was the same except for the arrival of J. Peter Lesley and Andrew A. Henderson and the departure of Sheafer who moved on to a career in coal mining operations as a consultant and mining engineer. Assignments were slightly altered, so that Holl and Trego were assigned to the first district; Whelpley and Lesley the second, Trego, McKinley, Henderson and Jackson the third; McKinney and Ward in the fifth, and Hodge tackled the sixth. No one was assigned to the fourth district, and very little work was planned for that area. Robert Rogers and Martin Boye continued as chemical assistants, and George Lehman was hired as draftsman.[30]

Of all of Rogers's assistants through the years, J. Peter Lesley would be with him for the longest period of time, and in 1874, Lesley succeeded Rogers as director of the second Pennsylvania state geological survey. Lesley graduated from the University of Pennsylvania the year preceding his appointment to the survey with the expectation of entering a theological seminary. Ill health made it impossible for him to do so, and Bache, who was a friend of Lesley's father, suggested a year of work on the survey for the sake of the young man's health. Although Rogers questioned hiring him for the rigorous work of the survey because of his health, he acquiesced to Bache's wishes, appointing Lesley to work with Whelpley. Lesley desperately wanted to succeed in his role with the survey and felt he was in a special position with regard to Rogers because of the way he had gotten the appointment. Lesley harbored a fear throughout his first year with the survey, however, that Rogers would ask him to leave if he showed any sign of poor health. He even cautioned his father, whom Rogers often saw in Philadelphia, to say nothing about the state of his health.[31] His fears proved to be groundless, and when Whelpley decided to leave the survey in June, Rogers put Lesley in charge of the second district, where he had been working under Whelpley's direction.

At 252 pages, the *Fourth Annual Report* is by far the longest and most detailed of any of the annual reports to that time. Rogers's larger group of assistants and three years of prior study had set the stage for the gathering of more information than ever before. Furthermore a full 50 pages of the report were devoted to chemical analyses of iron ores, no doubt intended as part of the report on iron requested by the state. The interest in iron notwithstanding, coal continued to be the first concern of the geologists as they mapped the great Pennsylvania coal basins. Lesley had taken over Whelpley's role as topographer, and in the *Fourth Annual Report,* Rogers called attention to the fact that "the map intended to exhibit, on a large scale, the topographical features of the several coal basins . . . is in a condition of considerable forwardness, though from the excessive complexity and truly wide extent of our anthracite region, much research will yet be necessary in order to complete it." In addition, he said, a series of at least 120 sections had been completed by Lesley, and more were planned.[32]

In response to Brodhead's criticism, Rogers elaborated on his plans for the cabinets in this report. There were, he noted, already nearly 8,000 specimens temporarily classified and arranged for study for the final report and in preparation for deposit in Harrisburg, the state capital. Only part of the collection was then in Harrisburg, he said, but eventually the cabinet there would be complete and accessible to the public. Rogers pointed out that it was impossible to say the specimens were nearing a final arrangement, since, as long as the survey continued, they would have to be rearranged each time a new one was added.[33]

This report differed not only in size and detail but in the way that the work of the assistants was handled, each receiving more direct credit for individual contributions than had been the case in earlier reports. Lesley later suggested that the assistants were disgruntled over the lack of credit for their individual achievements in the shorter reports of the first three years, and therefore, greater attention may have represented an effort on Rogers's part to appease them. There is also the possibility, however, that the differences in the length of the report and in the recognition of the assistants were the result of Lesley's assistance in writing the report. Lesley hinted at this in a history of the survey that he wrote in 1876.[34]

Rogers regularly dictated his report to Lesley, and it would not have been especially difficult for Lesley to write some portions of the report with guidance from Rogers. He was familiar with the areas that got the most attention and, just as Rogers would do, Lesley could refer to notebooks that the assistants turned over to Rogers at the end of each season, along with their own final reports of the season's work in their respective areas.[35]

When the *Fourth Annual Report* was received by the legislature, it generated discussion on several fronts. The number of copies to be printed was drastically reduced from the preceding year, reflecting an effort to save money, since the Bank of the United States was teetering on the brink of failure once again, and financial panic was about to engulf the state.[36] It was also clear that hostility toward the survey and Rogers was growing. When the report was not forthcoming in printed form soon after its submission, members of the house grew annoyed and assumed that it was Rogers's fault, although part of the blame rested with the printer, and action to rescind funding for the publication was started. Members of the house were also concerned about errors in the earlier reports and demanded that Rogers "be required to remain in Harrisburg during the printing of his report, for the purpose of reading the proofs and superintending its completion, or otherwise employ a competent person for that purpose."[37] They were justified in their concern, since the earlier reports had been riddled with avoidable errors. Rogers had probably dictated his early reports to someone unfamiliar with the general work of the survey and had then paid little attention to proofreading the final copy. The *Third Annual Report* was the worst in this respect. For example, Kittatinny Mountain on the eastern side of Pennsylvania was confused with Kittanning Mountain on the west, and the chemist Boye's name was consistently misspelled "Borje." The fourth report has few of these kinds of errors. Rogers either took the legislators' concerns to heart and did a better job of proofreading, or the improvement reflects Lesley's involvement in writing the report.

Although in each of the above cases, legislative motions to take action against Rogers were tabled and died, a potentially more damaging effort was also under way to repeal all the acts under which the survey operated, ostensibly because the governor had called for severe austerity measures in the state.[38] A motion to this effect was made on February 6 by Representative Strohecker. On February 10, a select committee was appointed to investigate the expenses of the survey and "to examine witnesses, as to the proper or improper application of the monies drawn from the State Treasury, for the purposes set forth in the Acts before referred to, by any officer or officers, employed in virtue of their provisions."[39] Neither effort resulted in any action, but the concern of the select committee about proper expenditures made it evident that legislative reaction against Rogers went beyond a concern for the state's general financial situation and implied that some representatives perceived a misuse of funds. There is no way today to reconstruct Rogers's expenses in order to determine whether there was anything questionable. Although some receipts have survived, they are not sufficient to give a complete picture.[40] Concern over a misuse of funds, however, may have been prompted by Rogers's iron furnace and its manager. Rogers hired John McKinney in 1838 as one of his geological assistants and promptly assigned him to the district where the furnace was located. From this time, McKinney managed the furnace. Although state geologists in other states sometimes used their positions to reap personal gain, such behavior was often questioned by legislators.[41]

The spring of 1840 left no doubt that Brodhead was not alone in his wish to be rid of the survey and Henry Rogers. It was, however, just as certain that the survey had friends, since an expenditure of $10,200 for the fifth year of work was approved.

His problems in Pennsylvania notwithstanding, Rogers was able to bring the New Jersey survey to a close in the spring of 1840 with the publication of the final report, but only after experiencing some difficulties there with the legislature.[42] The legislature had allocated $2,000 for the continuation of the survey in 1836 and again in 1837.[43] Rogers relied on Haldeman and one other assistant to do most of the fieldwork in 1836, and he employed an assistant to do chemical work during the winter.[44] It is not clear that Rogers had any assistant in 1837, or that he spent much time in New Jersey himself. He requested no funds after 1837, and no more fieldwork was done in the state. Rogers asked only for time to finish the report, the chemical analyses, and maps and sections.[45] The New Jersey legislature had expected the final report in 1838 and again in 1839 and had made plans for the distribution of copies.[46] When it was not forthcoming, the legislators began to suspect that something was wrong. Late in 1839, the governor, who was now William Pennington, called on Rogers for a report on his progress. Rogers responded that he expected to have all the work completed during the winter, and on February 18, 1840, he reported that the work was in press and was expected in mid-March.[47] The legislators did not trust Rogers's word, and agitated by the long delay, they called for an investigation of the survey.[48] Before the investigation got under way, however, the final report appeared in print, and no more was said of the survey or the investigation. The final report was a substantial summary of the geology of New Jersey, but it is very similar to the first annual report published in 1836 and thus shows that Rogers gave comparatively little attention to the state after that date.

A Capricious Master, 1840–1842   7

During the legislature's deliberations over the fate of the Pennsylvania survey, Rogers was preparing for the first meeting of the Association of American Geologists. When he came home from England in 1834, he thought that an organization like the British Association for the Advancement of Science was needed in the United States. He suggested the idea to Benjamin Silliman, but Silliman had urged caution, sensing that American science might not be ready for such an organization.[1] The lack of an organization as a forum for geologists, however, began to be widely felt as the state surveys increased in number. New York, for example, had authorized a survey in 1836, and the state was divided into four districts, each with its own chief geologist. In addition, there was a paleontologist for the survey and numerous assistants. This arrangement necessitated frequent meetings between all of the New York geologists during which they were able to discuss issues of common concern and coordinate their activities. Still, they were isolated from other geologists and had almost no contact with Rogers in spite of the fact that the geology of the border areas between the two states was important to both.

While it might be supposed that some discussion would have developed naturally between the geologists of New York and Pennsylvania, this was not the case. Rogers knew Timothy Conrad, who had started with the New York survey as geologist in its first district and had become paleontologist for the whole survey in 1837. It was he who attacked Rogers so vehemently in 1839. If Rogers knew any of the other principals in New York—Lardner Vanuxem, William Mather, Ebenezer Emmons, or James Hall—the acquaintance was casual. James Hall, who joined the

New York survey in 1837 as Emmons's assistant and then took over the fourth district, had no contact with Rogers until 1839, when they conferred regarding a possible meeting of geologists and when Hall wrote to Rogers in the hope of settling an old question in New York about whether there were, or were not, coal deposits beneath the surface. Hall asked Rogers's opinion on whether a conglomerate formation found in some areas in New York was the same as that associated with major coal deposits at Blossburg in Pennsylvania. Rogers correctly recognized that the conglomerate Hall described was not the same and thus put an end to speculation about profitable coal deposits in New York.[2]

The New York geologists were very much aware of Rogers's annual reports, and although Conrad was not favorably disposed toward some of Rogers's ideas, it was evident that an exchange of information was desirable with Rogers and with other state geologists as well. According to James Hall, Lardner Vanuxem in 1838 suggested a meeting of all the state geologists at least for the purpose of discussing nomenclature, which was a regular topic of discussion at meetings among the New York geologists.[3] Vanuxem's interest in expanding these discussions may have been immediately prompted by the appearance of Rogers's *Second Annual Report,* in which Rogers introduced a numerical system of nomenclature and argued against the assignment of names to formations such as was being done in New York and elsewhere.

Edward Hitchcock has also been given credit as the originator of the idea of a meeting of geologists and was a catalyst, even if not the originator. He had long felt that such associations had decided advantages, and in April, he wrote to both Rogers and Silliman urging a meeting.[4] Rogers took no immediate action, and because Hitchcock was concerned that a meeting would not be successful if Rogers was not a full supporter and participant, he went to see him in the fall to urge him to help in arranging a meeting.[5] He found Rogers hesitant about the prospect of a meeting limited to geologists. Rogers wondered instead whether it was not more advisable to wait until such time as a meeting of all scientists could be assembled, as he had suggested in 1834.

There were two movements afoot at this time to create a national scientific organization of broader character, more like the BAAS in concept. One, the American Institute for the Cultivation of Science, had been proposed by a prominent group of Bostonians; the other, the National Institute for the Promotion of Science, had been proposed in Washington by Joel R. Poinsett.[6] Rogers preferred to adopt a wait-and-see policy before deciding whether one or both of these proposed organizations better fit his idea of a scientific organization than an association limited to geologists. On the other hand, many of the people with whom he felt a scientific kinship, like Joseph Henry and Bache, were opposed to a national organization of a general kind at this time, fearing that American science might still be too colored by amateurism to present itself on a high level.

In spite of his hesitation, Rogers considered the possibility of holding a meeting for geologists in Philadelphia in the spring of 1839, but the winter passed without any attempt on his part to help organize such a meeting, in part because of his indecision about the nature of the association and in part because of the survey. In April 1839, Timothy Conrad, Lardner Vanuxem, and James Hall came to see Rogers, again to urge a meeting. Although he was still unsure about a meeting limited to geologists, this time Rogers felt he "could not refuse," remembering that he "was in this quarter the original mover," a reference to his 1834 suggestion to Benjamin Silliman that some kind of national organization was needed.[7] Rogers was asked to contact friends involved in state surveys about the meeting. While he may have done so informally, the invitations that have been preserved were sent to state geologists over the names of the New York geologists only.[8]

The first meeting of the new Association of American Geologists, as it was christened, was held in Philadelphia at the Franklin Institute. It was a three-day meeting that began on April 2, 1840. There were about forty geologists present, including the four Rogers brothers, several of Rogers's current and former assistants, and the New York survey geologists.[9] About half of those present were members of the Franklin Institute. Hitchcock presided, and although discussions and lectures covered a variety of topics, no formal papers on nomenclature were given, and nothing was recorded to suggest that this was a major topic at the meeting.[10] Rogers had adopted numbers for the Pennsylvania formations, and William had done the same in Virginia, but in New York, the geologists had named their formations in accord with the geographic location where each was most prominently displayed, grouping them together in larger units with names that reflected the geographical areas where particular sequences were most apparent. It was a procedure that Emmons later said was utilitarian, if nothing else, and whether it was better than Rogers's numbers or not, it was too far advanced by 1840 for any change.[11] Discussion on the issue of nomenclature was, therefore, an issue that was far less meaningful in 1840 than it had been in 1838, when the New York geologists first expressed interest in a meeting.

The meeting was pronounced a success by all who attended, and plans were laid for a permanent organization. The next year the organization changed its name to include naturalists, becoming the Association of American Geologists and Naturalists. This was undoubtedly done to recognize the presence of naturalists as part of some of the state surveys, like New York, that encompassed both geological and natural history surveys. In 1848, this association would become the American Association for the Advancement of Science.

As soon as the meeting was over, Rogers prepared to begin the fifth season of fieldwork in Pennsylvania. The members of the preceding year's staff returned, with the exception of McKinney, who resigned. He remained as the manager of Rogers's iron furnace, however, and his resignation supports the contention that his dual role was a subject of concern to Pennsylvania legislators. The only major shift in responsibilities for the season was the move of Hodge and Ward to the fourth district, a heavily wooded area that posed special problems for Rogers (figure 6). Rogers thought that Hodge had the best experience in dealing with such areas and gave him extra help to locate economic deposits in the area.[12] Rogers was more content with the progress of the survey during this fifth year than at any other time. Not only did he begin getting good information for difficult areas like the fourth district, but he was convinced that the accumulated experience of the years now made it possible to get a more accurate picture than ever before of the true relative positions of the various strata. Mapping proceeded smoothly. Although Rogers had been skeptical of Lesley's abilities as a draftsman when Lesley joined the survey, so much so that Lesley even thought about resigning, Lesley had proved himself to be highly skilled.[13]

While the scientific achievements had grown steadily more satisfying to Rogers through the years, his relationship with his assistants and the conditions under which they worked had grown steadily worse. Rogers was determined from the moment the survey began that it would be done as thoroughly and as carefully as the time and money at his disposal would allow. Nevertheless, as an administrator, he had many problems with his assistants that stemmed both from administrative inadequacies, particularly his general disorganization and frequent absences from the field, and from what seemed to the assistants an insensitive approach to their needs and desires. The letters of J. Peter Lesley, the best single source of information on survey operations after he joined the survey in 1839, show increasing frustration and irritation. Rogers was a demanding taskmaster, and his assistants complained freely about his disorganization. Indeed, during Lesley's first season with the survey, Lesley found him a "very capricious master who never knows his own mind 6 hours

FIGURE 6. Oil painting of a Pennsylvania survey field camp by William van Storkenborg from a sketch by George Lehman that was probably made during the late summer of 1840. At that time, Hodge, Townsend Ward (note initials on the tent), and Lesley were working in the Fourth District with a few helpers. The three top-hatted figures are almost certainly the three geologists. (Reproduced courtesy of the Pennsylvania Historical and Museum Commission, Collection of The State Museum of Pennsylvania.)

ahead." Lesley described what he considered an all too typical encounter in the field near Pottsville in 1839:

Rogers, after leading me, (for while he is here, I run at his coattail) into one mine after another, & like a schoolboy making a beautiful 2, running round & round to make a dot, shot off in a tangent yesterday to Mauch Chunk . . . [to see] Miss Kensington. . . . [although] he made a pretext, whatever the cause, of Sessman's [Lehman's] want of work here, & that he must carry a draftsman . . . to the Schy'l to give him employment.[14]

Rogers was careless in his attention to the needs of the men at work in the field, and salaries were often slow to be paid. Although this was sometimes due to the simple fact that the state was itself slow to appropriate actual funds, at other times the problem was Rogers's fault. For example, in September of his first season, Lesley learned that Rogers had not sent to the state word of his appointment, which consequently delayed payment of his salary.[15]

According to Lesley, it was general knowledge that there were problems between Rogers and his assistants, and soon after coming to the survey, he told his father that personal observations corroborated this and that Rogers played a "shameful" and a "cruel" game with his assistants.[16] Although Lesley did not make the nature of this game clear, it seems likely that he was referring to concerns among assistants that they were not getting proper credit for their work. Rogers took all of the field notes of the assistants and assembled them in one report, issued under his name. The assistants were named in each report and their contributions were acknowledged in general, but contributions were not normally attributed to individuals. It has already been suggested that this procedure may have contributed to Frazer's departure from the survey. Writing many years later, Lesley implied that this was a critical point of trouble between Rogers and his assistants. Whether Rogers's procedures irritated the assistants as much as Lesley thought they did will be discussed later, but it must have bothered them to a degree. They were all young men who were looking for careers, and some public acknowledgment of their contributions was potentially important for their own futures.

Rogers's frequent absences from the survey were another source of frustration to the assistants. In the early years, his time had been divided between Pennsylvania, New Jersey, and, to some extent, Virginia. Although the latter two received little attention after the first year of the Pennsylvania survey, they still took him occasionally away. Furthermore, as the years passed, he became more deeply involved in general studies of the Appalachians, including areas beyond the borders of Pennsylvania and Virginia. He and William had begun to plan what they liked to call their "general memoir" or "grand volume" on the Appalachians in 1836 and took every opportunity to pursue their studies.[17] They spent, for example, much of the summer of 1840 exploring the Berkshire Hills, tracing formations, and trying to work out the stratigraphy of that area.

Rogers's work away from the survey was not meaningless in the understanding of Pennsylvania's geology or the contribution to the greater field of geology that it permitted him to make. When the fifth season closed in the fall of 1840, Henry and William worked together on a paper describing their summer travels in the Berkshires, and Henry presented it at the January 1841 meeting of the American Philosophical Society. It was particularly significant because of the argument that Rogers made against a formation identified by Ebenezer Emmons in 1839 and named the Taconic. Emmons identified it as the oldest fossil-bearing formation in New York. Before this, the lowest formation had been assumed to be one called the Potsdam Sandstone. Although some of the New York geologists, especially Hall, were skeptical of the Taconic's place in the sequence of formations, as a group they accepted it. According to Emmons, the Taconic was especially well defined in the Berkshire Mountains, but after making their trip through the area in 1840, the Rogers brothers were convinced that Emmons was wrong and subsequently debated the point with him many times. In the paper for the American Philosophical Society, Rogers argued that the so-called Taconic rocks were only metamorphosed Hudson River Shales that lay above, not below, the Potsdam. Using his knowledge of overturned folds, Rogers said that the Taconic rocks had been confused with older rocks because the strata were inverted in order, so that the younger ones were beneath the older ones. The Taconic issue became a major one in American geology that was debated for much of the nineteenth century. Rogers was one of the most outspoken and earliest critics of the Taconic as defined by Emmons and the first to recognize the importance of structure in understanding the stratigraphic position of the formation.[18]

For all that might come from Rogers's travels, however, his assistants suffered, and their patience grew thinner with each absence. Without direction, their work sometimes came to a halt while they awaited Rogers's return. In addition there was sickness and at times a shortage of supplies. In the face of such problems, the adventures of the initial years of the survey vanished.[19] What had once seemed to be pretty ground now seemed rough and too hard for sleep. The yellowjackets were a constant source of irritation, "inviting themselves, without even the hint of welcome, to every meal regularly, begging for sugar." As Lesley succinctly summed up the situation in 1840, "one finds it very romantic at first, and very uncomfortable at last."[20]

The escalating difficulties were made worse as individuals vied with each other to achieve personal goals. Hodge planned a "future speculating and profitable business" with Rogers and tried to ingratiate himself with Rogers, going so far as to deliberately withhold money allocated for the needs of the fourth district crew in order to impress Rogers with his ability to run their camp on a small budget.[21] Hodge also tried to discredit his fellow workers Townsend Ward and George Lehman, even to the point of trying to get Lesley, who was transferred to the fourth district during the summer, to help him. These skirmishes were one more burden for the geologists to bear, and Rogers was no help to them.

According to the original legislation creating the survey, 1840 should have been its final year. By the time that year was over, the final report was to be completed. When Rogers had presented his report on the fourth season, however, he had argued that, after the five years were completed, at least six additional months of work would be necessary to write the final report, prepare its illustrations, complete the geological map of the state, finish the chemical analyses, and organize the cabinets.[22] By the time the fifth season of fieldwork ended in the fall of 1840, Rogers was convinced that an additional six months was woefully inadequate and that he would need a full year just to finish the fieldwork. He called attention to this problem when he presented the *Fifth Annual Report* to the legislature. Since the survey had met with escalating criticism in the legislature in the preceding two years and since money continued to be a problem for the state, Rogers did not expect to get an immediate hearing before the legislature or one without opposition.[23] He had, however, the support of Governor Daniel Porter, who, in his annual message to the legislature, emphasized how much had already been done, stressed the importance of completing the survey, and heartily endorsed the funding of another year of work.[24] His recommendations were referred to a select committee of the house, the purpose of which was to evaluate and report on all the governor's recommendations.[25]

Continuing opposition to the survey was immediately reflected in a move to expand the size of the committee and in the appointment of Brodhead to that committee. Brodhead introduced a resolution asking the state treasurer to present a statement showing the total amount paid out under all the acts of the survey and to specify the sum paid to each individual. This move suggests that Brodhead might have been behind the issue of expenses a year earlier, although he did not bring it up himself at that time. The request for the expense accounting was formally made on February 19, and on the same day the treasurer presented the requested information.[26] Rogers expected opposition and was thus pleased when it became obvious that the majority view of the select committee favored the continuation of the survey. On March 13, an act to supplement the survey to allow it to continue was introduced, recommending an appropriation of $12,000 to complete the survey. This act also rescinded the original provision for cabinets to be made available in every county in favor of three state cabinets, one in Pittsburgh, one in Harrisburg, and one in Philadelphia.[27]

A little more than two weeks later, Brodhead submitted the minority report of the select committee.[28] He demanded to know what was to prevent Rogers from asking year after year for an extension. Would the field researches ever be completed, he asked? He was critical in general of the annual reports because Rogers, in each one, had referred to details that would be saved for the final report. Arguing that Rogers was "bound by every principal of moral and legal obligation" to end the survey as originally scheduled, Brodhead again asked what would happen if the survey were not concluded in the additional time. Would the state have any recourse? He obviously felt that it would not. He emphasized the cost factor by pointing out that the money already paid or committed amounted to $64,285 plus the cost of printing, binding, and distributing the annual reports. Taking all of these expenses into account, a more realistic cost, he thought, was $100,000. Brodhead was angered by what he considered an intolerable expense and by the fact that the new act denied individual cabinets to each county. Without them, he argued, the residents would not know if any useful discoveries had been made. Brodhead's comments suggest that Rogers had been back of the move to reduce the number of cabinets, and he hinted that Rogers's reason for reneging on the county cabinets was because he had no specimens to furnish. While the move to reduce the number of cabinets was, no doubt, initiated by Rogers, Brodhead's assessment of the situation seems to have been unreasonable. There is no indication that Rogers did not have specimens, but the preparation of an individual cabinet for every county was a major undertaking, and since the survey fieldwork was still unfinished, the prospect of finishing it, writing a final report, *and* preparing many cabinets was formidable. Under such circumstances, Rogers cannot easily be faulted for looking for a way to reduce the burden.

Brodhead's attempt to quash the survey failed, and an appropriation of $12,000 for a sixth year of fieldwork was made. There was little reaction to the appropriation in the senate, but one senator cast his disapproval in the form of a sarcastic appraisal of an appropriation that made no provision for phrenology, physiognomy, animal magnetism, or water smell-ing.[29] The appropriation for the survey passed the house and senate and was attached to a general appropriations bill, which was not an unusual occurrence but in this case caused further problems for the survey. Be-cause of the unsettled financial situation of the state, Porter, in spite of his support of the survey, found it necessary to veto the entire bill. The setback was temporary, and the veto was overridden by the legislature. Rogers had spent many arduous days in Harrisburg lobbying for the survey appropriation and now, victorious, he prepared "to go home and engage my thought once more upon my professional business."[30] The legislative debate had been one more source of aggravation for the as-sistants because it had held up their last payment for the preceding season until the matter was settled.[31]

Fieldwork in 1841 resumed almost immediately after funding was as-sured, with essentially the same group of assistants. Lesley had become completely devoted to Rogers, and he was proud to note that Rogers had complete faith in him and entrusted him to make all of the arrangements and plans for the survey.[32] Rogers virtually left Lesley in complete charge, as he was again absent for part of the season. He and William, joined by Edward Hitchcock for part of the time, traveled extensively in a continuing effort to find the "true" relationship between the rocks of the main body of the Appalachians and their outcrops in Ohio and New York. They traced them "northeastward to the Mohawk, and thence westward through New York . . . and working round Lake Erie through Upper Canada and Michigan, [to] form a junction with the rocks of Ohio."[33] Tracing the formations by walking along them, could, they believed, give a satisfactory idea of the stratigraphic positions of these rocks and ultimately of the structure and the relationship of the whole of the Appalachian region to the rest of the continent. Rogers also spent some time with Charles Lyell, who was making an extended tour of the United States. He met him briefly in the late summer, and again in October when they were together for eleven days "scampering like the wind" on an excursion that took them through Reading and Pottsville to the Delaware River and Trenton.[34]

The Sixth Annual Report (published in 1842) was brief and largely given over to the plans for the final report, which, Rogers said, would require considerable time, energy, and money to prepare.[35] He planned to spend two years on it, completing it by 1843. Rogers promised that the cabi-nets, which he considered to be an illustration of the report, would be completed by the time the report was finished. The act of May 4, 1841, from which the sixth season's work had been funded, provided for an allocation of $2,200 for the preparation of the final report. Even though Governor Porter continued his support and assured the lawmakers that Rogers would complete at least part of the work before the legislature convened in 1843, the 1842 legislature would not release the funds for the final report, and opposition to Rogers surfaced again in more questions about the cabinets.[36] Nevertheless, Rogers set to work on the final report and employed Lesley to prepare maps, paying him from his own pocket. In 1843 the legislature approved the release of the $2,200 but signified continuing concern by adding to the authorization act a proviso giving the governor final control in the release of funds if and when he was convinced that Rogers would conclude the work called for by the state laws.[37] Since the governor continued to support Rogers, this did not create a problem.

Lesley used the materials from the assistants to produce a topographical and geological map of the state in color and thirteen long sections across the state by the 1843 deadline, not an easy task, since the information and drawings presented by the assistants varied from near perfection to "the rudest of pen sketches."[38] While Lesley accomplished his work in spite of the difficulties, Rogers did not fare as well. Rogers envisioned the final report as both the report on the practical findings of the survey and an exposition of the geological processes that had acted in the area. With regard to the first, he confronted a huge task that he could not complete by the deadline. Since the annual reports, with the exception of the fourth and fifth ones, were brief, Rogers had to rely largely on the raw materials from his assistants to complete the final report. Years later, Lesley accurately assessed the situation in which Rogers found himself when he said that

piles of annual reports of assistants now called for the authors to become editors; but these [the assistants] had all disappeared, leaving their unrevised manuscripts, their rough sketches, their unfinished maps, their uncatalogued, unstudied mineral and fossil collections behind. No matter which one of these reports the Chief Geologist might take in hand, he was sure to encounter on its first dozen pages a dozen nuts to crack for which he had no cracker. . . . To edit [the report] properly, intelligently, was to repeat in the office the work of the whole corps in the field. It was to task one man with the work of a number of men multiplied by the number of years they had been employed.[39]

The difficulties imposed on Rogers by this state of affairs were augmented by Rogers's perfectionism. He found it difficult to bring any work to an end because he always wanted to improve it. As a result, the final report was delayed sixteen years.[40]

His theory of elevation that had been maturing in his mind since 1837 might have met the same fate. Pressured by a series of events, however, Rogers presented the theory publicly in 1842.

8

A Theory So Much More
Satisfactory, 1842–1843

Rogers's theory of elevation, first written out in 1837, argued that elevation occurred because the molten rock beneath the surface of the earth's crust was set in motion. This event was triggered when tremendous pressure that had built up on the surface of the molten material tore open the crust, releasing the pressure. The sudden release of the pressure set the molten matter into a series of wavelike motions that lifted an elastic crust into similar wavelike structures. This theory of elevation, with elaboration and supporting arguments, was formally presented publicly for the first time at the third annual meeting of the Association of American Geologists and Naturalists (AAGN) in 1842. It was presented in a paper jointly authored with William entitled "On the Physical Structure of the Appalachian Chain, As Exemplifying the Laws Which Have Regulated the Elevation of Great Mountain Chains Generally."[1] The paper was divided into two parts. William read the first part, on structural features in the Appalachians that the brothers thought were critical to any theory of elevation that would successfully explain the mountains. Henry read the second part, which dealt with the dynamics of elevation.

The theory of elevation as given in this paper and as centered on the concept of wavelike undulations in the earth's crust was the unmistakable descendant of the paper Henry Rogers had written for the BAAS in 1837. In the intervening years, William and Henry had explored the Appalachians in detail and had often discussed the forces that operated there. Henry always referred to the theory as his and William's, suggesting that William's input had played a key role in refining the theory as presented in

1842. Since Henry Rogers had so clearly developed the idea earlier, however, he must be recognized as its principal author.

A year after entrusting his manuscript to the care of McIlvaine and Biddle in 1837, Rogers had called for the manuscript's return, probably to add to it and consider its publication.[2] Again, caution prevailed, although occasional traces of the theory appeared in the survey's annual reports, and Rogers discussed it informally with his colleagues. He expected to publish the theory as part of the final report of the Pennsylvania survey and/or as part of the volume that he and William planned to publish on the Appalachians. Since the latter was never published, the theory might not have been published until the final report appeared had Rogers not been alarmed on the one hand by his colleagues' failure to grasp his ideas and on the other by a fear that his ideas might be stolen. He also began to feel an obligation to make his ideas available because of increasing interest in mountain elevation.

In his 1837 manuscript, Rogers had discussed inversion and the frequency of folds showing a southeastern dip. In his *Third Annual Report* he pointed out that the southeast dip was more than just common—it was typical. Hitchcock had observed the same thing in Massachusetts. At the 1840 meeting of the Association of American Geologists, Rogers engaged in a discussion with Hitchcock and Emmons about this point, telling them that inversion and the southeast dip were the result of waves affecting the area.[3] When Hitchcock published his *Elementary Geology* later that year, he postulated that a single push had simultaneously affected and overturned the whole area, and he ignored Rogers's concept.[4] Rogers corrected Hitchcock in his address to the American Philosophical Society in January 1841 on the geological structure of the Berkshires (the paper in which he had denounced Emmons's Taconic).[5] Hitchcock made another brief reference to Rogers's theory in his anniversary address as retiring president of the Association of American Geologists and Naturalists in the spring of 1841. He discussed the dislocations and inversions of strata in the great mountain chains like the Alps and Andes, noting some similarities with the Appalachians but commenting that the causative agency was as yet unknown. Although he made passing reference to the fact that the Rogers brothers were at work on the problem of dynamics, his remarks made it clear to Rogers that his comments in casual conversation with his colleagues were having little impact.[6]

Rogers's concern about plagiarism was first awakened when Charles Lyell failed to give him credit for work he had done in the coalfields. When Rogers toured the anthracite fields with Lyell in the fall of 1841, he freely shared his work and ideas with him and felt that Lyell was sympathetic to his theoretical views. Lyell's observations and Rogers's comments were the basis of a letter Lyell sent to the Geological Society of London soon after the trip ended.[7] Rogers had understood that any information he shared with Lyell would be published under both their names, but it was not.[8] The letter discussed the structure of the anthracite coal region that they had visited together and made special reference to the confirmation found there of a suggestion made by William Logan that beds with the fossil plant *Stigmaria* were characteristically found immediately beneath the coal.[9] Logan, the future director of the Canadian Geological Survey, had himself toured the coalfields of Pennsylvania earlier that year but may not have noted this confirmation of his views. Rogers was aware of Logan's work and did not claim to have discovered the relationship between the *Stigmaria* and the coal, but he had pointed it out to Lyell and had wanted Lyell to give him some credit for this support of Logan's observations. When he did not get it, Rogers allowed himself to think that Lyell was taking advantage of him and might borrow more of his work without acknowledgment, perhaps even his theory of elevation. Almost immediately after Lyell's letter appeared, Rogers wrote to William, suggesting that they had better get a paper on the Appalachians ready for the 1842 meeting.[10]

Although at times Rogers seemed to have an almost paranoid fear that people would steal his work, he was not alone in his concern about Lyell. Many Americans shared his suspicion that Lyell was taking advantage of them. When it was reported that Lyell had arranged for publication of an American edition of his *Elements of Geology,* many geologists, including James Hall, concluded that if he did so, no American would be remembered for what had been done by Americans.[11]

Rogers's motivation to publish was further intensified by a growing general interest in mountain elevation, and he was anxious to see his theory placed in opposition to other theories. Shortly before the actual reading of the paper in 1842, Rogers expressed special concern about the popularity of the theories of William Hopkins, British mathematician, who approached the question of elevation on the grounds of mathematical, physical, and mechanical principles. Hopkins had read a paper in 1835 in which he tried to determine the effect of an elevating force acting from below on the surface of the earth. He concluded that the force was necessarily a vertical one affecting an elliptical area. According to his theory, such a force would first produce a series of longitudinal, parallel fractures and then a series of dislocations at right angles to them. He subsequently developed his theories in another paper on the Wealden District, an area in England noted for geological disturbance. Whether

Rogers was familiar with the 1835 paper or not, his attention was called to the last article by an address to the Geological Society of London by William Buckland in 1841. After reading Buckland's report, Rogers told William, that "our theory is I think so much more satisfactory that I am still more [intent] than I was that it should soon appear in print."[12]

The similarities between the 1837 and 1842 papers are remarkable, but there are some differences and a great many refinements that reflect the intervening years of work. Altogether the 1842 paper was a more mature statement and one that was a more clearly organized exposition of the theory and the structures that supported it than that which had been given in 1837. Two general considerations apparent in the 1837 paper were not included. First, the 1842 paper was concerned wholly with the process of elevation and not with the deposition that preceded it, which had been an important part of the 1837 paper. Second, in 1837, the theory was offered "for the purpose of throwing a strong light upon the structure of the country [rather] than of establishing a theory adequate to account for it," but the latter was, indeed, his main purpose in 1842, and so certain did Rogers feel about his theory that he suggested its application on a global scale.

Examination of the nature of the folds of rock after 1837 convinced Rogers that those with a southeastern dip were more typical than he had thought, but his view of gradation remained about the same, although somewhat sharpened. In the area of the initial rupturing of the earth's crust, broken and faulted folds predominated—the result, Rogers argued, of the violence of the event. Farther removed from the immediate center of disruption but still close to it, the folds were simply overturned but not broken. Still farther away, the force of the waves traveling across the surface of the molten interior created folds of rock that were neither broken nor overturned but displayed a gradation in the inclination of their axes (figure 7). Since the initial disturbance was in the southeast and the waves were pushed to the northwest, the types of the folds and the gradation within them represented the lessening of the force as it traveled to the northwest. For the same reason, the folds also became lower the farther they were from the source.

In 1837, Rogers had divided the Appalachian folds in Pennsylvania into five great groups. By 1842, the number of groups was increased to nine, and particular note was taken of the fact that the groups alternated between those with straight axes and those with curved axes. Rogers had noted the curving axes in the manuscript but had not elaborated on their meaning. He had also said that one group of waves influenced another, again without further elaboration. In 1842, however, he argued that as

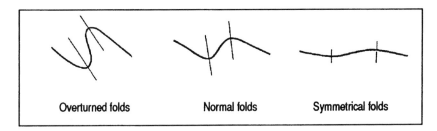

Overturned folds   Normal folds   Symmetrical folds

FIGURE 7. Schematic representation of folds. (From Henry Darwin Rogers, "On the Laws of Structure of the More Disturbed Zones of the Earth's Crust," *The Geology of Pennsylvania: A Government Survey* [Philadelphia: J. B. Lippincott, 1858], vol. 2, pp. 888–890.)

sets of waves traveled across the area, the folds with straight axes were produced first. As other waves followed, the already existing folds forced them to curve.

Rogers had talked about the lateral pressure of flexures already formed as sufficient to hold the folds in place in 1837, but in 1842, he replaced this idea with igneous intrusion as the supporting mechanism, an idea that he would maintain throughout his career.[13]

In the 1842 paper, Rogers offered an expanded statement on the nature of the force producing the waves by saying they were due to both vertical and horizontal motion. The waves that traveled across the molten surface represented vertical motion, but at the same time that the initial explosive event had set these waves in motion, it had pushed the entire landmass forward, giving the overall motion a horizontal component. He had first suggested this phenomenon in his *Fourth Annual Report,* and now he explained his reasoning by arguing that a merely vertical force "could only produce the same number of symmetrical anticlinal arches" and that a horizontal or tangential force could not "lead to any change in the position of the strata beyond an imperceptible bulging of the whole tract."[14] To understand fully the variations in the folds and their characteristic southeastern dip, both forces, he said, had to be taken into account.

A major addition to his theory in 1842 was the suggestion that the motion of the earth during earthquakes supported his theory. Some of Rogers's contemporaries, like Charles Darwin and his friend John Phillips, had already suggested a relationship between motion in a molten liquid beneath the crust and the motion of an earthquake.[15] Although there was strong opposition in the form of an argument that the motion felt during an earthquake was vibratory and resulted from a jar or fracture within the earth, Rogers concluded that the motion was, in fact, undulatory and that it therefore offered proof that waves regularly occurred in the molten interior of the earth.

Those attending the meeting of the AAGN not only heard William and Henry read their paper outlining the process of elevation but also were treated to a presentation by Henry on coal in which he discussed the origin of coal and how the theory outlined in the other paper accounted for coal formation. In this paper on coal, Rogers estimated the original area of the coal deposit at 900 miles long by 200 miles wide and reiterated his view as given in 1837 and his annual reports that although coalfields were now separated, they had once been joined as one continuous deposit.[16] He argued that the discontinuity of the present day coalfields had resulted when waters drained from the area after the great paroxysmal upheaval that lifted the Appalachian strata above the ocean. As the waters rushed away from the area, the coal deposits had been stripped away, and the coal was left primarily in synclines, or troughs, rather than in the higher anticlinal elevations. He again stated his opinion that the gradation in coals resulted from the dynamic processes that affected the area.[17]

There were many different theories about the formation of coal among Rogers's contemporaries.[18] It was clear that coal was derived from land plants, and yet it was interspersed with marine deposits. How was this to be explained? How had the plants arrived at the place of coal formation? Had they drifted there on currents in bays, or estuaries, or rivers? Had the coal been formed in situ? Why did fossil evidence indicate that the coal itself had been derived from fine, small plants, while rocks associated with the coal often contained large tree branches or even whole trunks?

To answer the questions, Rogers called attention to the regular nature of deposits associated with the coal and was one of the first persons to take into account certain specific formations both *above* and *below* the coal in explaining coal formation. One of the formations was the fireclay that appeared below the coal and was characterized by the presence of the plant *Stigmaria ficoides,* as Logan had noted. The other was a formation immediately above the coal, commonly called the "roof" by miners. Both, Rogers said, had a uniform relation to the coal that had been overlooked for the most part, although he pointed out that Edward Mammat had observed the relationship of the coal and fireclay in 1834 but had been ignored.[19] The fireclay was a relatively homogeneous deposit, suggesting to Rogers that it was deposited in quiet waters. The roof was another matter. Here there was a lack of homogeneity with layers of sandstone, grit, conglomerate, and slate commonly interspersed with each other. The slate was usually the first layer above the coal. While the plant fossils found associated with the slate were delicate, those in the other rock types were the larger fragments of stems and branches. This and the fact that these layers were composed of coarse sediments and sometimes bits of coal assured Rogers that their origin lay in a violent denudation or tearing away of the upper surface of the coal.

With these observations, he explained the origin of the coal as follows. The coal was formed in a series of marshy areas along the shore of a continent that stretched to the southeast of the present Appalachian area. *Stigmaria* grew abundantly in these marshes. Large numbers of a great variety of plants associated with the upper, or roof, formations, such as tree ferns, Coniferae, and Lycopodiaceae, grew inland. The decay of the *Stigmaria,* with the additions of other plants carried along the coastline by the natural movement of water, provided the material from which the coal itself was formed. This material accumulated over a period of time during which conditions were quiet. Rogers had determined during fieldwork for the survey that the paroxysmal movement that lifted the Appalachian Mountains above the water and caused the water to drain away from the area was only one of a long series of oscillations. Except for the one that followed his Formation III and lifted the White Mountains above the water, none before the end of the coal-forming period had been strong enough to lift the formations above the water. Periodic paroxysmal oscillations, however, offered a means of understanding some changes in the types of rock that characterized the area and, particularly in the case of coal, of explaining the consistency of formations above and below the coal. When these occasional paroxysmal disturbances occurred, water drained away from the marshes, and the plant fronds and twigs, along with the smaller plants, were spread over the surface of the marsh to form the slate. As the water rushed back into its basin, it abraded the land surface, knocking over forests and carrying the coarser materials that were normally found covering the slate. As the disturbances subsided, the water became tranquil again, and *Stigmaria* parts that were lighter in weight and had, therefore, remained suspended in the water, settled out to form the fireclay. The plants then regrew in the new marshes to start the process of coal formation over again. The presence of large tree trunks and branches in deposits associated with the coal but not in the coal itself had often puzzled geologists. According to Rogers, the fact that the *Stigmaria* grew in quiet conditions satisfactorily explained this absence from the coal, and the violent rush of the water after an oscillation accounted for the occasional presence of the larger pieces in the other formations.

It was obvious to Rogers that the deposition of the coal had occurred over a long period of time because of the thickness of the deposits, and it was, he thought, self-evident that the bottom of the marshy area in which the coal plants grew had steadily subsided to allow for such an extensive accumulation of material.[20] This same idea was found in Rogers's 1837 manuscript and in his later papers, but the 1842 paper on elevation failed to mention subsidence, since it was concerned primarily with the basic description of the area and how it was elevated rather than with how the sediment accumulated.

Rogers had one more opportunity to discuss his theory at the meeting, this time as part of a discussion on the drift. In the early 1840s, one of the critical questions that almost every geologist in the United States and in Europe addressed at some point was the reason for the large boulders and rocky debris strewn around the landscape, often with little relationship to the rocks of the immediate area. Indeed, it was often possible to show analogies between this rocky material and some far-removed place, indicating the source of the boulders and debris. In addition to these features, the rock in some areas often appeared to have been scoured, or deeply grooved and scratched, suggesting that a severe abrading process had occurred as the result of some natural phenomenon. While today these features are understood as the result of widespread glaciation, for centuries they were generally attributed to the action of water, thus the name drift, or diluvium, for many of the phenomena. In the 1830s, ice became a viable alternative to water. Lyell suggested that ice capping the north polar regions had been dispersed by volcanic activity and set in motion in the form of icebergs, and Henry de la Beche argued for an upheaval of the arctic ocean sending water and ice southward. In 1840, Louis Agassiz, a Swiss naturalist, proposed a different idea. Observing Alpine glaciers, he suggested that the features in question could be explained as the result of glaciers that spread over the land, carrying debris and boulders that were deposited on melting and that scoured and scratched the rock over which they passed while still embedded in the ice.[21]

Since most of the geologists attending the meeting of the association in 1842 worked in areas where glaciation had occurred, they were confronting the features almost daily and a discussion at the meeting not only recorded their concern but also reflected the fact that there was a controversy over the relative merits of Agassiz's theory versus Lyell's theory.

Most members at the meeting leaned toward Lyell's iceberg theory rather than Agassiz's glacial theory, although there were differences of opinion regarding the precise manner in which the icebergs traveled, or whether they were large enough to carry boulders. Some, like Hitchcock, whose studies in Massachusetts convinced him that the ice had played a role, thought it was more appropriate to talk of a glacioaqueous theory to reflect combined forces of water and ice.[22]

Lyell, who was at the meeting, discussed his concept of icebergs, and although Rogers was courteous enough to agree with him in principle, he noted that if the dispersal of the ice was "accompanied by an earthquake, rocking, or wave-like motion of the bed of the ocean," a mass of water would be thrown forward, like the rolling of a tremendous surf, carrying the ice with it.[23] His agreement with Lyell was perfunctory, for he would never accept ice as the explanation for these phenomena. During a later discussion at the same meeting, Rogers argued that scratches found at the bottom of valleys showed "all the local deflections which a body of moving waters would encounter," and that scratches and grooves were caused by "the friction of the overlying stratum of drift itself, urged into rapid motion from the north by one or more sudden inundations." The continuous breaking of the sea waves on land, "abrading and dispersing the fragmentary matter during repeated oscillations of the crust," also accounted, he said, for the widespread occurrence of coarse mechanical strata, such as the conglomerates, that was found in the Appalachians."[24]

Although drift and related phenomena are recognized today as fairly recent in geological terms, Rogers believed that the drift of the Appalachian area was deposited at the time of the great paroxysmal movement of the earth that lifted the Appalachians above the water after the coal was formed, and he argued that the climate was too warm at that time to support ice. He later acknowledged the possibility that drift in other areas was younger but argued that it had arisen in the same way as that found in the Appalachians except that the paroxysmal oscillations were of less magnitude in later times.[25]

There is no surviving record of the reaction to Rogers's comments on drift or to his paper on coal. The presentation of the paper on elevation was hailed as superb and eloquent, but the scientific reaction by the three or four dozen geologists who were present was generally negative.[26] Lyell was among those who objected. Only a few days prior to the paper's presentation, there had been a discussion about his belief in the gradual rise of continental masses. Rogers put his notion of a paroxysmal upheaval in opposition to Lyell's theory of a gradual rising of the continents and suggested that since fossils were not found uniformly distributed up and down mountain slopes, the land could not have been rising gradually through the ages. Lyell dismissed Rogers's theories categorically, and when he wrote about his visit to the United States a few years later, he summed up his view of them by saying that he could not "imagine any real connection between the great parallel undulations of the rocks and the real waves of a subjacent ocean of liquid matter."[27]

Comments surely came from those who already leaned to other theories, as Lyell did. James Hall and Timothy Conrad found uniformitarianism more appealing than the catastrophism embodied in Rogers's theory, and Conrad had already argued in another place that fossils showed no sign of the violent convulsions described by Rogers.[28] From other points of view, James Dwight Dana, who liked the theory of Elie de Beaumont, is likely to have made comments reflecting his preference. William Mather may have taken exception also, since he later argued that, although Rogers's theory might be correct in part, elevation could be fully understood only as the result of the gradual cooling of the earth *and* "the effects of gravitation and the rotation of the earth on its axis," an idea more closely allied to those of Elie de Beaumont.[29]

The most interesting thing precipitated by Rogers's paper was not a debate on the theory but a squabble that erupted between Rogers and Joseph Couthouy when Rogers accused Couthouy of plagiarizing his theory. At the AAGN meeting Couthouy read extracts from his personal journal, at the conclusion of which, he mentioned wavelike undulations of the crust. When Couthouy's remarks were subsequently published with the proceedings of the meeting that appeared in the *American Journal of Science,* Rogers referred to them as a "bold act of thievery."[30] Charles T. Jackson, secretary of the association, wrote to Rogers to apologize for what he called an error in publishing the remarks on the wavelike undulations, saying that he had understood Couthouy to be simply adding an observation to Rogers's paper.[31] He asked Rogers to clarify the situation by sending a note on the matter for the next volume of the association's proceedings or for the next issue of the *American Journal of Science.*

In the meantime, Couthouy and Dana had gotten into a quarrel over which one had been the first to note that temperature limited the distribution of corals, Dana claiming that he had told Couthouy about this phenomenon in 1840.[32] Couthouy rebutted Dana's claim in a paper to the AAGN in 1843. He took the opportunity to defend himself against Rogers's claim. He insisted that he did not know how mention of wavelike undulations had come to be included in the printed version of his comments and claimed that his comments referred only to recent volcanic action in Hawaii, which was "strikingly illustrative on a minor scale of those grand undulations of the earth's crust, so eloquently accounted for by Prof. H. D. Rogers." He continued:

I read nothing referring to undulations of this description by myself. My remarks were introduced at the request of Prof. R. on account of their bearing on his theory, as proving that effects similar to those described by him, were produced on a diminutive scale, by a less activity of the same agent, and also to prove the singular coincidence of expression between two observers of like phenomena.[33]

Whether Couthouy did or did not prematurely avail himself of Rogers's work will probably never be known with certainty, but it appears that the problem arose when casual comments made during the meeting were added to the official printed proceedings.

Almost immediately after Rogers had presented his paper to the AAGN, events compelled him to send his paper to the British Association for the Advancement of Science. On March 23, a month before the Association of American Geologists and Naturalists met, William Logan delivered a paper to the Geological Society of London entitled "On the Coal Fields of Pennsylvania and Nova Scotia." An abstract of the paper appeared in the *Athenaeum,* an English paper, in the issue of April 16, 1842.[34] Rogers first learned about the paper from this source after the meeting of the AAGN. Although Logan's paper dealt with the coal, there was enough in it to suggest familiarity with Rogers's work on the overall Appalachian structure. In his paper as read to the Geological Society, Logan cited his sources as Rogers, the Pennsylvania survey annual reports, and conversations with Lesley and McKinley (whose name he spelled as McKinnaly).[35] The abstract in the *Athenaeum,* however, which Logan wrote himself, failed to give specific credit. Rogers was mentioned, but the way the abstract was written made it sound as if the observations had been primarily Logan's own.

Lyell received a copy of the *Athenaeum* on May 18 while visiting Wheeling, West Virginia. Recognizing that Logan's report must be taken largely from Rogers's work, and, perhaps harboring a fear that Rogers might think he was responsible for giving information to Logan, Lyell immediately wrote to Rogers. He then tried to act as an arbitrator between Rogers and Logan, relaying messages in which Logan attempted to excuse his oversight in crediting Rogers in the abstract as the result of hurry. Nevertheless, Lyell shared Rogers's concern that his ideas were being circulated widely and without due credit. He urged Rogers to do something to give his theories greater distribution. Although Lyell believed Logan was wrong in his actions, he thought that it would be unwise for Rogers to attack Logan directly and suggested, instead, that

Rogers should either send a letter outlining his ideas to the Geological Society of London or send a communication to the British Association for the Advancement of Science, which was about to meet in Manchester.[36] Rogers chose to do the latter, sending an abstract of the AAGN paper to John Phillips in England to be read at the meeting.

Rogers's paper suggested that the theory was applicable to other mountain chains and that the Alps and the Jura Mountains of Europe manifested the same features that were seen in the Appalachians.[37] This idea guaranteed interest in the paper at the BAAS, and those in attendance listened carefully. Judging from an abstract of the meeting's proceedings that appeared in the *Athenaeum* for July, the reaction was much more spirited than it had been in the United States, and it reflected the many different views of mountain elevation that were being debated in Europe.[38] De la Beche argued that Elie de Beaumont's theory of tangential forces was sufficient to explain the phenomena and that Rogers had presented no evidence to suggest that another theory was necessary. Adam Sedgwick thought that the greater inclination of strata on the side farthest from the disturbing force, as Rogers had noted for many of the folds, was not in keeping with the origin of the folds as suggested and that the steepest side should be nearest the point of origin of the force. He also argued that the effects of the disturbing force in any given place were dependent on the kinds of rocks affected rather than on the degree of the force, some being more easily contorted than others. Although his comments seem not to have been as pointed as those of de la Beche, Sedgwick made it clear that he also thought Rogers had not given enough credit to the importance of tangential forces. In addition, he took exception to Rogers's notion of earthquakes, arguing that earthquake motion was vibratory and would not produce the kind of motion that Rogers said was necessary for elevation. The views of de la Beche and Sedgwick were typical of an overall negative reaction at Manchester, but support for the theory was not entirely lacking. John Phillips, ever diplomatic, wrote to Rogers:

I received your paper just before the Manchester meeting, at which my official occupations were excessive, and then commended it to the President of the Section, Mr. Murchison, Professor Sedgwick not having arrived. Against my expectation the paper was appointed for reading on the first day, when I was so entirely engrossed with my duties as not to be able to appear. Sedgwick, however, had arrived, and there was a lively and continued discussion. The opinions of the geologists present were apparently at variance with you, but I am very much of the opinion that, had your views been presented with large diagrams to show the mechanical reasoning, the result might have been different. There were several points in your argument which I approved and wished to advocate.[39]

Although he chose to say nothing about Rogers's theory at the meeting, Roderick Murchison emerged as his principal supporter in England. Murchison was president of the Geological Society of London, and in his "Anniversary Address" to the society, given at the beginning of 1843, which summarized the society's activities of 1842, he spoke favorably about the theory. Noting that America was mostly unfamiliar to Europeans, he shared Phillips's opinion that had they had illustrations or maps with which to follow Rogers's theories, the reaction would have been different. Having himself become better acquainted with it, he found it "a praise worthy effort." Furthermore, he thought the paper was perfectly in keeping with "the spirit of inductive philosophy," that it was backed by "a wide and successful field survey," and that it was "generously confided to the friendly feelings of the geologists of England."[40]

Rogers dismissed de la Beche's arguments as "feeble" and thought Sedgwick's objections that the steepest side of the flexure should be nearer, rather than farther from, the disturbances made no sense.[41] Sedgwick's comments on earthquakes, however, made him realize that it was important to try to substantiate the opinion that earthquakes were characterized by an undulatory, rather than a vibratory, motion. Consequently, he began to prepare a paper for the next meeting of the AAGN to be held in Albany in the spring of 1843. In early April, he told William that he had "been looking rather extensively into the subject of earthquakes, . . . and I think we may in a joint paper, as a sequel to our former one, greatly strengthen our positions."[42] A few weeks later he reiterated the importance of completing a paper on the subject, and at the meeting in May, Henry delivered a paper on earthquakes.[43] He cited a number of experiences and observations that he felt supported an undulatory motion in earthquakes, such as the rocking of houses and the opening and closing of great chasms. He also described an earthquake of January 4, 1843, that was felt from the East Coast inland at least as far west as Iowa. By comparing observations from various locations, he felt sure that he had proved the motion was like that of a wave crossing the area. As for the vibration that many thought was more typical of an earthquake than a rolling motion, Rogers argued that it was no more than the product of the undulation.

Shortly after delivering his paper on earthquakes to the AAGN in Albany, Rogers read part of the paper again before the members of the American Philosophical Society in response to a general invitation extended on the occasion of a special meeting celebrating the 100th anniversary of the society. According to an abstract of the paper, he dealt only with the theory of earthquakes in this paper and not with its application to his elevation theory, which he planned to give later in the meeting but never did. James reported to William that Henry gave a "truly masterful presentation" and "produced an impression on a room full of listeners that will not soon be forgotten."[44] While he gave an impressive presentation, Rogers won no converts to his position.

Rogers was elected president of the Association of American Geologists and Naturalists in 1843. When the end of his term came with the annual meeting of 1844, he, as was customary for the retiring president, delivered a paper reviewing the progress and current state of geology.[45] Rather than do a broad review, Rogers chose to concentrate on the work he and William had done on the Appalachians. He offered a brief summary of Appalachian structure, and having seen the importance attributed to illustrations as aids in understanding his ideas, he used a large map, measuring twelve feet by fourteen feet, to show the stratigraphy of the Appalachians. It was prepared for him by Russell Smith.[46] Smith, originally from Glasgow, had built a reputation in the 1830s as a theater scene painter and landscape artist. He was well known for his panoramic vistas, and a map of this size would certainly have been panoramic. In addition to summarizing his theory, Rogers expanded on his ideas about the drift and the support it offered for his theory.[47] Among other things, he argued that icebergs were impossible causative agents, since they were oceanic in origin and there was a lack of marine fossils in most of the areas covered by drift. There were, he noted, tongues of rock with marine fossils invading the overall area of the drift, but he argued that these represented places where the land had been depressed at some point, allowing ocean waters to flow into the valleys. He also attempted to dispel an argument that had apparently been leveled against him to the effect that water could not create a force strong enough to account for some of the features of the drift, especially the large boulders. Rogers was pleased to note in his address that William Hopkins had shown that a current moving at only twenty miles per hour could move a block weighing 320 tons and that such a current would result from a paroxysmal elevation of 100 to 200 feet. Upheaval generated by an undulating fluid was, in Rogers's view, more than adequate to produce the necessary force.[48]

Another part of his 1844 address to the AAGN concerned a nomenclature that he, with William's help, had developed for the Appalachian formations, one that he wanted to see adopted universally.

Names Make General Propositions Possible, 1843–1844

9

The classification and nomenclature of the Appalachian rocks that Rogers presented in 1844 reflected at least eight years of effort to isolate individual formations among the older deposits of the Appalachians, to group them into useful divisions, and to find names for the divisions and smaller units that would not lead to premature suggestions of relationship between the rocks of different areas.

Before Rogers began his work, very little was known about these rocks, particularly those that lay beneath the Carboniferous formations. They appeared to be without order and were often lumped together as Transition, or as Rogers preferred, Lower Secondary, rocks. During the first year of fieldwork in Pennsylvania (1836), Rogers and his assistants isolated twelve formations in the Appalachians. Rogers called the upper three (which included the coal) the Carboniferous system and the lower nine the Appalachian system.[1] He gave each formation a name that expressed some characteristic of the rock, like white fucoidal limestone. He began to refine this Appalachian stratigraphy in 1837 when he increased the twelve formations to thirteen.[2] Although his earlier names described the identifying characteristic of the rock, Rogers abandoned names altogether this year, replacing the names with numbers. As he had made clear in earlier arguments, he wanted a nomenclature that avoided the use of names that prematurely suggested relationship between areas.

Refining his understanding of the stratigraphy of Pennsylvania was a principal part of Rogers's work on the survey, and over the years that understanding underwent considerable change. As it did, Rogers became absorbed in the search for names to replace numbers, names that would adequately meet the needs of a successful nomenclature in describing groups of formations, the individual formations within the group, and the proper place of each group or single formation with regard to all others. By 1844, he was prepared to introduce his classification and nomenclature for the formations beginning with the coal and continuing to the bottom of his Lower Secondary. He proposed a division of the formations into nine parts. Each division was given a name signifying the passage of time, and this name was then incorporated in the name of each formation. The individual formation name tended to reflect the predominating mineral character, although Rogers did not rule out the possibility of a name based on paleontology. Figure 8 will be helpful in following the evolution of Rogers's classification and nomenclature from the thirteen numbered formations to this system.

When Rogers numbered his thirteen formations in 1837, he divided them into two very broad groups as he had done in 1836. The first eleven he called the Appalachian group and the other two (Formations XII and XIII) the Carboniferous group. Formation XIII was the coal itself, while Formation XII represented conditions immediately preceding its formation. Rapidly accumulating information from both William's and Henry's surveys soon began to call for further changes in these simple divisions. William seems to have been the one who took the first step beyond Henry's 1837 scheme when, early in 1839, he suggested that they isolate Formation XIII and make *three* groups of the formations below XIII.[3] In his scheme, William placed Formations I through VII in the earliest group, VIII and IX in the second, and X through XII in the third. Neither brother gave a reason for the division between groups 1 and 2, although the reason for the break may have been related to Henry's argument in his *Second Annual Report* on the geology of Pennsylvania that Formation VIII was composed of rocks and an abundance of fossils that clearly indicated a prolonged period of quiet deposition, giving it "marked peculiarities of . . . structure, colour, and composition."[4] The rationale for the third group was simply that the rocks of these formations (X–XII) "mark the gradual approach to the conditions necessary for the production of the great coal seams."[5] Thus each group represented particular physical conditions of deposition. The base of William's third group (Formation X) was maintained by Henry through nearly all of the following years as he realigned the formations into more and more complex divisions.

In joining the formations in divisions, the brothers were influenced by developments in England, where local formations below the coal were being grouped into larger units called the Cambrian, the Silurian, and the

William's first division April, 1839	Henry's division October 16, 1841	Henry's division October 29, 1941	Manuscript "System"	1844 paper	Manuscript "Textbook"	December 11, 1852 and final report	Rogers's numbers
Group III	Carboniferous	Hesperion	Hesperion	Seral	Seral	Seral	XIII
							XII
	Appalachian Evening	Post Meridian	Post Meridian	Vespertine	Vespertine	Umbral	XI
						Vespertine	X
Group II		Meridian		Ponent	Ponent	Ponent	IX
			Sandstone group	Post Medidial	Post Medidial	Vergent	
	Appalachian Afternoon					Cadent	VIII
				Medidial	Medidial	Post Meridian	
Group I			Meridian			Meridian	VII
			Older limestone group	Pre Medidial	Pre Medidial	Pre Meridian	VI
		Ante Meridian				Scalent	
				Levant		Surgent	V
	Appalachian Morning		Ante Meridian		Levant	Levant	
							IV
		Eoan	Eoan	Matinal	Matinal	Matinal	III
						Auroral	II
				Primal	Primal	Primal	I

FIGURE 8. The evolution of Rogers's nomenclature through 1844 and subsequent changes that were embodied in his final report on the geology of Pennsylvania. His thirteen numbered formations are shown on the right.

Devonian. These divisions, defined as global sequences, were based principally on paleontological markers, although the separation of the Cambrian from the Silurian relied heavily on structural evidence. Together, these divisions created a new framework in which to view the older rocks. The Cambrian and Silurian represented the old Transition or, in Rogers's definition, Lower Secondary rocks.

Murchison had identified the Silurian in 1833. At about the same time, he and Adam Sedgwick defined a still lower group, which Sedgwick named Cambrian. The Cambrian was not as clearly definable on the basis of fossils as was the Silurian, and Murchison and Sedgwick began to wonder if the two groups were separable. By the end of the decade they suggested that the two might be grouped under one heading. Later, however, Murchison abandoned this idea, convinced that the Cambrian had no identity distinct from the Silurian. Sedgwick became just as convinced that the Cambrian was distinct, and this difference of opinion led to more than a decade and a half of debate between Sedgwick and Murchison.[6] Murchison's definition of the Silurian, however, was widely accepted in Europe and in the United States.

The Devonian, which lay above the Silurian, was simultaneously the subject of intense study and debate.[7] By the end of the 1830s it had been recognized by Murchison, Sedgwick, Phillips, and others, as a system of rocks equal in universal importance to the Silurian. A formation called the Old Red Sandstone, considered by some to be Secondary and by others to be a distinct boundary between the Transition and Secondary, was made part of the Devonian sequence.

After reading Murchison's 1839 volume on the Silurian, William declared that he was "in raptures with the work," and noted that he recognized many Appalachian fossils in Murchison's illustrations and that there were other similarities with regard to inversion of the strata and the strike of slates in Formations I and II. "With what delight," he told Henry, "could I labour with you in comparing the results, so beautifully put forth in this work, with our own!"[8] He appears to have been ready to accept the Silurian as a designation for the older rocks of Pennsylvania, much as Conrad was doing for New York, and so was Henry, who noted in his final New Jersey report (1840) that the older formations of that state had closer affinities with the Silurian than with anything else.[9] When Henry Rogers was with Lyell in 1841, he was able to confirm with him the similarity of some of the Appalachian fossils with those of the Silurian of Great Britain, and when Lyell recommended a new work by John Phillips on the Devonian, Henry told his brother that with that and the Murchison volume "we shall have the whole subject before us."[10]

In mid-October 1841, while waiting for Phillips's book, Henry Rogers suggested a change in William's classification. Once again he grouped Formations XII and XIII together as the Carboniferous. He continued a triple division of the rocks below this but one that differed from William's. He suggested making rocks from Formation I to the top of V into one group, those from VI through IX into another, and those from X through XI into a third. With this suggestion, Rogers began an effort to bring his divisions into alignment with the Silurian as suggested by work in New York, particularly that done by Conrad. Conrad had, in 1838–1839, adopted the Cambrian and Silurian for New York but in 1840–1841 dropped the Cambrian designation in favor of the Silurian, dividing the Silurian into an upper, middle, and lower series.[11] Rogers made Formations I through V the equivalent of Conrad's Lower Silurian, while VI through IX were nearly the equivalent of his Middle and Upper Silurian together. This grouping differed from Conrad's only in that Rogers included Formation IX, whereas Conrad thought it was probably Devonian. Rogers assigned Formations X and XI to the Devonian but felt unsure as to whether Formation XI really belonged here or to the Carboniferous. It was a formation believed by many to be the equivalent of one in Europe, called the Mountain Limestone, that was usually considered part of the Carboniferous. Reasoning that it was a marine formation, while XII and XIII (the Carboniferous) were terrestrial in origin, Rogers decided to make it part of the Devonian.[12]

It was at this time that Rogers also began to think seriously about names for the Appalachian rocks. In New York, the geologists were naming formations in accordance with the local geographical area in which they were best represented. Rogers disliked the use of geographic names, because, when the time came that enough was known about formations in each locality to recognize true relationship between areas, such geographic names would not lend themselves to use in each area. He wanted names that would be exempt "from the difficulties of either numerical or geographical reasoning" and would prevent any implication of relationship between rocks in different areas.[13]

Rogers was not the only person at this time to be concerned with names. Names were critical to geologists because they expressed the ideas and philosophy behind the stratigraphic arrangement.[14] Although there were many sources of this concern, Rogers was supported and guided in his views by the work of William Whewell which, he had told William earlier, contained "many sound and broad views . . . in agreement with our own."[15] To Whewell, the whole process of giving things names to describe them was an inductive process, requiring the discovery of those characters that relate things on a permanent and real basis. "We must," he said, "impose our names, according to such marks." Only by doing so will it be possible to "arrive at that precise, certain and systematic knowledge, which we seek; that is, at science." Whewell went on to say that "the object, then of classificatory sciences is to obtain FIXED CHARACTERS of the kinds of things, and the criterion of the fitness of names is, that THEY MAKE GENERAL PROPOSITIONS POSSIBLE."[16] Whewell found fault with geological nomenclature because it was not based on fixed characters. Names based on local terms or local places were not good descriptive names, he argued, because the character on which the name was based might not be essential and could obscure natural marks of connection. The natural mark of connection must come first, before the classification and nomenclature.

Whewell worried that universal formations would be assumed without good evidence, and that names, if not carefully chosen, encouraged this tendency. He argued that strata must be studied and compared within their own country or area first so that their natural relationship with each other was understood before they were submitted to a broader, perhaps more universal, scheme of interpretation. Rogers had followed Whewell when, in his earliest statements on the subject of names, he had rejected the use of Lyell's names for the Tertiary formations and the name Cretaceous because they led people to believe that there was a relationship between areas that had not been proved.

When Rogers first used numbers to identify formations, he believed that the goals outlined in Whewell's philosophy would be well served. Ideally, he thought, the Appalachian rocks should be studied and placed in relative position to each other in their own context under his numerical designation. If geologists in other areas would do the same, as the relationships with those areas became clear, the confusion of a morass of different names referring to rocks of the same age and position would be avoided.[17] Nevertheless, it was always Rogers's hope that he would eventually be able to replace the numbers with names that would meet Whewell's criteria.

Since fossils could not be trusted as absolute guides in stratigraphy, they did not represent a "fixed character" to Rogers, but lithological characteristics, he believed, were "fixed." It was, therefore, important to him that nomenclature reflected these lithological characteristics whenever possible. It was also important that it reflected the chronological position of the formation, as his numbers did. His system of nomenclature evolved toward these dual ends.

Rogers's first step in developing a nomenclature was the designation of his three divisions below the Carboniferous as the Appalachian morning, Appalachian afternoon, and Appalachian evening, chronological designations each of which he hoped to identify with a suitable Greek word. Less than two weeks after suggesting the divisions and these names, Rogers offered another revision, increasing the three-part division of the rocks below the Carboniferous to a four-part division and maintaining the Carboniferous as it had been. He substantially changed the points at which the divisions were made, and each division was now given a Latin name to signify its place in the passage of time:[18]

Eoan	dawn	Formations I–II
Ante Meridian	morning	Formations III–VII
Meridian	noon	Formations VIII–IX
Post Meridian	afternoon	Formations X–XI
Hesperion	evening	Formations XII–XIII

The Carboniferous remained segregated as the Hesperion, while Formation XI (the equivalent of the Mountain Limestone and the affinities of which he had felt uncertain) along with Formation X remained together as the Post Meridian. He now went back to William's grouping of Formations VIII and IX, convinced that Formation IX and at least some part of VIII represented the Devonian, noting that the fossils of the upper part of VIII and IX were essential "as forming the groundwork with Conrad for his American Devonian."[19] Although Rogers was convinced that Formation IX should be included in the Devonian, just how much of Formation VIII belonged to the Devonian was a question that neither he and William nor apparently Conrad could settle with certainty. Had they but collected a better suite of fossils from the rocks above VIII the preceding summer they might, Henry mused, have been able to solve the problem. About three months later, when as a member of the publications committee of the Academy of Natural Sciences of Philadelphia, Rogers refereed a paper by Conrad, he pointed out to William that Conrad believed that the Devonian had begun about the middle of their Formation VIII, but Rogers was not ready to yield on this point. The question is, he said, "shall we take the bottom of VIII, or shall we divide VIII [for the bottom of the Devonian]?"[20] The Eoan and Ante Meridian represented the Lower and Middle Silurian of Conrad.

In an unpublished and incomplete document entitled "A System of Classification and Nomenclature of the Paleozoic Rocks of the United States," Rogers made further refinements in his divisions.[21] Internal evidence suggests that this manuscript was written in 1842 or 1843. Rogers returned to divisions more similar to the ones he had suggested first in

1841. He did, however, extend the Eoan to the top of Formation III, and although his reason for making this change is not stated, he had made frequent references to a major disturbance between Formations III and IV and the fact that Formation IV rested unconformably on III. The new Eoan equated with what he would later accept as the Cambrian System in the United States. The Ante Meridian extended upward to the top of Formation V only, the point at which the original Appalachian Morning ended. The Meridian began with Formation VI and ended at the top of Formation IX, before the advent of the conditions that led to the coal formation. He did distinguish two parts of the Meridian, maintaining the division at, or near, the boundary of Formations VII and VIII. The upper division was a sandstone group with a limestone group below, coinciding with Formations VI and VII. This may have been a concession to lithology that allowed him to keep some limestones above and below the boundary between VII and VIII in the same group.

His criteria for nomenclature were developing rapidly at this stage from the general to the specific, and he began to associate his division names with the specific formations within each through a binomial nomenclature introduced in the "System of Classification":

First it [the nomenclature] should comprise in a symmetrical form of terms all the wider as well as the more restricted groups of strata, presenting in due subordination the great systems of rocks and the several subdivisions into which their individual members group themselves. Secondly, it should possess such pliancy as to admit of expressing by some simple adjunct all the modifications of type exhibited by the minor subdivisions in different or distant regions. Thirdly, the primary idea suggested by the names of the great divisions should be that of their order in time. This in the subordinate divisions is connected with characteristic mineral or paleontological features. [22]

Each division name, like Eoan, became the basic part of a stratigraphic name. Thus the Eoan was composed of the Eoan Conglomerate, the Eoan Sandstone Slate, the Eoan Vitreous Sandstone, and the Eoan Ferriferous Slate. Together these were members of the Eoan Sandstone group. Following this group came the Eoan Limestone group composed of the Eoan Magnesian Limestone and Eoan Fossiliferous Limestone.

Rogers wanted to present the nomenclature formally in a paper to the Association of American Geologists and Naturalists in 1843, and the unpublished "System of Classification" may have been prepared toward this end. He hoped that, after judging the reaction to the nomenclature, he would be able to send it to the British Association for the Advancement of Science for reading. [23] The paper was not presented, possibly because William was unable to be there, but more likely because Henry was anxious to present his earthquake paper that year and thus decided to forgo the nomenclature. It was a year later at the meeting of the AAGN that his ideas were first heard as part of his presidential address. By then, he had made some striking changes.[24]

Although still denoting time, Rogers changed the names, realigned formations, and increased the number of divisions to nine. His scheme was as follows:

Primal	dawn	Formation I
Matinal	morning	Formations II, III
Levant	sunrise	Formations IV, V, part of VI
Premedidial	forenoon	part of Formations VI, VII
Medidial	afternoon	part of Formation VIII
Postmedidial	sunset	part of Formation VIII and just beyond the boundary of Formation IX
Ponent	evening	Formations IX, X
Vespertine	twilight	Formation XI
Seral	dark	Formations XII, XIII

He incorporated his binomial system in his address, noting, much as he had in the earlier "System of Classification," that each name was

composed first, of the name of the period to which it appertains, and secondly, of a word or words descriptive of the *ruling* mineral *character* of the rock; and to these is appended, when we wish to specify the type under which the formation is referred to, the name of the district or place where it is so developed. . . . The well characterized formation called in the New York survey the Marcellus shales, is named by us the *Postmedidial* older black slate. . . . and a member of the Clinton group of New York . . . we propose to call the *Levant iron sandstone*.[25]

Although at the outset, Rogers had been concerned primarily with making his groups match the Silurian and Devonian of other geologists, he had found these divisions too large to have significance. While they would continue to offer a broad general framework for him, his groups, which he called "series," offered greater refinement and a more immediately meaningful structure in which to view the formations in their totality. His series were intermediary between the broader divisions and local formations. Although New York and other states used the group or series concept, Rogers was certain that his series, based primarily on lithology, were more accurate and that his names would have universal applicability because they reflected time as well as the principal characteristic of the rock rather than geographical location.

Rogers thought that the nature of his names and the binomial structure would make relationships between areas absolutely clear, no matter how complex those relationships might be. His determination of the place of a blue limestone found at Cincinnati, Ohio, provides a good example of how he thought this system could work. In 1841, he and William had undertaken an extensive field trip through parts of Ohio, New York, Michigan, and Canada to "set in a clear light some essential points in Lake Erie geology" on the basis of lithology rather than paleontology. As the result of these studies, Rogers had decided that the blue limestone was the equivalent of the New York Niagara limestone. He subsequently changed his mind, first thinking that the blue limestone was more properly equated with the Hudson River group of formations in New York and still later that it was the equivalent of both the Trenton Limestone and shales in the Hudson River group. He was probably influenced in this change by the fact that the New York geologists believed that the blue limestone was the equivalent of the Trenton. In any event, his binomial system provided a means of expressing this dual relationship that would not mislead anyone. He called the Trenton Limestone the Matinal Newer Limestone and the Hudson River shales the Matinal Newer Shales. He named the Cincinnati blue limestone the Matinal Newer Limestone and Shale.[26]

Like his theory of elevation, the nomenclature met a less than enthusiastic reception. The New York survey had been completed and the final reports published in impressive quarto volumes in 1842 and 1843. The geographic names assigned by the New York geologists established criteria for nomenclature. Rogers was unable to convince other geologists that his names were more adequate than ones with which they had become familiar. Although Rogers continued to use his system, altering it once again in the early 1850s, it became nearly instantly a relic.

A Mind and a Heart with Scope to Unfold, 1843–1845   10

The years immediately following the close of the survey's fieldwork represented a transitional period for Rogers when events in his professional and personal life led him to forsake Philadelphia and to seek a new beginning in Boston. The constant problems he had experienced with Pennsylvania legislators in the last years of the survey had tried his patience. The enormity of the task he faced in preparing a final report was overwhelming, although he continued to work on it steadily to the point of near completion in 1845. The lack of interest in his scientific ideas was discouraging, and Rogers experienced a growing isolation from the scientific community, a situation largely imposed on him by several members of the Franklin Institute who had once been his friends. On top of these troubles, he suffered a severe financial loss when his iron furnace failed.

Trouble had brewed for a long time between Rogers and Bache's circle at the Franklin Institute. It began with Frazer's departure from the survey and continued with the argument over the Delaware Breakwater and Rogers's disagreement with Bache over the institute high school.[1] By 1842, Rogers had few, if any, friends left at the institute, and he gradually withdrew from activities there. The final closure on his relationship with this scientific body was signaled when his name was dropped from the Committee on Geology and Mineralogy, on which he had served for many years.[2]

His change in standing among his onetime friends at the institute was apparent when Trego published *A Geography of Pennsylvania* in 1842 that drew, in part, on Rogers's work. Trego, who had done so much to help the survey win legislative approval, had been an assistant on the survey until 1841, when he returned to the legislature. He was a close friend of Frazer and of Bache, and his attitude toward Rogers reflected the difficulties that these men had with Rogers. In addition, he was angry with Rogers because, after spending a substantial portion of his early legislative career promoting the survey, he now helplessly watched as Rogers failed to produce a final report. He questioned Rogers's intentions with regard to the final report and doubted that it would ever be written. Therefore, when Trego published his *Geography,* he included a section on geology, underscoring the point that its inclusion was intended to make up for the lack of any other published source on the subject. Although his chapter on geology included a brief discussion of Rogers's thirteen formations and his numbers, he deliberately omitted any mention of Rogers's name, and in the same vein, he failed to call attention to Rogers and the work of the survey throughout his book, although some of the information included was based on the survey.[3] In the years to come, Trego, as a member of the legislature, would do all that he could to prevent the state from making it possible for Rogers to finish the survey report.

There was an ever widening gulf between Rogers's views on science in America and those held by Bache and his friends, particularly on the issue of a national scientific organization that embodied all the sciences. If there was any doubt in Rogers's mind about just how great the disparity in their views was or how seriously it was taken by the men from the institute, it was dispelled at the 1843 meeting of the American Philosophical Society, the same meeting at which he read his paper on earthquakes, when Rogers became involved in a controversy over the National Institute for the Promotion of Science.

The National Institute was founded in 1840. A debate over its purpose ensued, with some arguing that it should be a national museum and others equally vociferous in proclaiming that its purpose should be to provide a forum for scientific papers.[4] With regard to the latter, it was seen by many as the potential equivalent of the BAAS, which was the kind of organization Rogers had advocated ever since returning from England in 1833. A few months before the meeting at the American Philosophical Society, the National Institute had invited members of the society and of the AAGN to join in its 1844 meeting, to be held at Washington at the same time that the AAGN was scheduled to meet there. Publicly, both organizations expressed opposition to a joint meeting because the National Institute advocated a link to the federal government through which financial help for its activities would be available.

There was strong feeling that such an arrangement would foster government control in science. Privately objections were focused more directly on the concern that American science was not mature enough to support a broad national organization with the same level of participation that characterized the BAAS. Rogers, however, argued that such concerns were unjustified and that the National Institute held great promise for science in the United States.

Concern about the National Institute ran especially high among the members of the Franklin Institute, and many of them were, or had just become, members of the American Philosophical Society, including Frazer, Booth, Trego, and Franklin Peale as well as Patterson, who was vice president, and Bache, who was secretary.[5] James Rogers, who attended the meeting at the Philosophical Society, told William that it "was manifest throughout the session that it was the design to keep him [Henry] if possible in the background" while bringing others forward "who could more easily be used for party purposes."[6] The effort to keep Rogers in the background reflected the hesitation of Bache and the others on the issue of a national organization, but if the effort to keep Rogers quiet was planned, as James implied it was, then it also reflected the depth of Rogers's alienation.

As the gulf between Rogers and the others grew, his onetime friends at the institute spread the word that Rogers's actions were self-serving. A case in point concerned Rogers's failure to help James Hall with some maps. In 1843, Hall was working on two maps for his report on the survey of New York and another map of the United States. Although Rogers provided information on northern Pennsylvania and New Jersey for a map of New York to accompany Hall's survey report, he failed to provide information on Pennsylvania that Hall had requested for the other map to accompany the report or for the map of the United States.[7] Why he failed to send the information for the survey map is not entirely clear, although he may have felt either a lingering hesitation to provide such information before the official Pennsylvania report was completed or a fear that his work might be used inaccurately or without proper credit.[8] Rogers had a more specific reason for refusing to provide information for the map of the United States. He and William were working on a similar map. When Hall learned this, he wondered whether he should stop his work on the United States map and wrote to John Frazer for advice. Frazer, not content to acknowledge Rogers's position, suggested that Rogers's refusal was no more than a self-serving act "made merely to stop you, and without any *immediate* intention of carrying into effect its purpose" [emphasis his].[9]

The positions and attitudes of the institute members isolated Rogers from the scientific community of which he had been a part for more than ten years. Unfortunately, his personal life offered little comfort, because he suffered a critical financial setback caused by the failure of his furnace. The first two years of the operation had gone along without much trouble, but the economic condition of the country worsened between 1840 and 1842, and the iron industry faltered.[10] Rogers's furnace produced pig iron, much of which was sent to rolling mills in Pittsburgh. The economy forced the rolling mills that purchased Rogers's pig iron to close. Without a steady income to balance expenses, the furnace was heavily in debt by the spring of 1842, and several creditors began to demand payment. The principal creditor was A. Collwell (sometimes spelled Colwell), from whom McKinney, the manager of the furnace, purchased most of the supplies for the furnace and its workers. McKinney described Collwell as a "skinflint" and "the terror of the country people," and he pleaded with Rogers to stay on good terms with Collwell "for as long as people see that he is satisfied, knowing that we are indebted to him, they think well of our concern."[11] The furnace had been financed from the start by Nicholas Biddle, either directly or through Biddle's borrowing power. Rogers was forced to borrow nearly $1,800 from Biddle in May 1842 to forestall trouble from Collwell and other creditors, but even so, notes were coming due that could not be paid. Lawsuits were threatened.[12]

Roswell Colt, Biddle's agent, visited the furnace in late May or early June and remained optimistic in spite of the financial trouble. Although he arranged for the sale of thirty tons of metal at a reasonable $22 per ton, the income was used as a repayment of money that Biddle had invested in Rogers's furnace and therefore failed to help the daily operation of the furnace. Nevertheless, Colt thought the furnace capable of developing sufficient profit to pay all the debts incurred for it by Biddle no later than the fall, with an $8,000 per year profit after that time.[13] Colt's optimism was misplaced. Two hundred tons of unsold metal were still on hand. Worse for the continued operation of the furnace was the fact that the company store was low on stock and McKinney was having difficulty finding anyone willing to restock the store on credit. Since the workers at the furnace were usually paid in goods from the store rather than money, a well-stocked store was essential. The rolling mills were still closed, and if these were not sufficient troubles, the lower part of the furnace wall fell in on the hearth, necessitating a shutdown. There was not enough money to repair it immediately, and as creditors learned that the furnace was closed, their demands for payment increased.[14] At about this time, ownership of the furnace shifted from Rogers alone to Rogers and Colt, probably in an effort to increase borrowing power. By September, McKinney was desperate, and Biddle managed to get him $500, money, he told Colt, "drug dollar by dollar from the bones." Biddle maintained a slight sense of humor over the situation and told Colt that: "If by accident you should find a pocketbook with 2 or 3 thousand dollars going down the falls stop it—or if by any other . . . means you could raise even a thousand dollars, I wish you would." His real feelings were more aptly summed up when he told him that "few things have annoyed me more than this engagement."[15]

McKinney became suspicious that Collwell and others were purposely making it difficult to raise money for the furnace because they wanted the furnace, but in late 1842 when Collwell demanded payment of $1,000 due him, and Biddle was able to give him a draft for only $500, Collwell accepted it without argument.[16] By mid-December, McKinney had the furnace back in full operation with a small foundry, probably with a loan from Jacob Christopher Painter who was already active in the iron trade in the area and who would soon emerge as the furnace's major creditor. In spite of the reprieve, efforts made in the spring of 1843 to meet payments and restock the store failed, and the workers refused to work, forcing the furnace to close completely before year's end.[17] At this point, Colt urged Rogers to sell the property and furnace to Collwell or to Painter. Rogers rarely made visits to the furnace, but he went in November of 1843 to see the situation for himself before making a decision on the sale. He found things as bad as he had been told, and when he totaled all the outstanding debts, he found the furnace in debt for about $13,000, which was owed principally to Painter. In addition, Biddle had borrowed and invested money on which he had not seen a return. This amounted to another $10,000. Exactly what happened next is unclear, but it appears that the courts intervened, for the furnace was scheduled for sale at a sheriff's auction the following spring, and Rogers reported to Colt that Painter and Company "have for the purpose of keeping the Furnace in order until the day of sale agreed to rent it for [$]700–800." The furnace was purchased by Painter, Peter Graff, and Reuben Baughman for $7,200.[18]

The transaction seems to have satisfied the creditors but not Colt. Rogers considered the money raised by Biddle and Colt as an investment in a partnership and therefore assumed that Biddle would absorb the loss on the money he had invested. Colt did not understand the arrangement this way and threatened to hold Rogers personally responsible for all the money invested by Biddle. Since Rogers was sure Biddle saw the arrangement as a partnership, he thought this an unlikely event, writing to

Colt that "if you *could* substantiate a claim against me and against the firm, certainly Mr. Biddle in *honor* could not permit it."[19] Biddle died in 1844, however, and whether he might have protected Rogers or not, Colt pressed the issue, and Rogers spent the next ten years paying off the several thousand dollars that had come from Biddle and Colt.

Rogers was able to escape from his mounting concerns when he accepted an invitation to give a series of lectures in Boston late in 1843. It was the beginning of a long association with Boston where he hoped to find the comfortable stability and place in the community of scientists that had escaped from him in Philadelphia. Boston, like Philadelphia, was a center of scientific and cultural activity in the United States, and many things about the city appealed to Rogers. Harvard was there and was already a distinguished center of scientific education. The American Academy of Arts and Sciences and the Boston Society of Natural History provided scientific forums not unlike those available in Philadelphia.

The invitation to lecture in Boston came from George B. Emerson, president of the Boston Society of Natural History, who, like Rogers, was interested in the practical application of science in daily life and whom he had met at meetings of the AAGN.[20] William was optimistic about his brother's success in Boston, advising him to "entertain the fullest confidence in your entire success, and suffer no misgivings to damp for a moment that animating sense of power which is your right to feel."[21] The coming lectures were announced in the *Boston Daily Advertiser*. Rogers intended to cover the elementary principles of geology and to give more detail about areas in the vicinity of Boston that his listeners could visit to make their own geological observations. He also planned to show specimens and to illustrate his lectures with "an extensive set of geological drawings, prepared by a skillful artist, and believed to be much the best ever exhibited in this country."[22]

The lectures were to begin on Friday, December 1, and continue on the following Tuesdays and Fridays, but the first was postponed until December 5 because the illustrations, being done by Russell Smith, were not ready or had not reached Boston. Partly as an apology for the postponement, and partly to encourage attendance, Emerson sent the *Daily Advertiser* a superb testimonial about Rogers. His letter, which appeared in the November 29 edition nearly on the eve of the original starting date for the series, cited Rogers's magnificence as a speaker before the AAGN and commended his broad learning. Emerson tried to convey the excitement of geology and the interest that it would hold for young and old alike once they had taken advantage of the opportunity to learn from a man like Rogers. He pointed out that no other person was expected to lecture on geology to the public that winter in Boston.

Although attendance at the lectures was small, they were well received by what a local newspaper described as "many of the most cultivated minds in this city."[23] Shortly after he had completed the lectures, Rogers returned to Philadelphia to give his course at the University of Pennsylvania and to prepare his address as retiring president of the AAGN. James was moved to note that Henry's return to teach his course would "gratify his friends and show his enemies that he is above their petty malice."[24]

In the short time he was in Boston, Rogers had made many friends, and not long after his return to Philadelphia, he received an invitation to return to Boston to give a series of lectures on geology at the Lowell Institute.[25] The Lowell Lectures had started in 1840, established by the will of John Lowell, Jr., as a means of providing educational opportunities for the citizens. These popular lectures were intended to demonstrate the ways in which science had practical application and were recognized as among the most important in science in the United States. The lecturers, paid handsomely, received as much as $1,200 for twelve lectures. Rogers was an obvious choice. His belief in the importance of practical science was attractive to the officials of the institute, his credentials as the director of the state geological surveys attested to his importance, and his commanding style as a speaker assured that the audience would be satisfied.

Rogers delivered his Lowell lectures early in 1845 to a friendly and receptive audience.[26] Joseph Lovering wrote to William:

You have probably heard before this of your brother's success in Boston. He has only left a few days since, much to our regret, and I hope not without some long and lingering looks on his part, back upon the city where he has made so many sincere friends, and where he is so much loved. I heard half of his lectures, and should have been glad not to have missed any. He found a docile and attentive audience, and a large one, too; and it is not strange that his winning style of lecturing, his calm eloquence, his chaste and beautiful language, and his comprehensive views of his vast subject, should have riveted the attention that was freely offered to him, and stormed hearts that were by no means closely sealed against him.[27]

The detailed contents of his Lowell series are not preserved, but the topics may have been the same as those which he planned to give in Portsmouth, New Hampshire, immediately after leaving Boston. These lectures were to cover the nature and cause of earthquakes and their effect (prefaced with remarks on "the nature of strata, and the mode of investigating geological appearances"), fossils, the classification of rocks, geologic time, former conditions of the world, coal (its origin, nature, and meaning), and, finally, the various theories of drift and the nature of grooves on boulders, including the iceberg theory and his own theory. In short, Rogers presented the fundamentals of the science of geology clothed in his own theories. The first Portsmouth Lecture, given in mid-January at the People's Lyceum, a public hall, was to be followed by three more at the "Temple" during the next week and a half, given for a "private class."[28] The Temple was used by the Washingtonian Temperance Society as a lecture site from 1844 to 1876. This, apparently, was what Rogers meant by "private class." Although the public lecture drew about 1,000 people, the three private lectures were never completed because they were not popular. A local newspaper commented, "Had a sufficient opportunity [been] offered for making our citizens acquainted with the merits of the lecturer, the value of his illustration, and the importance of the science, the Temple would have been crowded, instead of being compelled prematurely to close its doors."[29]

As Lovering's comments suggest, Rogers found an amiable and helpful group in Boston that included Emerson; John Amory Lowell, a Harvard trustee and administrator of the Lowell Institute; Amos Binney, a zoologist and the founder (in 1830) of the Boston Society of Natural History; Charles Sumner, a lawyer and later a United States senator; Benjamin Peirce, a mathematician and Harvard professor; George Ticknor, an educator, author, and former Harvard professor; and George Hillard, a lawyer and author. Although his friends included members of the scientific community, they were predominantly literary and political, many associated with Harvard and most of them known to each other. Furthermore, these friends were nearly all involved in antislavery and abolitionist activities. Rogers was never as overtly involved in these activities in the United States as Sumner, Hillard, Ticknor, and the others, but he was as strong in his beliefs that slavery was a cruel institution that denied the basic human rights. All in all, Rogers's friends in Boston were an influential group in the city and at Harvard.

In spite of the cancellation of the Temple lectures, Rogers's overall reception in the Boston area and his newly developing friendships renewed an enthusiasm for his work that he had lost in Philadelphia, and when he returned there, he lamented to Robert, "Things here look dismal enough to me after Boston. There my mind and my heart had scope to unfold in; here, like a frightened coral, I draw myself within my stony shell."[30]

In his "stony shell," Rogers immersed himself in work and spent the spring in Philadelphia preparing a paper on slaty cleavage for the 1845 meeting of the AAGN, held in New Haven.[31] Cleavage is the tendency in some rocks like slate to break along a precise surface as the result of the way in which the crystals that make up the rock are aligned. It was a subject that had interested Rogers for some time. He had commented on it for the first time in his *Second Annual Report* when he discussed the economically important slates of eastern Pennsylvania, pointing out that the dip of the cleavage planes, like the axis planes of the folds, was invariably to the southeast. He returned to the question of slaty cleavage in his final report on New Jersey to say that it was an "interesting problem" the cause of which he hoped to connect to "views concerning the elevation of our primary chain." He did not return to this issue in his 1842 paper, although he told William that they should include it as part of their overall work on structure.[32]

Cleavage, confused with original bedding until Sedgwick clarified the difference in 1835, was a subject of great interest to geologists, particularly with regard to the mechanism of its origin. Because it is a phenomenon found predominantly in mountainous areas, elevation and cleavage had long been associated. Whether heat or pressure was the most important cause of cleavage, however, was the subject of debate. By the time Rogers wrote his paper, the majority of geologists had decided in favor of pressure as the causative factor.[33] Rogers, following a suggestion by Sedgwick that cleavage resulted from a chemical (rather than a mechanical) rearrangement of the crystals, had no doubt that cleavage was caused by heat.[34] After noting in his paper that cleavage was found primarily in the most disturbed areas of the Appalachians and that the dip of the cleavage plane was invariably to the southeast and "nearly parallel in direction and steepness to the *anticlinal and synclinal planes,*" he argued that the penetration of the severely broken rock by hot steam and gas from the molten interior had resulted in layers of alternating heat and cold in the rocks. This distribution of heat was, he said, "precisely that which ought to impart through the new polarities it would awaken in the mass, a corresponding symmetry and parallelism in the planes of maximum and minimum cohesion or in other words the planes of cleavage."[35] Once again on the less popular side of an argument, Rogers's comments received little attention and had no discernible effect on the majority view.

Rogers had been contemplating a trip to Europe after the meeting of the AAGN in 1845 but decided instead to spend the summer traveling with William to study the White Mountains of New England, later going on to Lakes Champlain and Superior to report on mineral deposits as a paid consultant.[36] It was on this trip that Rogers decided that the White Mountains (as mentioned in Chapter 5) were metamorphosed sedimentary rock rather than primary rock as other geologists believed. Because the strike, or orientation, of these mountains was different from that of surrounding ranges, Rogers concluded that they had been elevated at an earlier period than the main body of the Appalachians, specifically, during the disturbance that occurred after the close of his Formation III.[37] Evidence for an extensive disturbance at this time existed elsewhere in a widespread unconformity, but the White Mountains were the only ones that he thought were lifted above the water at that time.

Rogers's trips to Lakes Champlain and Superior were probably made for the Franklin Copper Company, although he named no particular person as his employer this time. Rogers did several jobs for the company in the next two years, helping to locate copper in Michigan and acting as an intermediary for the company with agents in Michigan who were acquiring public lands for the company whenever such lands were made available by the government.[38] This was a period of copper mania in the area, and companies were vying with each other for potentially profitable lands. Geologists were much sought after for their expertise in locating areas likely to be the most rewarding and profitable. Rogers was also employed during the summer by Horace Gray, merchant in Boston, to survey a district near the Hudson River with a view to the practicality of starting iron furnaces there; and he was hired by Charles Jackson, to provide information on iron mines in New Jersey.[39]

Rogers's thoughts were never far from Boston, and before the year ended, he returned. He planned to stay for a few months with George Hillard, who had become a close friend, while looking into the possibilities for permanent residency.[40] There was a growing sense in the United States of the mid-1840s about the importance of the sciences for practical pursuits, and schools were looking for new ways to extend educational opportunities in the so-called useful arts.[41] Rogers hoped that Boston might provide the proper environment for the school he and William had often discussed since their days at the Maryland Institute. If not, he at least hoped to find a teaching position there. It would not be long before Rogers considered Boston his home and was deep in his quest for a new career.

Faithful Labours Cruelly Repaid, 1846–1848

The thought that Boston might provide a site for the school of practical science that he and William hoped to open someday was uppermost in Rogers's mind when he went to Boston. Consequently, one of the first things he did after returning was to approach John Amory Lowell, head of the Lowell Institute, about the possibility of attaching such a school to the institute, and he asked William to put their ideas on paper for Lowell.[1] William responded with a long plan for a polytechnic school, a school that would "embrace full courses of instruction in all the principles of physical truth having direct relation to the art of constructing machinery, the application of motive power, manufactures, mechanical and chemical, the art of engraving with electrotype and photography, mineral exploration and mining, chemical analysis, engineering, locomotion and agriculture."[2] Although Lowell considered the plan, nothing came of Rogers's hopes for a school at the Lowell Institute.

Nevertheless, Rogers did not consider this a particular setback, for he felt that the atmosphere and time were right in Boston for him and for William to start such a school themselves, independent of any other organization. William, just as enamored of Boston as Henry, later wrote that "ever since I have known something of the knowledge-seeking spirit, and the intellectual capabilities of the community in and around Boston, I have felt persuaded that of all places in the world it was the one most certain to derive the highest benefits from a Polytechnic Institution."[3] In spite of the brothers' enthusiasm for Boston and for their own school, William was not ready to leave the University of Virginia, and Rogers sought an academic post for himself.

139

When Rogers arrived in Boston, Harvard's Rumford Professorship of the Application of Science to the Useful Arts was vacant. This professorship had been established in 1844 by a bequest from Benjamin Thompson, Count Rumford. Lectures by the Rumford Professor were entirely separated from those offered as part of natural philosophy and aimed at providing the kind of information needed by engineers, surveyors, metallurgists, and manufacturers.[4] Jacob Bigelow, the first Rumford Professor, equated such a practical education with democratic egalitarianism and actively pursued efforts to make the kind of education embodied in his professorship a greater part of the Harvard curriculum. Bigelow left the job in 1827, and a new appointment was not made until 1834, when Daniel Treadwell became the Rumford Professor. Treadwell, an inventor, continued the tradition established by Bigelow, but whereas Bigelow was active in efforts to expand practical education, Treadwell was passive.

New and active interest in developing the teaching of the practical sciences at Harvard came from longtime faculty member and mathematician Benjamin Peirce, who in 1843 recommended a consolidation of all the endowed chairs in science into one school to train engineers and scientists. The Rumford Professor would head the new school. Treadwell was about to resign because of poor health, and the Rumford Professorship became the focus of Rogers's search for a teaching position in Boston, and the only opportunity for him to have a direct hand in a school of applied sciences.

Rogers enjoyed the strong support of his new Boston friends as well as many old friends who wrote to recommend him for the position. They included Hillard, Emerson, Binney, Sumner, probably Lowell, Henry Vethake, Matthew Fontaine Maury (head of the National Observatory), Edward Hitchcock, Samuel George Morton, Franklin Bache (Alexander's first cousin), Daniel B. Smith (principal at Haverford School), McIlvaine, C. C. Biddle, J. W. Bailey (professor of natural philosophy and chemistry at West Point), and Joseph Henry.[5] Together, they were an influential group, and in the early stages of the search for the Rumford Professor, Rogers was considered a leading candidate by the college. When the appointment was finally made, however, the position went to Eben N. Horsford, a young chemist recently returned from study with the noted chemist Justus Liebig in Germany. The story of Rogers's failure to get the position at Harvard is one of innuendo, fundamental differences on scientific questions, and hostility generated on the part of one individual who personally preferred another candidate.

Soon after Rogers made his interest in the Rumford Professorship known, rumors began to circulate in Boston that he was an unscrupulous, self-serving, and insincere opportunist. The nature of the rumors is clear in a letter written by Boston anatomist Jeffries Wyman to his brother Morrell, who was himself considering a bid for the position. After expressing complete surprise that Rogers was pursuing the post, Wyman attacked both Henry and his brother James in the harshest terms, claiming that "the two Rogers go hand in hand in everything & are universally regarded by scientific men as 'snakes' (it is no expression of mine) who go about prowling among the nooks & corners of this land, ever on the look out for a warm spot in which they may locate themselves."[6]

Why Wyman thought that James and Henry spent their time looking for jobs is not clear. James had experienced considerable difficulty in finding a place where he was comfortable, but since 1840, he had found security as a lecturer at the Philadelphia Medical Institute and at the Franklin Institute. Henry had been able to support himself reasonably well through his lectures, through teaching at the University of Pennsylvania, and through his growing consulting activity. Although he wanted to establish himself in the academic field as a teacher, there is no evidence to suggest that he was as consumed in his effort as Wyman's comments would lead one to believe.

Admitting that he knew of nothing "decidedly outrageous" about Rogers and acknowledging that he was entitled to respect as a geologist, Wyman went on to say that Rogers's "unpopularity results from his plausibility of manner & his utter insincerity of character . . . [that] overturns all respect which he has acquired." He further claimed that no one at the Boston Society of Natural History had "the slightest respect" for Rogers. In a confidential letter to his sister, Elizabeth, Wyman continued his attack on Rogers's character by suggesting that Rogers often tried to win a position through an appropriate marriage and that he was currently pursuing a young lady of influence identified only as "Miss M." Wyman assured Elizabeth that Miss M would not be "fool enough" to take Rogers seriously.[7]

Wyman's allegations are puzzling. As he was wrong about an incessant job hunt by Rogers, he was wrong on his other charges as well. Rogers had many friends at the Boston Society, including Amos Binney, its founder, and Emerson. He was a regular participant at the society's meetings then and for years afterward. These were meetings to which he looked forward eagerly as times to see friends. On the subject of Rogers's interest in an opportune marriage, it is an unlikely scenario. While Rogers would like to have contemplated marriage, the financially disastrous end to his iron furnace kept him from entertaining such thoughts.

He took the large debt on his own shoulders and spent ten years paying it off. He could support himself on his income and pay the debt in regular fashion, but there was no money left to support a family. When he finally did marry in 1853, he confided to a good friend that he had been unable to consider such a move until the debt was paid.[8]

Wyman was known to his contemporaries as a modest, nearly saintly man, who achieved recognition among his peers for his quiet dignity as well as for his skill in anatomy, comparative anatomy, and zoology.[9] He was not a man given to attacks on others, but he was determined that science and scientists should demonstrate the highest ideals. Although Wyman had attended some meetings of the AAGN, he did not really know much about Rogers other than what he heard, and what he had heard came in the form of rumors. According to Rogers's friends, these rumors originated in Philadelphia among people who did not like Rogers and were serious enough, as Wyman's mistaken view of Rogers indicates, to cause Rogers's friends to mount an all-out effort to counter them. They did this both by suggesting that the rumors sprang from factions who opposed Rogers on their own self-serving grounds and by calling attention to Rogers's excellent personal and intellectual qualities.

Hillard told Harvard trustee B. R. Curtis that Rogers had enemies in Philadelphia and the objections to Rogers that came from that city were all due to cliques and factions that demonstrated "all manner of unkindness and bitterness among them." In a ten-page letter, Hillard stressed Rogers's dynamic qualities as a lecturer, his scientific achievements, and, because a major part of the Rumford Professor's job was chemistry, his study with Turner in England. He wrote, "In the name of those friends [of the college] let me beg you to select a man in whose veins the warm blood of intellectual life freely runs." McIlvaine, one of Rogers's longtime friends from Philadelphia, wrote to Lowell along the same lines, without mentioning the city, saying that if Rogers was not praised by everyone, it was because "he has been pure in his office, and would not lend himself or the weight of his name, to promote the selfish interests or dishonest designs of individuals or of cliques in some of the societies with which he has been connected."[10] The references of McIlvaine and Hillard to cliques surely referred to Rogers's enemies at the Franklin Institute—Frazer, Peale, Patterson, and especially Trego, who at about this same time was working hard to discredit Rogers and the final report of the Pennsylvania survey with the state legislature.

Equally serious in Rogers's defeat was his support of the idea of organic evolution, a concept that the majority found to be atheistic. Rogers had lectured on the subject of evolution from time to time, and soon after giving his lectures in Boston, he was viewed in that city as a spokesman for the theory. Like many of his contemporaries in geology, Rogers recognized a succession of life in the fossil record, at least of the major classes of animals. Like other catastrophists, he believed that the appearance of each class was associated with a particular geological period and coincided with pronounced geological events.[11] Such a view did not necessarily embody the principle of organic evolution, for God could ordain the change, and most other geologists were content to leave it at that. But Rogers found the idea perfectly plausible that a natural process of change in the animal classes was prompted by geological change. While some of those who attended his Boston lectures were impressed with his arguments, evolution remained for most a highly questionable theory and one that denied God's active presence in the appearance of life forms.

Evolution was far from a new concept in 1845. It can be found in eighteenth-century studies, and it is probable that Rogers's interest in the idea began with an introduction to Erasmus Darwin.[12] Darwin believed that species could change, and he entertained an idea about the cause of that change that was similar to one made famous a few years later by Jean Baptiste Pierre Antoine de Monet, chevalier de Lamarck, and called the doctrine of acquired characteristics. Darwin's statement on evolution appeared in a single chapter of his *Zoonomia; or, The Laws of Organic Life,* published between 1794 and 1796, and had limited impact. Nevertheless, Rogers is likely to have been introduced to the concept by his father, who was such a great admirer of Erasmus Darwin that he gave Henry the middle name "Darwin."

Unlike the theory of Erasmus Darwin, that of Lamarck received widespread attention. It was the most complete evolutionary idea to date, but the doctrine of acquired characteristics was unacceptable scientifically, and the notion as a whole was theologically outrageous. More than twenty-five years later, Robert Chambers, a Scottish journalist, attempted to support the general concept of evolution (but not the doctrine of acquired characteristics) by gathering together all the evidence in favor of evolution. In an anonymous publication entitled *Vestiges of the Natural History of Creation,* Chambers argued that species continuously changed in response to changing environmental conditions, progressing from lower to higher forms of life. Although Chambers presented evolution as part of a divine plan, he met more severe criticism on religious grounds than even Lamarck had met. Chambers believed that evolution might occur in accordance with some natural law that was part of the divine plan, but the idea of an event occurring without the direct supervision of

God left too much to chance in the mind of the average person. Furthermore, Chambers included man in the evolutionary scheme and by so doing ensured an immediate and hostile reaction to his work. If evolution was to have any chance of a favorable hearing, man had to remain spiritually and physically aloof from it.[13]

Reaction to Chambers was at its height as Rogers began his quest for the Rumford Professorship. Rogers had read Chambers's book shortly after its appearance, and he told William that it contained "many of the loftiest speculative views in Astronomy and Geology and Natural History, and singularly accords with views sketched by me at times in my lectures."[14] There is little to suggest the full extent of Rogers's evolutionary thought or of his agreement with Chambers beyond the basic premise that some process of organic change had led to new animal classes, but he may have shared the sentiments of his friend Samuel Haldeman with regard to natural change at the species level. Haldeman argued that there was so much variation within species that it was often difficult to define a particular species and that, given this kind of variation, it was entirely possible to think that certain variations could cross the line and become new species.[15] Whether Rogers was willing to go so far as to include man in the evolutionary scheme is not evident in any of his remarks, but his youthful religious liberality suggests that he was not as constrained by formal doctrine as others were in the mid-1840s, leaving his agreement with Chambers on the issue of man a possibility but one that is unsubstantiated.

Rogers's discussion of evolution in his lectures in Boston was his undoing in the eyes of many people. Although his eloquence and power as a speaker persuaded some people, including Benjamin Peirce, that evolution might be an idea worthy of consideration, his stand raised questions about his moral and religious values. The importance of the issue with regard to the Rumford Professorship was apparent in that the letters addressed by his supporters to the Harvard trustees tried very hard to counter any suspicion about Rogers on these very counts. McIlvaine stressed that Rogers's lectures were always moral, reflecting strong Christianity. George Emerson wrote to Harvard treasurer Samuel Eliot that although the orthodox of all sects might place Rogers "in the same rank with most of those who reject the popular doctrine of theology," he was confident that Rogers was interested in the moral condition of mankind and "certainly holds to the morality of the Gospel."[16]

Because of Rogers's stance on evolution, Asa Gray, a botanist and a powerful member of the Harvard faculty, became his severest opponent and did everything that he could to prevent Rogers from getting the

Rumford Professorship. Although twenty years later Gray would become an ardent defender of evolution, he did so only after its scientific credibility was established by Charles Darwin, grandson of Erasmus, in his work *On the Origin of Species by Means of Natural Selection,* published in 1859. Before then, Gray found no saving grace in Lamarck or Chambers and was determined to see that the heresy embodied in their arguments was not allowed to spread to Harvard. Gray's public attacks on Chambers began soon after *Vestiges* appeared, and they worked against Rogers. By the spring of 1845, Gray was able to report to John Torrey, his friend and fellow botanist in New York, that he had quelled a tide of support for Rogers's views and that he had persuaded Benjamin Peirce to stop supporting Rogers's position on evolution.[17] Gray may have thought his attacks had quieted Rogers, but they had not, and Rogers continued to talk about evolution. At the AAGN meeting in New Haven in 1845, a paper by Samuel Webber on attraction and polarity was read in which a reference was made to *Vestiges*. An argument followed during which Rogers urged the association to take up the question of generation, referring to Chambers as putting forth "sublime and glorious views . . . of creation."[18]

Gray also misjudged his own effect on the Harvard trustees, for they were not dissuaded from their interest in Rogers for the Rumford Professorship. By late January 1846, it was clear that in spite of negative opinions about Rogers and in spite of Gray's attacks on Chambers, Rogers might be given the post. When Gray realized this, he took immediate action directly with the Harvard Corporation to stop Rogers.[19] The powerful Gray rallied his many allies at Harvard to fight Rogers's appointment, including Peirce and Louis Agassiz, who had recently arrived in the United States.

The Swiss-born Agassiz came to the United States with impeccable credentials in zoology and geology. Although he came to give a series of Lowell Lectures, it was expected that he would become part of the new scientific school at Harvard. Rogers met Agassiz shortly after he arrived in the United States and found him "amiable, engaging and philosophic," someone from whom he could "draw new power and impulse," but the belief that he had found a new scientific colleague was short-lived.[20] Agassiz's vehement antievolutionary stance made him a natural ally of Gray's and an opponent of Rogers.

Peirce, who had given birth to the idea of the scientific school at Harvard, became nearly as powerful an opponent as Gray and Agassiz. While he had initially looked with favor on Rogers's arguments for evolution, he yielded to Gray on the issue. Furthermore, he had his own candidate for the job in the person of Morrill Wyman, and he took up Jeffries Wyman's argument that Rogers's work was inaccurate and that he was untrustworthy both as a person and as a scientist. Seeking to refute favorable comments coming to the trustees from Rogers's friends, Peirce solicited a statement from Alexander Bache. Bache had become a nationally prominent figure in science, and Peirce knew that an unfavorable statement from him would carry weight with the trustees.[21] Rogers's friend Hillard was so worried about this possibility that when he wrote to Curtis in support of Rogers he warned him not to give heed to any statement from Bache.[22] Bache did not support Rogers for the Rumford, but his response to Peirce was not as damaging as Peirce might have hoped. In spite of their difficulties, Bache had remained convinced that Rogers was an excellent geologist. Bache had prepared a speech for the National Institute two years earlier and, in what was undoubtedly a reference to Rogers's work on mountain elevation, said that with it, geology in the United States had reached a desirable level of "bold generalizations."[23] Now he told Peirce that Rogers's career in geology had "been a very distinguished one" and that Rogers was a brilliant lecturer whose theories often "captivated" him.[24] "For a Professor of Geology," Bache continued, "I would take him in preference to any one, perhaps." Nevertheless, Bache refused to support Rogers's candidacy because he did not think he was well enough prepared in other areas for which the Rumford Professor was to be responsible. Although not a damning statement, Bache's refusal to support Rogers was important, and Peirce saw to it that his letter got to the Harvard trustees.

While Gray, Peirce, Wyman, and others were fighting against Rogers's appointment, a third flank of opposition to Rogers was forming. The central figure was John White Webster, Harvard's professor of anatomy, who wanted his young friend Eben Horsford to have the Rumford Professorship. Webster was known to many of his colleagues as a man of questionable temperament. Many people disliked him, and there was even a current of suspicion swirling around him in regard to his competence in the study of anatomy. Whether or not Webster sensed the discontent with his performance, he wanted Horsford there to relieve him of some of his teaching responsibilities, and Rogers's position as a leading candidate for the job sent Webster into a frenzy of activity.[25]

Horsford was in Germany, studying chemistry with Justus Liebig when he received a letter from Webster proposing "that I [Webster] continue the usual course of general and elementary lectures to the *undergraduates* and *medicine class*—that you be associated with me as adjunct professor of Chemistry, Mineralogy and Geology—you to take the

Rumford Chair and be the head of the scientific school—to give all the practical instruction in that—and to cooperate with me in the instruction in mineralogy and geology." Since it was near the time of his return to the United States, Horsford looked eagerly on this prospect.[26]

Horsford had been an active participant in science in the United States, had been a candidate for a position with William Rogers on the Virginia survey in 1837, had served with James Hall on the New York survey, had taught natural and mathematical science at the Albany Female Academy, and had been a candidate for a professorship of chemistry at Philadelphia in 1844. He lost the latter opportunity to John Frazer, who was backed by Joseph Henry and Alexander Bache. Although Horsford had considerable backing for the Philadelphia position, including that of Henry Rogers, he lost, according to one friend, because Joseph Henry's word was "law in Philadelphia."[27] Horsford was not widely known, however, certainly not in Boston, and this made him a decidedly secondary candidate for the Rumford. Webster saw his role as having two parts. On the one hand, he had to discourage support for Rogers; on the other, he had to build support for Horsford. If he could do both at the same time, all the better.

Webster solicited recommendations for Horsford far and wide. James Hall liked Horsford and replied to Webster's request with strong support, as did many others. Webster told Horsford that Hall had met privately with a trustee of the Harvard Corporation and had informed him of the "real standing" of Rogers among scientific men, "which you must know is very low in the scale."[28] Whether such a meeting ever actually took place is questionable. Webster was not above embellishing a story for his own purposes, and although Rogers and the New York geologists frequently disagreed with each other, and Hall had failed to get information for his maps from Rogers, he remained friendly toward him. In fact, at about the same time that Webster was writing this to Horsford, Hall wrote to a correspondent, saying that he had recommended Horsford for the position, but had no personal ill feeling toward Rogers.[29] There were two reasons why Hall did not recommend Rogers. First, he was initially under the impression that Rogers was being considered for some other professorship, and second, like Bache, he did not believe that Rogers was closely connected with the subjects taught by the Rumford Professor.

With all the action against Rogers, the situation over the Rumford chair grew ever more complex, and by the end of February (1846), trustee Thomas Eliot suggested to President Edward Everett that the best plan "will be to defer action upon it for some time."[30] Webster, however, continued to solicit recommendations for Horsford. He wrote to William

Norton at Delaware College (then also called Newark College and now the University of Delaware), asking him to forward letters in his possession that had recommended Horsford for the professorship at Philadelphia. Norton, who was Horsford's cousin, complied but was concerned that the use of the letters might cause some reaction against Horsford. The letters included the one from Henry Rogers, endorsed by William, and Norton felt that it was not right to use this letter in Horsford's support, since Rogers was a candidate for the Rumford chair. Rogers's letter should certainly not be used, he told Webster, unless Rogers tried to demean Horsford's abilities. Webster prevailed on Norton to allow him to decide the matter, and unmoved by Norton's concern, sent all the letters to Everett, emphasizing the inclusion of a letter from Rogers.[31]

By the end of May, Webster was concerned that Everett was leaning toward Rogers and that the fact that Horsford was relatively unknown in Boston was working against him with the Harvard trustees. His efforts to get support for Horsford escalated. He implored Hall to write to the trustees, this time stating that there was some question whether Rogers had really done the chemical analyses for which he claimed credit, thus casting his abilities as a chemist into doubt.[32] There is no evidence that Hall ever responded to this request. Webster urged Horsford to prepare a paper for publication so that he would become better known in Boston; then he solicited funds from various persons, including Hall, to aid in its publication.

Over the summer Horsford grew concerned about Webster's unrelenting efforts to discredit Rogers. He became so upset at one point that he was willing to withdraw his candidacy for the Rumford Professorship. In a letter to Hall, he wrote that he was particularly "mortified" over Webster's efforts to raise money, noting that he sent the paper only because "my friend desired something that might be published."[33] He also noted that he had talked with Liebig about the affair and that Liebig had advised him to withdraw his candidacy for the Rumford chair and to offer himself "for a post in Chemistry if one should be created in the new organization [the scientific school] at Cambridge." He told Hall that he was further encouraged in his plan to withdraw by information he received that there was an opening, or soon to be an opening, in physiology at Harvard that he thought more desirable. Horsford sent a letter of withdrawal to Webster advising him to hand it to Everett if Hall determined that the opening in physiology would materialize.[34]

When Webster learned that John Collins Warren, professor of anatomy and surgery at Harvard, had received a letter from a friend favoring

Horsford with the request that it be handed to Everett, Webster seized the opportunity to make sure that Warren was in Horsford's camp as well. After talking with him, he sensed, however, that Warren leaned toward Rogers, and in an effort to sway his opinion, he informed Warren that several Europeans, including Liebig, supported Horsford, along with Americans Hitchcock, Silliman, Morton, Vanuxem, and Espy. His list of Americans was not accidental, since it included prominent geologists who might be supposed to support Rogers if the latter's work was creditable. Since Hitchcock and Morton had both written to recommend Rogers, however, Webster's list shows how far he was willing to go to stop Rogers.

Since the Rumford Professor would probably head the proposed new scientific school, it was essential that the person selected be respected in science and capable of working harmoniously with others. "Should," Webster told Warren, "a gentleman, who has been much talked of, but whose chief or only qualification is in geology, be appointed, it will defeat the desirable end."[35] He went on to say that a prominent member of the faculty had refused to come into the new school if Rogers was there and that his loss might prove fatal to the total project, a reference to either Gray or Agassiz. Rogers, he said, "has been unsuccessful as a Professor in another city, was very much disliked by his pupils and became obnoxious to his scientific brethren." Furthermore,

It was owing to accident that he gained the approbation of some gentlemen in Boston, who were not in a situation to form a correct estimate of his real standing as a man of science. He was said to have an European reputation but it was not known to them that this reputation was founded upon an article, one that was valuable in which was the labor of others adopted & thus used without acknowledgment & which contained chemical analyses promulgated as the writer's own, but are actually made by another!

The determined Webster continued his accusations with an attack aimed directly at Rogers's work in geology:

Since then a flimsy geological paper has been published by the same person in which, starting with a presumption. . . . the writer has made a tremendous occasion of displaying logic in its application to geological reasoning. A most thorough geologist, to my knowledge, went over this paper with Mr. Lyell, and the judgment & decision of both in regard to the conclusions were that they were "merely wind & words."

Webster's remarks reflected little knowledge of Rogers's activities. Part of his accusation was based on the attack by Harlan so many years earlier, an attack that the final publication of Rogers's paper by the BAAS had dispelled. Webster's reference to chemical analyses done by someone else and for which Rogers took credit was probably to chemical analyses in the annual reports, since there are none in the paper for the British Association. The chemical assistants were given credit, however, and thus the criticism carries little weight, and although few people felt that Rogers's elevation theory held much promise, only Webster pronounced it "flimsy." The only message that Webster wanted to give was that Rogers was incompetent and incapable of any significant work of his own.

Webster was ecstatic when he heard that Joseph Henry was leaving Princeton to go to the Smithsonian Institution, because he thought an important element favoring Rogers would thus be removed.[36] At the same time, someone tried to help Rogers by attacking the credibility of the school at Geissen where Horsford studied with Liebig, an attempt revealed in a letter that James E. Teschemacher, Boston businessman, mineralogist, and active participant in the Boston Society of Natural History, wrote to Horsford in September. Teschemacher also indicated that efforts were under way to replace Webster because the latter had been unsuccessful in developing the chemical department of the college, and it also indicated a growing concern over Webster's activities as Teschemacher acknowledged "the difficulty of what to do with Dr. W."[37]

Of Rogers's detractors, the motivations of Webster are the least clear because of the extent to which he went to defeat Rogers. They are more easily understood as the result of a disturbed mind than as an earnest effort to bring Horsford to Harvard. Many people considered Webster unstable, and their views prevailed a few years after the Rogers episode when, in 1849, the state accused Webster of the murder and dismemberment of his onetime mentor and friend Dr. George Parkman, to whom he owed $2,000 and with whom he argued over the debt. After a trial, the jury found the unfortunate man guilty and agreed that the state must hang him for the crime.[38]

Early in 1847, after more than a year of deliberation, the Harvard Corporation finally made a decision in favor of Horsford. It has been argued that a recommendation for Horsford from Liebig persuaded the trustees to hire him, and Asa Gray contended that it was primarily his intervention and the arrival of Agassiz that defeated Rogers.[39] It was, rather, a combination of things. While his opponents varied in their motives, each contributed to his defeat. Together they created so much confusion over Rogers that the Harvard trustees were hesitant to give the job to him even though they liked his qualifications and the recommendations had come from reputable persons to whom they were willing to listen. They also liked Rogers's popularity as a lecturer. He was repeatedly asked to speak to public gatherings, and when he opened the season of Lowell Lectures in October 1846 with a course on geology, he addressed an audience of nearly 2,000 persons.[40] It was a difficult decision for the Harvard trustees, but when they finally had an opportunity to meet Horsford, they found him to be well informed and amenable to their goals. His credentials were good, and they could at last feel at ease in putting aside the questions about Rogers.

Although Rogers was disappointed at not receiving the appointment, the defeat strengthened his and William's resolve to start their own polytechnic school. Neither thought the new science school at Harvard, although generously endowed by Abbott Lawrence, would develop along the lines they wanted.[41] Throughout the fall and winter of 1847–1848 Henry and William planned for their school, and early in 1848, William decided to resign from his position at the University of Virginia and come to Boston.[42] Intense opposition to his resignation by students and faculty at Virginia, however, persuaded him to stay, and the brothers' dream of a polytechnic school was put off once more. It was a dream that would not be fulfilled with Henry's participation, but their ideas succeeded in 1861 when William played the major role in the development of the Massachusetts Institute of Technology.

Although disappointed over his failure to achieve his goal, Rogers had more than enough to do. He continued to work on the Pennsylvania survey report, and his consulting business was flourishing, so much so that he would soon begin to identify his occupation as that of "practical geologist" in reference to the application of his knowledge through his consulting work to utilitarian ends.[43] He also spent a considerable amount of time preparing for another course of Lowell Lectures to be given in the fall of 1847, discussing their content at length with William. He intended to talk on something other than geology in these lectures, possible subjects including ventilation, heat, and light. Hoping to give a second set of lectures later, he explored the possibility of devoting them to the subject of ethnology. He also traveled to gather information for the geological map of the United States, on which he had continued to work and which Lesley was drawing for him. He went to New Brunswick and Rhode Island to settle points for the map, and by the spring of 1848 he was anxious to get William's Virginia map, because Lesley was ready to add its data.[44]

During this same period, Rogers was also occupied with preparations for the first annual meeting of the American Association for the Advancement of Science (AAAS). A year earlier members of the Association of American Geologists and Naturalists had voted to change the organization into the broad-based one that Rogers had so often advocated. In 1845, Rogers had made a motion to expand the AAGN to "an association for the promotion of science," to be called the American Society for the Promotion of Science.[45] His effort bore little fruit, however, until Louis Agassiz addressed the issue at the 1847 meeting of the AAGN in Boston. Joseph Henry was also at the meeting and added his voice in support of expansion because he found the AAGN potentially useful as a platform from which he could discuss the goals of the Smithsonian Institution, of which he was now secretary.

With Agassiz and Joseph Henry giving encouragement, the seed planted by Rogers when he returned from England in 1833 finally sprouted and the association voted to become the American Association for the Promotion of Science. It quickly changed its name to the American Association for the Advancement of Science.[46] Rogers, Benjamin Peirce, and Agassiz were made a committee at the 1847 meeting to formulate a constitution and by-laws for the new organization. The committee's charge was to revise and modify the AAGN constitution and rules, and Rogers and the others worked throughout the early spring of 1848 to accomplish the task. The new constitution was published as a circular to be distributed at the annual meeting to be held in the fall.[47]

Rogers added several "explanatory" pages to the circular that embodied his view of the need for a society that would bring the sciences together:

Much of the business of research can only be successfully performed through combined exertion; for many of the most important investigations are of a kind, so necessarily laborious in their details, and so complex and divergent in their bearing, as to require the cooperation of many minds with very various powers and talents. . . . Individuals, however insulated their efforts, may at all times do something for the cause of knowledge; but associated and wisely organized, they can give a wider scope and a higher direction to their exertions, and accomplish incalculably more.[48]

Rogers felt that he had done his job well and that the constitution was "democratic, federal, flexible and expansive, progressive, with all the true conservatism these features imply."[49] Unfortunately, Rogers was unable to attend the first meeting of the new AAAS because he decided to go to Europe in July, nearly two months before the meeting was to be held.

Rogers's decision to go to Europe rather than to attend the first meeting of an organization he had so long advocated was based on two factors. He continued to hope that he and William would soon be in a position to open a school and felt that, if they succeeded, he would then have no time for travel. The more important consideration, however, was his health.[50] Although Rogers characterized his state of health wholly in physical terms, he was depressed by events that occurred in the spring of 1848 concerning the survey. He had completed the work, and early in 1848, the legislature approved a resolution for its publication. The same legislature failed, however, to appropriate the necessary funds. James Rogers blamed Trego for the defeat, and William was prompted to tell Henry that his "faithful labours" and "unbending uprightness" had been "cruelly repaid," a reference directed at Henry's enemies in Philadelphia who had worked first to discredit him in Boston and now to prevent the survey's final conclusion.[51] In spite of William's efforts to encourage him, Henry's spirit was broken, and his physical health suffered. By the early summer of 1848, he was "unable to do more than half a fair day's work of any kind without fatigue and much nervous and neuralgic pain, and I have a nervous and sick head ache every four or five days, as often in other words as I urge myself beyond a very moderate amount of exertion."[52] An escape to Europe was essential.

12 A Spirit Oppressed, 1848–1851

Rogers had not visited Europe since his trip there in the early 1830s with Owen's group. On that trip, he had decided on a career in geology. Now he looked forward to a trip as a time to renew old acquaintances as well as an opportunity to renew his health. He also relished the opportunity to study some of the geological features of Great Britain and the continent in the framework of his theories.

He lost no time in his quest for new geological information. He arrived in Scotland in early July and spent several days traveling in the Highlands, where he focused his attention on the Parallel Roads of Glen Roy, striking parallel terraces that were once the shore lines of an ancient lake. Their origin had been debated by geologists for a long time. Most of those who had observed them believed that they had originated in standing water, but Rogers was true to his theory and interpreted them as the result of floodwaters pouring in from the Atlantic during a paroxysmal upheaval. He told William scarcely a month after his arrival that he had solved the problem of their origin. He intended to give his solution at a meeting of the London Geological Society within a month or so, but it would be many years before he made his idea public.[1]

After reaching his conclusions about the Parallel Roads, Rogers undertook a whirlwind trip to Edinburgh, Glasgow, and finally London, where he arrived in late July or early August. From there, he went to Swansea for the annual meeting of the British Association for the Advancement of Science. Murchison, Sedgwick, and Lyell were not at the meeting, but Rogers was glad to find de la Beche and Richard Owen there. He asked Owen, England's leading paleontologist, to identify several fossil reptiles that he had brought with him from the Greensand formation of New Jersey, fossils that Owen identified as the remains of mosasaurs.[2] Rogers gave two short papers at the meeting, one on the geology of Pennsylvania and one on earthquakes, both no more than summaries of his ideas.[3] September found him in Switzerland studying the geology of the Alps and the Jura Mountains and finding what he believed to be confirmation of his elevation theory. In October he was in France where he met several persons whose work interested him, among them, Dominique François Arago, Philippe de Verneuil, and Elie de Beaumont.[4] Arago, the perpetual secretary of the Academie des Sciences, whose work centered on the physical phenomena related to electricity, had studied earthquakes; Elie de Beaumont was the architect of the elevation theory discussed earlier; Verneuil had visited the United States in 1846 to study geology, but he and Rogers had not met at that time. Verneuil had nevertheless become acquainted with Rogers's work and referred to Henry and William's "beautiful researches" on the debituminization of coal.[5] Rogers was pleased with the reception he received, noting in particular that Elie de Beaumont seemed "right glad to see me." Because he found them familiar with his theory, and because they greeted him enthusiastically, Rogers expected his theory of elevation to "meet a prompt reception by the French geologists, even while many of the English may hesitate."[6] It did not, and the expressions of interest in France proved to be little more than ones of courtesy.

Although the English hesitated to accept his theory, Rogers nevertheless felt at home in the British atmosphere, just as he had during his first trip. He enjoyed a camaraderie and intellectual stimulation there that exceeded any he found at home. He told William that he had so many invitations that he could not possibly accept them all and that he was enjoying the "rare sport" of exciting intellectual conversation.[7]

Rogers had always believed that his elevation theory was universally applicable, and making use of some of his new observations on the Alps, he was eager to show his doubters how his theory explained the complex features of this mountain system. Therefore he arranged to give a paper comparing the Alps and the Appalachians at the November meeting of the Geological Society of London.[8] He did not submit his paper for publication in the society's journal because of ill health and an excess of work when he returned to the United States, but his argument can be determined from the abstract of a similar paper presented the following year in the United States at a meeting of the AAAS and from comments on the Alps made in his final report on the geology of Pennsylvania.[9]

The Alps are a complexly folded group of mountains stretching across

Europe. Close folds, intense disruption, and extensive inversion of strata, causing the younger rocks to plunge below the older ones, are among the things that make them so complex, and geologists had been attempting to find a satisfactory explanation of their origin and structure for many decades. It was generally believed that the elevation of the Alps began with the eruption of a single igneous core, which precipitated the lifting of the strata in all directions outward from the core.

Rogers thought he detected two central cores in the Alps rather than one, each characterized by massive disturbance and igneous intrusion. He concluded from his examination that the structure of the Alps was a folded one not unlike that of the Appalachians albeit more complex, and he concluded that the steeper sides of the folds were always away from one of these central cores. He reasoned from this that a violent disturbance had occurred in each of the two areas, spawning the characteristic wavelike motion of the molten interior of the earth. In contrast to the formation of the Appalachians, where there was a lateral push in one direction only, waves had moved out from both sides of each of the centers of disturbance in the Alps. As many as four main belts of waves were associated with the Alps at any one moment during their formation. With this concept, Rogers explained the origin of some particularly complex areas within the Alps as the result of the intersection of waves between the two areas of disturbance. The emanation of waves in two directions also helped to explain a fanlike arrangement of the rocks that had long puzzled geologists. It was, Rogers said, the visual result of the dip of the axes of the folds augmented by the cleavage lines that, as he had argued in an earlier paper, always paralleled anticlinal and synclinal planes.

Many geologists assumed that the Jura Mountains, which lay to the northwest of the Alps, were the result of the same event that created the Alps. Rogers reached a very different conclusion. Calling attention to the appearance of steeper folds on the side of the Jura Mountains facing the Alps, he argued that the Jura had been lifted by forces farther to the North, propelling the rock forward and steepening the advancing sides that faced the Alps. Thus, the Jura did not originate in an outward push of rock as central Alpine axes were elevated. Had that happened, the steeper flanks would be, not where they were actually found, but on the opposite sides.

Rogers's Alpine theory met with little enthusiasm in England, even prompting Henry de la Beche to attack Rogers's entire theory by arguing that the structure of the mountains could not sustain the idea that tension had built within the earth of sufficient force to give rise to waves on repeated occasions. Had it happened, he argued, the molten matter injected into the rents created by the explosive release of the tension would have driven the once continuous formations farther and farther apart, with the intervening space filled up each time by igneous matter.[10] This was not, he said, borne out by any observation.

Failing to gain new support for either his general theory or his Alpine theory, Rogers was heartened when Murchison once more came to his defense, this time supporting his principle of inversion. Murchison had given more support than others to Rogers's general ideas on elevation when presented to the BAAS in 1842 and had continued to show interest in the theory in a work on the geology of Russia, published in 1845. He argued that part of the Ural Mountains strongly suggested wavelike undulations which, he said with a reference note to the Rogers's 1842 paper, was a "view which American geologists have so admirably worked out in the Apalachian [*sic*] chain."[11] Murchison had tried earlier to explain inversion as the result of the collapse of rock into cavities created by the eruption of igneous matter, but dissatisfied with this idea, he introduced Rogers's theory of inversion in his work on Russia. Although by 1848 when Rogers gave his paper before the Geological Society, Murchison had become unsure of the wave theory, he still thought that the lower, less complicated Appalachian range would provide essential keys to understanding mountains that the Alps and other European ranges would never provide, and he was more interested than ever before in the theory of inversion.[12]

Murchison probably heard Rogers's paper at the Geological Society, but just two weeks later the two were together at a meeting of the Royal Society where they discussed Rogers's theory and Murchison showed Rogers sketches of several Alpine geological sections with inverted strata.[13] A day later, Rogers sent Murchison a letter and several diagrams illustrating his theory and the kinds of faults (displacements of rock that result from fractures) found in closely compressed and inverted flexures.[14] Part of Rogers's letter and his sketches of inversion were incorporated in a long address Murchison gave at a meeting of the Geological Society that was read on December 13, 1848, and January 17, 1849, and was subsequently published[15] (figure 9).

The work and contacts during his trip revived Rogers's spirits, and he returned to the United States early in January (1849) in good health. Another series of Lowell Lectures, for which planning had started before he left for Europe, began on March 6.[16] Their preparation, the refinement of the paper on the Alps and the Appalachians for the AAAS meeting in August, and the preparation of a second paper for the same

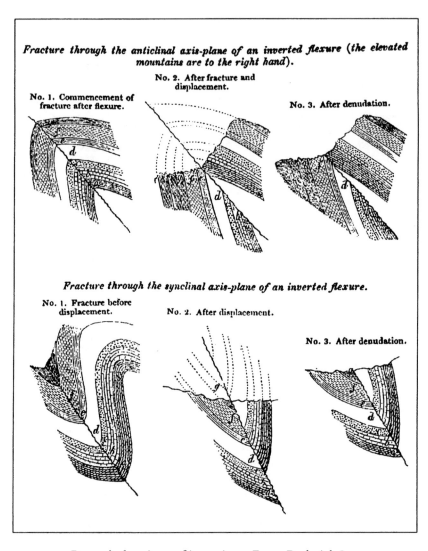

Fracture through the anticlinal axis-plane of an inverted flexure (the elevated mountains are to the right hand).

No. 1. Commencement of fracture after flexure.

No. 2. After fracture and displacement.

No. 3. After denudation.

Fracture through the synclinal axis-plane of an inverted flexure.

No. 1. Fracture before displacement.

No. 2. After displacement.

No. 3. After denudation.

FIGURE 9. Rogers's drawings of inversion. From Roderick I. Murchison, "On the Geological Structure of the Alps, Apennines, and Carpathians," *Quarterly Journal of the Geological Society of London* 5 (1849): 251.

meeting, "Analogy of the Ribbon Structure of Glaciers to the Slaty Cleavage of Rocks," extracted a toll.[17] It was not long until Rogers's state of health was as poor as it had been before he went to Europe, forcing him to give up his work and spend the next several months traveling. By August, his health was much improved once again and he was able to deliver his two papers at the meeting of the American Association for the Advancement of Science that month.

The paper on the Alps and Appalachians has already been discussed. The paper on ribbon structure represented an attempt to develop new support for his theory. Since the early 1840s when James D. Forbes, Scottish physicist, and Louis Agassiz had called attention independently to alternating bands of blue and white ice in glaciers and had suggested that they had something to do with the glacier's forward movement, this so-called ribbon structure had attracted interest.[18] Forbes led the way in suggesting that the ribbons were the result of internal pressure in the ice that caused the ice to act as a viscous, or plastic, substance and to flow in a manner similar to water. He argued that the surface moved faster than the lower portion and the center faster than the sides, the ribbon structure occurring where one part of the ice dragged past another part.[19] With variations in detail, the pressure theory remained popular. Rogers, however, explained the ribbon structure as the result of heat, arguing that the ribbons were veins paralleling the surfaces of higher temperature that were the walls enclosing the glaciers, just as cleavage paralleled zones of higher temperature in rocks.

Rogers had observed the analogy between ribbon structure and cleavage during his study of the Alps in 1848 and may have been the first person to see this similarity, although he said that Forbes had cautiously hinted at it.[20] By the mid-1850s the similarity between ribbon structure and cleavage had been noted by others and, with the widely held belief that cleavage was caused by pressure, was used to support a pressure theory of glacial flow and ribbon structure that had been proposed by John Tyndall and others. It differed from Forbes's pressure theory in suggesting that the ribbons were caused when brittle ice fractured under pressure and then underwent regelation.[21] Whether Rogers's suggestion of the similarity had any effect in calling the attention of Tyndall and other later investigators to the analogy is impossible to say. On the one hand he may, indeed, have discussed the idea in England, and a summary was available in the *Proceedings* of the AAAS; on the other hand, since his was a theory of heat causality and few agreed with the idea that cleavage was caused by heat, it may have had no impact.

No reaction to the paper on ribbon structure was recorded at the

AAAS meeting, but the reaction to the paper on the Alps was predictable and critical, with Agassiz and his friend, Arnold Guyot, leading the challenge. Guyot, a colleague of Agassiz in Switzerland who had come to the United States in 1849 to give a series of Lowell Lectures, was expressly concerned about Rogers's conclusions regarding the Jura Mountains. Although he agreed with Rogers that a lateral pressure was involved in the origin of the Jura rather than a vertical one, he saw that force in the context of Elie de Beaumont's theory, not Rogers's. He felt that the preponderance of evidence mandated that the originating force for the Jura lay in the Alps, and he argued that the reason Rogers thought folding had occurred independently of the Alps was the complexities in the formations that resulted from at least three upheavals that affected the Jura. Agassiz and Guyot provided formidable opposition to Rogers's efforts to explain Alpine structure, but James, who attended the meeting, assured William that Henry "acquitted himself with his usual ability in the face of much opposition."[22]

The skirmish between Agassiz, Guyot, and Rogers reflected an alienation of Rogers and Agassiz that had been growing since Agassiz had arrived in Boston in the midst of the efforts to keep Rogers from the Rumford Professorship. On scientific matters, there was little on which the two could agree. Not only were they on opposite sides of the evolution question, but Agassiz's principal claim as a geologist was his glacial theory, which Rogers firmly opposed. In the few weeks that Rogers had spent in the Alps, he had found nothing to support Agassiz's theory and much, he thought, to support his own diluvial view. When Agassiz decided to remain in the United States, it was a foregone conclusion that he would be the professor of geology and zoology at the new Lawrence Scientific School, but nevertheless, his actual appointment in 1847 must have galled Rogers.[23]

The worst problem between Agassiz and Rogers was, however, personal rather than professional. When Agassiz came to the United States, he was followed soon after by Edward Desor, who had been his associate and assistant since 1837. With considerable disregard for his long association with Agassiz, Desor tried to undermine him and win a reputation for himself by claiming that embryological studies for which Agassiz was well known were in large part his own. Desor went so far as to intrude in Agassiz's personal life with charges that linked Agassiz to an Irish servant girl and suggested that Agassiz's lack of concern for his wife had led to her death in 1847.[24] The height of Desor's attacks on Agassiz occurred in 1848 and 1849. Sometime in this period, Desor and Rogers became friends, ending any chance of cordiality between Agassiz and Rogers for many years, and for years to come Rogers and Desor lent support to their mutual dislike of Agassiz.

Not long after the AAAS meeting, Rogers fell into another state of depression. Whether it was directly related to the reaction to his paper, to continuing concern over the survey, or to some combination of factors, it was deep and prolonged. William was his only comfort, and he poured out his feelings to him:

I thank you for your very kind letter of the 11th, and for your sincerely affection-ate words. These are ever to me a source of cheerfulness and consolation and they seem at this time of double value, coming when my spirit is oppressed with an unwonted sense of loneliness and of life's disappointments. In all hours of trial, in all time of need, your love has given me strength. The faith that some turn of fortune may bring me again to live, as in earlier blessed days, with you and our generous and gentle Robert has for a long while past been to me the one calm star of hope that, when all other beacons have gone out, has never once grown dim. Daily do I take counsel with my heart that it may keep itself worthy of a companionship out of which, if pure, it will derive a peace such as is not in store for it from any other earthly source. That Heaven may shed upon you both, my dear brothers, its sweetest blessings is my never ceasing prayer.[25]

William had married in the spring, and he and his wife were in London, where he wrote to Henry, commenting on events there that he felt sure would lift his brother's spirits. He had attended the meeting of the BAAS at Birmingham, where he reported on his own work in Virginia and discussed informally the theory of elevation. He told Henry that the Brit-ish held him [Henry] in high regard and noted that Murchison, Lyell, and de la Beche had commented very favorably on the importance of the theory. It was, William said, "a real triumph and joy to hear them suc-cessively declare that our development of the great law of flexures was one of the grandest contributions to geology ever made, and to find that they gave us the entire and exclusive credit of having thus furnished a clue to the most difficult problems in European geology."[26] This was undoubtedly a reference to the delineation of various kinds of folds and particularly to the principal of inversion, rather than to the wave theory itself. While the British geologists did not think any more highly of Rogers's wave theory of elevation than the Americans, they, as exemplified by Murchison, ad-mired his work on structure.

In spite of William's encouraging words, Rogers's depression did not end. He lost interest in many of his activities and found others no more than lackluster duties. He joined the Warren Club, an informal group founded by John C. Warren, the surgeon who had favored Rogers for the

Rumford Professorship. The club met twice a month from December to April and was partly social in nature and partly a place for discussion of common interests in the scientific and literary. Nevertheless, Rogers found their meetings "dull enough."[27] His appointment to the chairmanship of a committee for the ventilation of the rooms of the Boston Society of Natural History prompted him to tell William that "for a man of any brains whatever, Boston has no peace or quiet, all is restless excitement and unproductive change of thought and of pursuit." The brain is overworked "without the fruits of intellectual labour."[28]

In December 1849, Rogers's spirits improved momentarily when he spent some time in Providence, Rhode Island, with the president and trustees of Brown University, who were contemplating a change in their school's educational plan. Francis Wayland, president of Brown, was one of the nation's most progressive educational leaders. His interest in a meaningful and utilitarian education intrigued Rogers so much that he asked William to send information on their proposals for a school of arts to Wayland.[29] The visit held no suggestion of a place for Rogers at Brown, but it was stimulating and rekindled his desire to find some way of developing a school with William. Nevertheless, Rogers's general despondency remained, and his depression grew so great that for a time in the spring of 1850 he turned his attention away from his usual pursuits entirely and wrote a short paper on the improvement in the intonation of the organ, which, he said, contained philosophy and mechanical information on the evolution of musical tones.[30]

This diversion worked, his depression began to wane, and Rogers resumed his activities in geology with regular participation in scientific meetings, lectures, and consulting work. In May, he gave the first in what was to be a two-part series of lectures on geology at the Smithsonian Institution. A year earlier Joseph Henry had laid plans for a lecture series at the Smithsonian, intended to show the best of American scientific activities. The Smithsonian was the product of a large bequest by James Smithson, and Henry was determined, as its secretary, to shape the institution for the betterment of science in the United States. He had written to William Rogers in May 1849 about his planned lecture series, explaining his intent to invite only "those who have distinguished themselves by original research" and assuring William that he and Henry were among those to be invited.[31] Rogers's May lectures were on geological dynamics, and he expected to give a set of lectures on paleontology the following December.[32]

As heavy development continued in the anthracite coal lands of Pennsylvania, opportunities for Rogers to serve as a geological consultant grew. To a lesser extent, his work as a consultant on iron mining also grew. Most of his consulting work was done for companies run by Charles P. and William L. Helfenstein. The latter, a lawyer and former judge of the Court of Common Pleas in Dayton, Ohio, was the moving force behind almost all the development of Pennsylvania's western middle coalfields from Mount Carmel to Trevorton in the late 1840s and early 1850s. Two major companies were incorporated in 1850 for this development. The first was the Zerbe Run and Shamokin Improvement Company; the second was the Mahanoy and Shamokin Improvement Company. In order to transport the coal, Helfenstein established several railroads to connect the mines with the rest of the region.[33] Rogers was hired by the Helfensteins to report on the relative merits of the coals in the area.

Rogers's former assistant, Peter Sheafer, was also working as a mining engineer for these companies, and the two were soon joined by Lesley, whom Rogers called for help when he suffered yet another round of poor health in the summer of 1850. After returning to work for Rogers from 1846 through part of 1848, Lesley had gone back to his theological studies in 1848 and had become pastor of an orthodox Congregational church near Boston, but by 1850 he had run into some serious difficulties with the church over their "ultra calvinistic logic," and his "geological rationalism." According to Lesley, the church felt that "no practical harmony could be established between Science and Religion," which Lesley said meant "between his science and their religion."[34] Although he continued his ministry until 1852, he was only too glad to rejoin Rogers and get back to work in geology.

The result of Rogers's work for the Helfensteins was three published reports in late 1850 and 1851 based on work done mainly in 1850. The working conditions for the preparation of these reports were probably as ideal as Rogers might ever have wished for in his years of surveying. He was given a large group of miners, a corps of geological assistants, and as much time as he wanted, and he claimed that his studies were "the most minute, elaborate and systematic hitherto authorized by capitalists in the development of any coal territory in the limits of the United States."[35]

Together the reports constituted an extensive geological survey of the anthracite areas, taking twelve months and covering some six thousand acres. Lesley later said that some of the first and most important instrumental measurements were done at this time, resulting in a hypsometrical map (one measuring heights with reference to sea level) of the Shamokin coalfield.[36] Rogers's first report on the area was a *Report on the Coal Lands of the Zerbe's Run and Shamokin Improvement Company*, completed and presented in August of 1850. The second was *Reports on the Combustible Qualities of the Semi-anthracites of the Zerbes's [sic] Run Coal Fields,* for which he had hired Dr. Augustus Hayes to assist him with chemical analyses. Hayes owned the Roxbury (Massachusetts) Chemical Works, which had been in operation for about twenty years. His job for Rogers was particularly to analyze the coal so that the relative commercial value of all coals in the vicinity of Zerbe's Gap at the west end of the Shamokin Basin could be assessed. The third report, the *Report on the Coal Lands of the Mount Carmel and Shamokin Rail Road Co.,* was probably a prospectus for the developer, but the Mount Carmel railroad was never built.[37]

The consulting work took an enormous amount of Rogers's time, but it was lucrative, for he told William that he had managed to put aside $1,000 from his earnings, with which he bought one bond of the Michigan Central Rail Road, a substantial investment in 1850. In 1851, he arranged to be paid $3,000 for further work in the area but was so concerned that information on his income might go beyond his private correspondence that he cautioned William to "keep very close in regard to all my pecuniary matters."[38] He may have been expressing no more than an interest in privacy, but since he was still repaying the Biddle debt, his concern may have had something to do with that.

In August of 1850, shortly after finishing his first report on Zerbe's Run, Rogers took time to attend the annual meeting of the AAAS at New Haven, where he gave three papers. One paper was intended as a refutation of an earlier paper by Isaac Lea in which Lea had suggested that footprints found in Devonian rocks were reptilian.[39] Reptiles had been commonly associated with the later Carboniferous period, and Rogers, in keeping with his belief that animal classes appeared suddenly and that their appearance was associated with specific geological conditions, was unwilling to believe that any reptilian form appeared before that time.

He also gave a paper on coal, repeating views he had stated elsewhere, and one on salt. Salt, Rogers argued, was leached out of volcanic rocks by rain. If the lake which received the runoff had no outlet, salt built up and created a salt lake. In arid climates, evaporation exceeded rainfall, causing extensive deposits of salt along the shore lines of the salt lake. Using this principle, Rogers argued that it ought to be possible to judge past climates by examining ancient seabeds and their salt deposits. Rogers had given the part of his paper that dealt with the creation of salt lakes, but not that dealing with the association with climate, at the Boston Society of Natural History earlier in the year. Shortly thereafter he had learned that Charles Darwin and others had already attributed the origin of salt lakes to the lack of outlet, but Rogers considered his statement about rain and evaporation and the application of the idea to an interpretation of ancient climates a new contribution.[40]

Rogers's comments on climate pleased Agassiz, who put aside his true feeling about Rogers long enough to compliment him on providing a means of assessing ancient climates. Spencer Baird, permanent secretary of the AAAS and responsible for its publications, wanted Rogers to send the paper for publication in *Scientific American,* where information on the association's proceedings regularly appeared. Rogers did not send it, but Baird saw to it that the same abstract that appeared in the AAAS Proceedings appeared in *Scientific American* along with Agassiz's comment.[41]

From time to time, Rogers was called on as an expert in law cases involving questions of geology, especially those having to do with the mining of coal and iron. In July 1848, for example, his testimony was sought in a court case involving an iron deposit in Lebanon County with which he was familiar because he had studied it in the course of the survey.[42] In November 1850, he nearly became a part of a different sort of lawsuit that involved several geologists and a chart. He was asked to come to Albany to testify for James Hall and Agassiz in a lawsuit brought against them by James T. Foster, a schoolteacher and the author of a geological chart published in 1849 for use in the public schools. Hall, Agassiz, and others thought Foster had no credentials in geology and denounced the chart for gross errors in portraying American geology. They were particularly upset by the inclusion of the Taconic. In addition, Hall felt a proprietary interest in all the results of the New York survey and resented Foster's attempt to use them as part of his chart. Hall was also preparing a chart of his own for the New York schools.[43] Late in 1849, Eben Horsford had sent, or taken, the Foster chart to Charles Jackson, James Dwight Dana, Benjamin Silliman, and Rogers for their opinions. Rogers responded immediately that it was little more than a gross plagiarism of a chart published by Brongniart in Paris some sixteen years earlier, but added that, in spite of its shortcomings, any response to it was more likely to promote the piece than to discredit it.[44] The others pressed their opposition.

Because the Taconic was included, Emmons, who had first proposed the formation, became a principal defender of Foster's chart and even tried to get Hall to endorse it. He also approached William Mather, but when Mather looked at the chart and recognized it for what it was, he refused to have anything to do with it. Emmons claimed, however, that Mather endorsed the chart.[45] This added more fuel to the growing battle over the chart, which continued throughout 1850, leading to the lawsuit against Hall and Agassiz. Although he was not sure whether his other obligations would permit a trip to Albany, Rogers was willing to come if possible. He told Hall that he had in his possession an English translation and a facsimile of the Brongniart chart, which had been published in Glasgow in 1837, and that he would also attempt to find an original French copy that he believed was in William's possession in Virginia.[46] He planned to leave Boston in late November for Albany for Hall's defense, go on to New York to meet with a Helfenstein representative, and then go to Washington, where he was to begin his second series of lectures at the Smithsonian in early December. He had also received an invitation to give a series of twelve lectures in New Orleans, for which he was to receive $1,200, and he was contemplating a lecture tour with stops in Cincinnati, Mobile, and Charleston on the way to New Orleans after leaving Washington.[47] The hearing, however, was postponed until March 1851, and Rogers was unable to participate. He was thankful not to do it because he continued to think it was better not to call further attention to Foster in spite of the fact that he felt no sympathy for him.[48]

Because the hearing was postponed, Rogers was able to go directly to New York. He arrived on December 1, 1850.[49] He suffered a serious illness on the way that was diagnosed as erysipelas, a streptococcal infection characterized by intense local inflammation of the skin. He was able to conduct some business in New York in spite of it and attempted to travel to Washington for his lectures, but the illness forced him to turn back. He immediately contacted Joseph Henry to reschedule the lectures. Henry did not reply until late December and then suggested that Rogers postpone his lectures until the following season. He told Rogers that Benjamin Silliman would deliver a series of lectures in the winter instead.[50]

Throughout the late fall, Rogers had grown impatient with Henry's slow response to efforts to confirm arrangements for the lectures.[51] Against a relationship already strained, the letter from Henry irritated Rogers. He thought Henry was being unfair and that his action in substituting Silliman was a deliberate effort to slight him. Silliman wrote to Rogers as the new year began, trying to persuade him that Henry had acted in good faith and had not realized that Rogers might recover in time to give his lectures in the current season. Nevertheless, the cordial relations that had existed throughout the years between Rogers and Henry were altered, and Rogers never gave the lectures. Silliman, who had not formally accepted the invitation to lecture in Rogers's place, also declined to deliver any lectures because of the circumstances.[52]

Throughout December, Rogers continued to conduct his business with representatives of the Helfensteins in New York, transmitting maps from

Lesley and making arrangements for future work, but he gave up the idea of going to New Orleans because, he said, so many were seeking his help on coal and because it was rumored that there was cholera in the city.[53] By January 3, 1851, Rogers had settled back in Boston, where he would soon see the Pennsylvania survey begin again.

13 In Pursuit of a Great Objective, 1845–1852

During the period when Rogers was making his transition to Boston in 1845 and 1846, the survey had remained a constant concern. Trego had been reelected to the legislature and was determined to harass Rogers as much as possible. In mid-February 1845, he introduced a resolution demanding that the secretary of the commonwealth present all available information on the current state of the survey, particularly the state of the final report.[1] In response Rogers noted that he faced a nearly insurmountable task in correlating the diverse materials that came to him from the assistants. His response also shows, however, that he had continued to work on the survey and that the work was in an advanced state by 1845. In spite of meager finances, Rogers said that he had succeeded "in so far completing the description of the survey as to make it highly expedient that the Legislature should forthwith authorize its publication." He continued:

The three state cabinets have been labeled and catalogued and deposited, under directions from the Secretary of the Commonwealth, in the places assigned by the Legislature to receive them, namely, in Harrisburg, Philadelphia and Pittsburgh.

The geological maps, and a very extensive series of sections, diagrams, pictorial views, and other drawings essential to the elucidation of the report, are also finished and only wait an appropriation to go into the hands of the engravers.

The chief portion of the text of the final report is written, and the remainder, at present necessarily in outline, will be completed long before the engravings and the numerous wood cuts can be sufficiently advanced, to make it practicable or judicious to begin the printing.[2]

Arguing on the basis of what he estimated to be the nearly $76,000 spent on the survey, on its general importance to the state, and on the fact that it was so nearly completed, Rogers urged the legislature to allocate the funds necessary to undertake immediate publication.[3] He expected that the final publication would cost the state about $15,000, and offered assurance that no part of that was to compensate him for his time. Although later questions suggest that the cabinets were not as fully prepared as Rogers indicated, his report temporarily quieted further questions about the survey, if indeed anyone but Trego had questioned it, but his arguments for more funding fell on deaf ears. It would be another year before the legislators returned to the survey question.

In March 1846, Trego was appointed with two others to a legislative committee to report to the house on the survey.[4] Trego would do nothing to help Rogers, and the committee reported that, in their opinion because of the financial situation of the state, the $15,000 needed for the publication of the report could not be allocated. Furthermore, the committee argued that enough practical information had been presented in the annual reports so as to make the final report wholly unnecessary. But as useless as they suggested a final report would be, the committee members were concerned by the fact that Rogers had in his hands the raw data, a nearly completed final report, and a catalog for the cabinets. The committee members implied that Rogers was intentionally withholding all the material and that he should be forced to surrender it.[5] No one moved to take any action against Rogers, but it was evident that as long as Trego had power, money to complete the survey was unlikely to be forthcoming.

Undaunted by this obstacle, however, Rogers was willing to continue work toward the full completion of survey report without state financing, and he was able to rehire Lesley late in 1846, at his own expense, to assist him. Lesley gave up what had become nearly a full-time commitment to the ministry and took up residence in Boston where, in return for his work on maps and sections, Rogers offered him $400 for four months' work and free attendance at his lectures.[6]

The survey found strong support the following year when, in his annual message to the legislature in January 1847, Governor Francis Shunk ignored Trego's committee completely and spoke out for the survey's completion, calling attention to the production of a final report as the main purpose of the survey. This and the money already spent made the matter "of sufficient importance to secure legislative consideration."[7] As was customary, the governor's suggestion was referred to a joint committee of the house and senate for consideration. Trego was not a member of the committee, and unlike the report of the house committee of 1846, this report was favorable. It emphasized the practical importance of the survey, Rogers's cooperation with the state, and his long months of work without compensation. Rogers presented this committee with a partial manuscript for the report, some 1,400 foolscap pages in length. Noting that the final cost of the report might be cut by eliminating some of the illustrations, the committee spoke against any such deletion as being detrimental to the total work. In answer to the comments of the 1846 committee that suggested that Rogers was deliberately withholding information and papers, the committee assured the legislature that Rogers "still retains the possession of the work, but only for the purpose of maturing it, and is ready, at any moment, to deliver the same to the Secretary of the Commonwealth, printer, or any other person the Commonwealth may designate to receive the same."[8]

By the end of 1847, a completed final report of nearly 2,000 pages had been submitted to the secretary of the commonwealth. Governor Shunk, in his annual message on January 5, 1848, once again encouraged its publication, and it was submitted to the house and senate for study.[9] The whole survey had become little more than a burden to Rogers, but now, he thought, the end was in sight.[10] He described his plans for it in detail to Samuel Haldeman. There would be 600 wood engravings, 70 to 80 large plates, and 2 large maps. Two thousand copies were to be printed at a cost of about $20,000. "The superintendence will of course be given to me, a labor I do not covet, for so numerous and various are the illustrations, between two and three years will be required to pass the whole through the press."[11]

His expectations of finally completing the report did not seem misplaced when a resolution favoring the publication was introduced in the house and senate along with another about the cabinets. The resolution for publication met with little resistance and was approved.[12] The resolution concerning the cabinets was a more difficult matter as house and senate committees went back and forth over the issue. Judging from the final resolution accepted in April 1848, concern focused on deciding whether an additional cabinet was warranted and in what institutions some of the cabinets were going to be permanently placed. According to Rogers's earlier report to the legislature, cabinets had already been put in Harrisburg, Philadelphia, and Pittsburgh. Rogers was working on a fourth one (which was to be a second one in Pittsburgh) to be made up of specimens taken from each of the three other collections. The final resolution directed that the latter be deposited with the Western Pennsylvania University in Pittsburgh *only* if the collections in Philadelphia and Harrisburg were unharmed by its creation.[13] To ensure that the collection deposited in Philadelphia, which was then in a building at the University of Pennsylvania, would be available to the public, the state insisted that the city see to its deposit in a public building.[14] The city designated the Franklin Institute as the repository for this collection, and a $5,000 bond assuring that the specimens would be properly arranged and exhibited was posted by the institute. The collection was probably transferred to the institute in the summer of 1848.

In spite of the legislature's approval of the resolution for publication, it did not approve the appropriation for the publication. Once again the completion of the Pennsylvania Geological Survey was stopped. Rogers did not himself name a person responsible for the defeat, but James contended that it was Trego.[15] Trego had not been a member of any of the committees that studied the resolutions in 1848, but having been so critical of Rogers, he would be expected to lobby against it. As a long-time legislator and the year before a member of the Ways and Means Committee, he wielded considerable influence without sitting on any committee. In addition, opposition to the survey in 1851 from William F. Packer, who was speaker of the house in 1848, suggests that Trego may have marshaled his support to help defeat the appropriation in 1848.

Before the momentum for publication in 1848 developed, Rogers had thought about taking all the material for the final report to England for publication.[16] Therefore, when he went to Europe in 1848, after the defeat of the survey appropriation, he took some of the materials with him to explore avenues of publication.[17] Nothing came of his effort immediately, but in the long run his visit to England had a direct impact on the eventual publication. While he was in England, Rogers negotiated with Alexander Keith Johnston, geographer to the Queen and noted publisher of atlases, for the publication of an earthquake chart. Although such a published chart has not been located, the alliance between Rogers and Johnston established a relationship that eventually resulted in the publication of the final report.[18]

Apparently the specimens deposited in the prescribed locations had not been arranged fully. Early in 1850, therefore, the Pennsylvania legislature asked Rogers to complete the arrangement of the cabinet in Harrisburg so that it would be available to the public. Rogers told his brother, James, that he thought that the request probably "originated with Trego's love of fossils, if he is at Harrisburg."[19] But he argued that the cabinets were illustrations of the final report and should not be completed and available until the final report was published. Even if the legislature insisted, he told James emphatically that he would not do it unless he was paid for his time. There is no indication that he did anything with the cabinet in Harrisburg, but at about the same time, he did send a catalog to the Franklin Institute so that members could organize the cabinet that had been placed there by the city of Philadelphia. The state of the cabinets is indicated by the fact that they experienced a problem in matching specimens to the catalog and that when the task was finished, nearly 1,000 specimens were left unidentified.[20]

Perhaps because the legislators wanted to see the cabinet in Harrisburg arranged for the public and because Rogers insisted that the cabinet illustrated the final report and was relatively meaningless without it, a motion was made in the house of representatives in the spring of 1850 to have the Ways and Means Committee look into publishing the final report. This small glimmer of hope faded as quickly as it had come, however, when the motion was amended to remove any chance of funding for the publication.[21]

The turning point for the Pennsylvania survey came in 1851 when the legislature approved funding for the publication of the final report, and fieldwork was reopened in order to bring information up to date. General interest in the survey's final report had been on the increase because of the continuing rapid development of Pennsylvania's coalfields. Mining developers needed geological information, but there was virtually no place where they could get it unless they commissioned private studies such as Rogers did for the Helfenstein group. In 1849, for example, when the Pequa Railroad and Improvement Company was arguing for the development of the western end of the southern coalfield, only the limited information available in Rogers's *Second Annual Report* could be cited in support of their argument.[22]

When the 1851 legislative session opened, the governor told the assembly that "the true interests of the State and a just appreciation of her character for enlightened enterprise" demanded the publication of the final report and the completion of the cabinets. This time the governor would not be alone in his efforts, for as he spoke, the foundation for renewed legislative interest in the survey was being laid by William Parker Foulke.

Foulke was an ardent promoter of scientific activity and was concerned with development in the coalfields. He was a member of Philadelphia's Academy of Natural Sciences, where there was much discussion of coal, discussions that were encouraged by members like Samuel Haldeman and Richard Taylor, who were deeply interested in the coal resources of the state. Haldeman was a longtime friend of Rogers, and Rogers and Taylor had known each other for years. Taylor had written a major study of coal in 1845 in which he referred favorably to Rogers's work and called Rogers "one of the ablest geologists of the day."[23] Foulke was sympathetic to the needs of the coalfield promoters and was convinced by such comments of Rogers's abilities and the need to complete the survey. Foulke, who was well known for his work on behalf of prison reform, had many influential friends, including several state legislators. He knew how to work with the state legislature and how to bring existing support for Rogers together in an effective package.

Foulke expressed his first interest in the survey in September 1850 when he wrote himself a memo in which he identified Rogers and briefly sketched the reasons why the final report had not been published, as well as why it should be. Among the reasons for the failure of the legislature to allocate funds for the final report, he listed disappointment in the annual reports by people who wanted detailed studies of their area and failed to find them, generally unreasonable expectations of the survey, and the actions of persons hostile to Rogers.[24] These were, in Foulke's mind, unimportant when compared to the benefits to be derived from the publication of the final report.

The first weeks of the legislative session were filled with feverish activity to reinstate the survey and publish the report. Early in January, Foulke met with Rogers and encouraged him to initiate efforts to solicit support from groups such as the American Association for the Advancement of Science, but Rogers was hesitant, telling Foulke that he wanted the support of that organization to be voluntary and not prompted by him. Rogers told him that the AAAS had appointed a committee in 1849 to encourage the publication of the final reports of all surveys, many of which, like that of Rogers, were waiting for funding but that the committee had forgotten about it. He sent Foulke the names of the people he thought were members of the committee in the event that Foulke wanted to contact them. They included Agassiz as chairman, Bache, Joseph Henry, Silliman, and William C. Redfield of Connecticut, a committee that Rogers thought spoke for the importance the AAAS gave the final report.[25] Rogers seems to have forgotten that he was himself a member of the committee, but he was right in saying that the committee had done nothing.

Foulke pursued a much more fruitful line when he enlisted the help of his friend George S. Hart, representative from Philadelphia and member of the house Ways and Means Committee. Hart had been a strong supporter of the survey in 1847 and 1848, but he had been no match for Trego. Although Trego was no longer a member of the house, his adamant opposition to the survey and concern about the cabinets, along with various complaints about the survey's lack of thoroughness had created a current of hostility in the legislature that was directed at Rogers personally. Foulke and Hart had to deal with this problem as judiciously as possible, and Hart decided that the most prudent thing was to keep Rogers away from Harrisburg.[26] Instead of having Rogers speak with legislators about the survey, Foulke worked with the legislators primarily through Peter Sheafer, Rogers's former assistant, who was now a mining engineer working in the Shamokin coalfields. Sheafer had become politically astute in the years since he left the survey, and he was able to act as an aid to Foulke, as a contact with the legislature, and as a lobbyist for the survey.[27] He was willing to work without pay for the sake of getting the final report published and with the hope of getting a job with Rogers should the survey be reopened.

At the same time that Foulke began his campaign with the legislature, Rogers hired Charles Hegins to lobby in behalf of the survey. Hegins, who had been a supportive member of the state legislature when the survey was in its early years, was now a member of the Helfenstein group. Sheafer attempted to work with Hegins, but according to Sheafer, Hegins did little, leaving him with the whole burden. Sheafer spent a great deal of time in January working on behalf of the survey, all without pay, and because of his own efforts, Hegins's failure to contribute much began to irritate him. Foulke smoothed the ruffled feathers by assuring him that Rogers would surely give him a job if the survey resumed.[28] Aware of the problem with Hegins, Rogers decided to get rid of him and in mid-February tried to hire Foulke as his formal agent with the legislature, offering him a financial arrangement similar to one he had with Hegins. Foulke refused, content to do the work without the formality.

By early February, Sheafer reported that things looked good and that Thomas Bigham, author of the favorable report on the survey in 1847, was going to adapt the report for presentation to the current legislature. Bigham, however, wanted to stress the practical nature of the survey by adding information on the importance of geology to agriculture.[29] In the meantime, Foulke had asked Rogers to prepare a summary of arguments favoring publication to give to Hart.[30] Since he was still preparing his arguments for Hart when Bigham's request for more information was received, Rogers included the information on agriculture in the report, which he sent to Hart on February 11, 1851.

These early efforts by Foulke and Sheafer got the attention of the members of the house and senate, and on February 12, Sheafer reported that the whole issue of the survey and its continuance had been referred to a combined committee of the house and senate. Favorably disposed toward continuing the survey, the committee immediately began to discuss the amount of the appropriation that would be necessary to bring it to a final conclusion, taking into account both the cost of publication and the fact that Rogers felt it was necessary to update the survey results with more fieldwork, especially in the coal areas. Hart suggested a budget of $30,000. Sheafer and Rogers estimated the cost at $35,000, however. Sheafer was alarmed by the potential shortfall in funds if Hart's suggested figure of $30,000 prevailed, but Foulke reminded him that $30,000 was better than nothing.[31]

In the meantime, Foulke had been gathering together testimonials in favor of the survey's completion, including ones from Samuel George Morton, Booth, Frazer, and the American Philosophical Society.[32] Frazer and Booth may not have liked Rogers, but they wanted to see the information from the survey made available. In an effort to encourage support for the legislative committee, Sheafer gathered together all the testimonials and other information that he thought were important to the deliberations of the joint committee and, impressed by its scope, began to contemplate writing what he called a "state paper." He proposed to prepare the report secretly but to let the committee issue it as their work. Hart agreed, and after some hesitation, so did Bigham.[33] Sheafer intended to prepare his paper without any help, but he did want to meet with Rogers in Philadelphia so that they could discuss the details of the final report, including style of the maps and reports, the cost of publishing, and the appropriation for each year. Thus, Foulke sent a telegram to Rogers on February 17 to come to Philadelphia, but ignoring Sheafer's comments that he (Sheafer) would write the report, Foulke told Rogers the materials for writing the report were all in Philadelphia and that the committee wished to have Rogers write it to suit himself.[34]

The following day (February 18) Foulke received word from Rogers, who said that his experiments on anthracite combustibility for the Zerbe's Run survey, then under way with Hayes in Roxbury, would make it impossible for him to leave Boston for nearly two weeks. Alarmed, Foulke sent an urgent letter to Rogers making it clear that his presence was needed immediately, particularly for the preparation of a budget for the survey. To emphasize the importance of the situation and the need for Rogers's presence, Foulke sent a telegram to Rogers almost as soon as the letter left his hands.[35] Reluctantly, Rogers put aside his work to come to Philadelphia, but his enthusiasm was gone. "I care little," he told William, "for the issue matters may take at Harrisburg, but I feel it to be due to myself and those who have extended themselves in the Survey to exert myself at this."[36]

Rogers arrived in Philadelphia on February 20 and met with Foulke and Sheafer the next day. They began work together on the report, and by the end of February, it was nearly completed. Rogers had wanted to go to Harrisburg as soon as he met with Foulke and Sheafer in Philadelphia, but Foulke, knowing that some people felt Rogers's presence there would do more harm than good, had persuaded him to stay in Philadelphia. By early March, however, with the report in an advanced state of preparation, Rogers did go to Harrisburg, no doubt to meet with the members of the joint committee to discuss a draft of the report.[37] As a result of this meeting, committee members unanimously agreed that the report should be printed and distributed.

While all this was going on, informal bids for the publication of the final report were being solicited by the legislature, possibly to help the committee make a determination on the amount of the appropriation request, or possibly to establish a sense of its cost independent of the committee. Rogers made it known that he wanted the Boston publishing house of Little, Brown to publish the final report because he could maintain closer contact with them than he could with a Philadelphia firm. Rogers projected a publication cost of $20,000, and Little, Brown submitted a bid of $16,000. The lower bid was nice, but once again Sheafer was alarmed because he feared that if the bid was eventually accepted, the legislature would use it as an excuse to lower the $30,000 allocation that Hart had suggested.[38]

With news of the committee's favorable stance and word that the state of Virginia was seriously considering publishing William's final report, Rogers's confidence that success was at hand prompted him to admonish Lesley, who was then working on the coal land maps for Rogers's independent surveys and on the long-planned map of the United States, to rest his eyes for the survey work. Foulke instructed Sheafer to send several hundred copies of the committee report to influential persons in Virginia, probably hoping to encourage the publication of William's report, which in turn would add more support for the publication of the Pennsylvania report.[39]

On March 7, the committee's report was presented to the house and the senate. It summarized the history of the survey, noted that other states and other countries had given great emphasis to geological surveying, pointed out that Pennsylvania was so well endowed with natural resources that their proper understanding through a survey was incumbent on the state, and discussed the general contents of the final report. In keeping with Bigham's earlier charge to stress the practical, the report emphasized mining, manufacturing, and agriculture. Underlining the obligation of the state to make its richness fully known to the world, the report closed by emphasizing the state's responsibility to industrial interests and to commercial prosperity. The state's geological riches, it said:

have been bestowed on us to use for ourselves to the fullest measure of our abilities to understand and enjoy them, but likewise to transmit and to diffuse. . . . Only by so doing, can Pennsylvania show that she appreciates and is worthy of the high career in the history of the world's progress to which she has been called; only by so doing can she show that she possesses the qualifications of a pure ambition, a liberal enlightenment, and a sense of duty to fit her to accept that exalted station which the Great Creator has assigned her amid the communities of men.[40]

Sheafer's fears that a low bid from Little, Brown would cause a reduction in the final appropriation recommended by the committee ended when the committee report recommended an appropriation of $32,000, of which $15,833 was for the publication itself and the remaining $16,167 for two seasons of fieldwork and two more years to see the work updated and through the press.[41]

Sheafer lobbied hard in Harrisburg for the appropriation. Three days after the report was read, Hart presented a memorial from the manufacturers of iron in the city and county of Philadelphia, asking for an appropriation to complete the report. Because he realized that the iron manufacturers spoke from a powerful economic base in the state, Foulke had especially wanted this memorial, so much so that he wrote it himself and then had it signed by nearly thirty of the manufacturers, representing the controllers of the principal ironworks east of the Allegheny Mountains.[42]

Neither Foulke nor Sheafer expected much opposition to come from the Pennsylvania senate, but when the report was presented, opposition was, indeed, forthcoming from Senator William F. Packer of Lycoming County. Packer, former speaker of the Pennsylvania House of Representatives, had been elected to the senate in 1849 and was a powerful member of the senate's finance committee. Packer was concerned about the reopening of the survey, fearful that it was economically motivated and a self-serving ploy on the part of Rogers. He questioned whether Rogers would bring the survey to a creditable conclusion. Foulke directed Sheafer to emphasize in discussions with Packer that the reopening of the survey meant only a limited resurvey of some areas and that Rogers wanted nothing more than to bring the survey to a conclusion. Point out, he told Sheafer, that Rogers "has already suffered much disappointment and mortification from the delay of publication" and that his "chief expectation of reputation benefits is connected with one survey. [Therefore] it *must be the* great work of his professional life—He can never engage in one as important to his standing among scientific men." Foulke also thought it worth stressing that the Virginia report was then pending and that the publication of both reports would "govern the future geologists of the U. States," establishing a standard reference for position and nomenclature.[43]

On March 17, the senate passed an amendment to the house resolution, reducing the total expenditure to $20,000.[44] This action would seem to have been due to Packer, because he was adamantly opposed to reopening the survey fieldwork. Foulke, in Harrisburg regarding a pending bill on prisons, visited Packer and tried to persuade him that the report should be updated. By the time they parted, Packer had agreed to talk again about it.[45] Foulke also talked with Senator Malone from Bucks County, who he felt was favorable to the bill. Malone thought the answer to the problem was to rephrase the appropriation request so that it still asked for $32,000 but called only for publishing the final report without mention of reopening any fieldwork. Malone felt no qualms about this even though he made it clear to Foulke that Rogers could still use part of the money for fieldwork. It would simply be a silent part of the preparation of the final report.[46]

The new phrasing was introduced, but it did not escape the attention of several senators, especially Packer and Timothy Ives, who immediately introduced an amendment specifically stating that no money appropriated for the survey was to be used "for any re-surveys, explorations, or other out-door operations whatever."[47] Foulke and supporters of the survey had, however, done their job well, and this amendment was defeated by a margin of six votes. The resolution was adopted in the senate on March 27.

After the bill passed the senate, Rogers was optimistic about its success if it could reach the house in time to be put in the general appropriation bill. It arrived too late, however, and Rogers's mood turned once more to despair and disgust. He told William he no longer had any desire to do more work for a state he had come to despise.[48] Things looked dark, but Malone fought to attach the appropriation to another bill. He succeeded, and on April 14, the governor signed into law act 341, directing the publication of the final report. The legislature put the power of choosing a publisher in the hands of Hart and Abraham Olwine from the house, Malone and Henry Muhlenberg from the senate, and the secretary of the commonwealth. The state was so confident of the imminent appearance of the final report that orders were accepted for copies.[49] Rogers was just as sure and put a crew in the field immediately. Lesley was already work-ing for him on his private coal surveys, and they were joined by Peter Sheafer, Edward Desor, and also John Sheafer.[50] In spite of the optimism, it would be seven years before the report was published, and any notion that the process would proceed smoothly during this time was pure illusion.

Difficulties began at once, when the Treasury Department balked at the appropriation, saying that there had been no specific appropriation of $32,000 made in addition to the state's general appropriation and that therefore no money was available. Rogers, concerned that the fieldwork just begun would be terminated, decided to keep his crew in the field at his own expense, hoping for eventual reimbursement.[51] He spoke with the governor and members of the publication committee appointed by the legislature and received a favorable hearing, but he did not receive funds during 1851. Nevertheless, he kept the work going at his own expense and was reimbursed the next year.

Rogers went to Cincinnati in May to attend the meeting of the AAAS, but most of May, June, and July was spent in Philadelphia, meeting with the publication committee and haggling over potential publishers. The only thing Rogers enjoyed about this was that he stayed with his brother James and had a chance to visit with old friends. Since the legislature strongly opposed the use of a firm outside Pennsylvania, Rogers decided it would be prudent to gather information on possible publishers in the state, even though he wanted the work done by Little, Brown of Bos-ton.[52] Two Philadelphia publishers considered the job, Sherman (who had been one of three publishers of Rogers's 1844 presidential address to the AAGN) and the firm of Hogan and Thompson.[53] The law required that the publisher give security bonds for the work, and both Sherman and Hogan and Thompson asked Rogers to post the bond for the field-work. He refused. Sherman offered an alternative in the form of a part-nership with Rogers, with the firm paying for the bond. Rogers refused this also, and as a consequence Sherman may never have submitted a formal bid.

Hogan and Thompson tried to dodge the problem by suggesting that another geologist (presumably one who would post the bond) be hired to supervise the work. In spite of the fact that Rogers did not look forward to the production labor, he certainly was not willing to let someone else do it. Fortunately, Foulke and Hart stopped the move, and the matter was settled when Rogers agreed that if Hogan and Thompson were awarded the contract, he would find others who would post the bond for him.[54]

Hogan and Thompson underbid Little, Brown by $225, and the firm was awarded the contract for publication.[55] In spite of Rogers's wishes to the contrary, he was reasonably content with the decision, but as will be seen later, the mandatory awarding of the contract to the lowest bidder sent the survey hurtling toward failure once again when Hogan and Thompson went out of business before the year was over.

Once matters in Philadelphia were settled, Rogers returned to the field only to leave again in August for another meeting of the AAAS. The still young American Association for the Advancement of Science was struggling with challenges and dilemmas, and Rogers felt that his attendance was critical. Bache, Joseph Henry, Agassiz, Peirce, Frazer, Torrey, and others had continued to grow as a powerful scientific clique in the United States. They were "men committed to promoting American science through research and institutions for its support," their every action encouraging a high level of professional science in the United States.[56] They were determined to see that the AAAS developed as an acceptable scientific forum for their purposes, but in doing so they alarmed many of the older members who had found the original Association of American Geologists and Naturalists attractive either because it was an association only for geologists and naturalists, or because it had not made serious distinctions between those persons who met the criteria of Bache and his group and those who did not. Henry and William Rogers were outspoken advocates of democratic scientific organizations that allowed every person to have his say, and Henry had taken care to see that the constitution of the AAAS was, in his words, "democratic." But Bache and the others feared that such an organization might fall under the domination of amateurs who would prevent it from reaching their goals. Rogers had tried to ward off the problem in 1849 by proposing two classes of membership, one that would include the researchers, who would control the association, and the other the "amateurs." The effort failed, and it seemed to him and many others that there was a force at work bent on taking over the association for its own purposes, which, as far as Rogers and others could tell, were the interests of the United States Coast Survey, which Bache headed, of Henry and the Smithsonian, and of Agassiz and the Lawrence Scientific School. Too much power, Rogers thought, was being put in the hands of a limited few, who he later referred to as the Mutual Adulation Society, or as did others, the "Holy Alliance."

By 1851, it seemed clear enough that those whom Bache and his group considered the less "professional" members of the group were being squeezed out by the standing committee of the AAAS. The standing committee was composed of three current and three retiring officers, chairmen of the sections, and six elected members. It was charged with making all decisions and arrangements between regular meetings. Since the committee controlled the papers to be presented, it was able to guide the organization in preferred directions. The committee included Rogers as well as Agassiz, Peirce, and Bache, and Rogers felt that he arrived in Albany "just in time to act as the voice of the dissenters." He told Desor that the standing committee

has been usurping control of absolutely everything perpetuating committees in behalf of the Coast Survey etc. etc. . . . , and doing all sort of irregular things such as filling its own vacancies in spite of the requirement of the Constitution which demands the election annually of 6 members by Ballot. Agassiz seemed to have no thought of rules or parliamentary courtesy and when I spoke to a point of order calling him back to a certain rule, he acted like a very child, huffed and proposed to resign the chair.[57]

Rogers expected that at the following year's meeting (1852) in Cleveland there would be a "rare contest" over the function of the committee, but the contest never occurred, because the meeting was postponed for a year. Ostensibly the postponement was because of a cholera threat, although many of the members living west of the Alleghenies saw it as a deliberate attempt to exclude them from the organization. The Cleveland meeting was finally held in 1853, but neither Rogers nor Agassiz was there. When a local publication of the meeting's papers was suppressed in favor of an official one that omitted a paper the standing committee thought was too amateurish, the fears of Rogers and others were realized.[58] Thereafter Rogers grew less and less interested in the AAAS.

Rogers had been worried enough in 1851 about what was happening to the AAAS to think about starting a new society more along the comfortable lines of the old AAGN. When, in the early winter of 1851, Rogers and Desor met with Josiah Whitney and John Foster (not to be confused with James Foster of the geological chart controversy) in Pottsville for the purpose of discussing the anthracite deposits, the subject came up. Whitney and Foster had a wealth of experience in geology and had been longtime members of the AAGN. Foster had served on the Ohio geological survey, and Whitney on the Iowa survey. Together, they had directed a government land survey in the Upper Peninsula of Michigan. They, too, were concerned over the policies of the AAAS, and they took advantage of their meeting in Pottsville to discuss the possible formation of a distinctly geological society that would have a library, a research collection, and a journal devoted solely to geology. Rogers did not pursue the idea, although others, including Whitney, Foster, and Charles Whittlesey, continued to talk about it for a few more years.[59]

In spite of his involvement with the structure and power struggle within the AAAS, Rogers gave three papers at the meeting: on the "Infra-Carboniferous" vegetation of Pennsylvania, on the age of reptilian footmarks, and "On the methods of investigation adopted in the geological survey of Pennsylvania."[60] His paper on the reptilian footmarks was a further defense of his belief that reptiles had not lived as early as the Devonian, a defense apparently so lively that Lesley told his wife that Rogers was "struck with a sort of amusing terror [at the thought of reptiles appearing so early], saying that if so then all the geological rules were overturned."[61] Whether indicative of his preoccupation with the survey or general disillusionment with the AAAS, he sent none of them for publication in the *Proceedings,* and no abstracts appeared in print.

While Rogers was in Philadelphia and Albany, the men were at work in the anthracite area, using Pottsville as a base. In August, the Sheafers, Lesley, and Desor were joined by Leo Lesquereux, whom Rogers hired to work on the coal fossils, and then, or shortly after, by John Colter.[62] Lesquereux, a classmate of Guyot in Neuchâtel, had attracted Agassiz's attention in 1844 because of work he had done on the fossil plants found in the peat bogs of the Jura Mountains. He came to the United States in 1848 with Agassiz's encouragement and became one of the earliest authorities on fossil plants in the United States. His work on the Pennsylvania coal plants would form an important essay in the final report as well as an independent publication.[63]

Although still unwilling to put too great an emphasis on fossils, Rogers had become more sensitive to their importance and was eager to have more fossil studies for the final report. He tried to find additional assistants who could, like Lesquereux, do these kinds of studies. Duties were to include "collecting organic remains, defining their position and limits by measurements, and working out in detail various paleontological boundaries which I am now studying."[64] His search for such assistants was fruitless, and Lesquereux remained the only member of the survey who studied fossils to any extent.

Before the fieldwork began, Lesley had agreed to work in the field for four months each year of the survey and to work an additional three months drawing maps. As the work began, however, Rogers decided that more extensive topographical work was necessary and that he needed to have Lesley in the field for at least five months. In late June, Rogers wrote to Lesley about the change and also asked him to meet him in Philadelphia to discuss various matters concerning the survey. There was a misunderstanding about the date they were to meet, and Rogers was upset when Lesley failed to arrive when he thought he should. He wrote an angry letter to Lesley.[65] The letter triggered a hostile response that marked the beginning of the end of their long friendship.

Lesley had carried a heavy burden as Rogers's principal assistant before 1841 and had repeatedly returned to work for him in the following years. The tedious work of drawing maps, which he frequently complained about, had taken a toll on his health and his patience. Field conditions in the renewed survey were hard, and Lesley felt that Rogers did not pay sufficient attention to the needs of his field assistants. When he received Rogers's letter, it was the last straw. Lesley began to act in an erratic fault-finding manner toward Rogers. He wrote to remind him of their years of friendship and the fact that his affection for Rogers had been the reason for his agreement to resume working for a commonwealth that he (like Rogers) found less than desirable. He assured Rogers that the only reason he had continued to work on the survey was his feeling for Rogers, but he also told Rogers that he [Rogers] had a recurring problem with other people that stemmed from an all-consuming passion for understanding the geology of the state. "In the pursuit of a great object," Lesley said, "you are too apt to neglect . . . your best friend." He went on to say that this kind of selfish behavior had earned Rogers enemies in the past "and alienated one after another of those whom you need to lend you support."[66] He accused Rogers of paying him too little and complained that Rogers had not sent him a promised agreement outlining his responsibilities, an agreement that he wanted in case Rogers died. Although he still felt personally devoted to Rogers and said he would continue to help him, Lesley felt that their relationship had to be restricted to business or to friendship and that it could no longer be both.

Rogers was genuinely concerned, and even baffled, by Lesley's letter.[67] He sent the desired agreement to Lesley, and when Lesley returned to Pottsville in mid-August, he was warmly received by Rogers who, Lesley said, insisted "on attending to my luggage."[68] For a while it was as if nothing had happened between them, and things were again peaceful as Susan Lesley joined her husband in Pottsville and described a typical evening in their parlor: "Here, Rogers and Shaefer [*sic*], Colter and Desor set of evenings, and talk of coal and conglomerate, and all sorts of geological matters. . . . They arrange their plans for the next day, and suggest to each other different arrangements."[69]

Fieldwork continued into the early winter, and Lesley prepared a map of the coal area, a tedious and complex job best described in his own words:

I undertook the task of mapping on a large scale, and representing in light and shade the surface aspects, the outcrop terraces, the sandstone ribbing of the gaps, the varied erosion of the crests, and the relationship which the opened gangways bore to these. The Southern Basin, from ten miles east of Pottsville to twenty miles west of it, was cross-barred in parallel lines, one or two thousand feet apart, running from the crest of the Sharp Mountain, about N. 25 W. to the crest of the Broad Mountain, and in some cases to the summit overlooking the valley of the Mahanoy. These lines, measured, leveled, and staked, were tied together by transverse staked and leveled lines, one set running along the hill tops, another along the valley bottoms. To this system all the railroad surveys were tied; the branch roads, gangways, trial and air shafts were located properly; and in a few indispensable cases, where beds had not been opened, new trial shafts were made.[70]

As the result of such work, Lesley's health gave out in November, and he ended his work in the field earlier than planned. He offered to fulfill his obligations to the survey through "office work" (a euphemism often used by Lesley when he meant the actual construction of the maps and sections from his field notes and measurements) and to go into the field in the spring in order to finish the map and to make up time he had been unable to work in November because of illness. Thereafter, he said, he wanted to end his involvement with the survey. Rogers agreed.[71]

Rogers expected Lesley to follow this proposed schedule, but when spring came, the relationship between Rogers and Lesley deteriorated rapidly. The two had reached what appeared to be a satisfactory agreement on the completion of Lesley's work, and Rogers was cordial and solicitous of Lesley. Lesley became increasingly nervous about details of his work, however, particularly his salary. After Rogers's problems in getting his funds from the state the previous year, Lesley wanted absolute assurance that the financial arrangements for his work in the spring were secure. Rogers tried to reassure him but felt that he could not guarantee anything until the money was actually in hand. Nevertheless, Rogers hoped that Lesley would consent to work even beyond his commitment and would spend up to five more months in the field.[72]

For reasons that are not clear, Lesley suddenly refused to go into the field at all. As a consequence, Rogers fired him.[73] Rogers believed the problems that had developed between him and Lesley were the result of Lesley's mental state rather than of any particular actions he had taken. Indeed, Rogers thought Lesley was insane, telling Desor that Lesley's "malady of Temper, I fear of *insanity* is evidently growing upon him."[74] Rogers was relieved to be rid of him but lamented to his friend Desor that the episode was more evidence that "I am ever the victim of ill fortune in this ill-fated survey." He was determined to finish the survey, but so many things had gone wrong through the years, he no longer believed that it would be "what I once fondly dreamed I should make it."[75]

The anger Lesley had begun to show toward Rogers turned to bitterness very quickly, and as the years passed, he would take every opportunity to denounce Rogers. In turn, Rogers would retaliate by pointing out what he considered to be the great inaccuracies of Lesley's work. He became so convinced that Lesley had done a poor job that he had nearly all the sections and maps that Lesley had done redrawn. Of all the people Rogers had worked with, he had the easiest and longest relationship with this man who would one day be his successor as the head of the Pennsylvania survey. They had accomplished much together. Desor, although not considering the break detrimental to the personal happiness of either, noted that "for the sake of science" it was, indeed, an unfortunate event.[76]

Rogers not only lost Lesley as an assistant in 1852, he lost Desor when he returned to Switzerland early in 1852 because of family illness. Not long after Desor left, Peter Sheafer also left, and John Sheafer probably left, too, since he is not mentioned in later survey records. Rogers missed Desor's presence on the survey, but he also missed him as a friend. There were few, other than Desor, Lesley, the Hillards, the brothers George and Charles Sumner, and his own brothers, with whom he had ever felt close, and in letters written to Desor over the next few years, he never failed to talk about their friendship and how much he wished Desor would return to the United States. Separation from William, and particularly the death of his brother, James Blythe, on June 12, 1852, deepened his sense of isolation.

Lesquereux nearly left the survey as well. Hired to work for three months in 1851 for $500, Rogers wanted more of his time in 1852 at an increase of $100 in salary. Lesquereux, however, felt put upon by the arrangement and threatened to leave. He changed his mind, although the course of his relationship with Rogers grew stormy over the coming months and years.[77]

Lesquereux's attitude toward Rogers worsened in the fall of 1852. By then, he had had a chance to become familiar with the earlier work of Rogers's assistants both by reading the annual reports, most of which he saw for the first time in the summer of 1852, and by corresponding with Lesley. He also met and had frequent conversations with R. M. S. Jackson, another former assistant. As a result of the things Jackson and especially Lesley told him, Lesquereux began to regard Rogers as a poor director and to believe that the assistants had done an overwhelming amount of work that had gone to waste under Rogers's direction. His relationship with Rogers was altered by his new ideas about the earlier years of the survey, and it worsened in the winter as he and Rogers argued over his salary and about the relative importance of fossils in identifying coal formations. As the months passed, Lesquereux became increasingly bitter toward Rogers, calling him jealous, unjust, vain, ego-tistical, ignorant, and incompetent.[78] He thought Rogers was afraid his work would be stolen and thus too protective of it, but in a strange twist, Lesquereux began to worry that Rogers would steal his work on the coal fossils. Lesquereux became so paranoid over the issue of protecting his own work that he threatened to go to the state with his concerns. When Rogers visited him at his home in Columbus, Ohio, in the late fall of 1852 to review some of his work with him, Lesquereux was sure Rogers's only purpose in coming was to claim the work as his own.[79]

In spite of his concerns, Lesquereux decided to finish his work and stayed with Rogers until 1854. When he handed his completed study to Rogers in the spring of that year, he found that his concerns had been for nothing. Rogers praised his work and drawings, guaranteeing that his name would appear with them, which it did, in the final report. In addition, Rogers wrote a flattering introduction to an article on fossil plants from the Pennsylvania coals by Lesquereux, which was published by the Boston Society of Natural History the following August.[80]

Rogers had difficulty finding replacements for Desor and Lesley but hired A. A. Dalson, a Finlander who, with Henry A. Poole, a land surveyor, completed the maps on which Lesley had been working at the time of his departure.[81] Dalson and Poole were with Rogers throughout 1852, and they were joined by James Rogers's son William, who was often called Willie by family members, in July. Dalson left at the end of the 1852 season, and Poole went to Nova Scotia to superintend mining operations in Pictou, leaving Rogers with only Willie to help him.[82]

Few to Take an Interest in My Volumes, 1852–1855

14

In spite of his skirmishes with Lesley and Lesquereux, the first few years after the survey resumed were relatively good ones for Rogers. His personal life was filled with moments of great joy occasioned by his marriage and by William's move to Boston. His consulting work continued to expand, and his intensified geological studies in the framework of the survey allowed him to make refinements in his classification and nomenclature that would be used in his final report.

When Rogers decided to keep his geologists in the field at his own expense in 1851, he did so with the certain conviction that he could easily support himself with his consulting work, which, he told William, was going very well.[1] Even when money for the survey expenses, including his salary, was finally available, he made the decision to continue to put his salary back into the operation of the survey. Therefore, he spent as much time as he could in his private consulting work in order to support himself. He saw no real conflict in this procedure, nor did he feel that it detracted from his survey responsibilities in any way, for the information that he gained from these private surveys was added to the information for the survey report.

Rogers prepared a number of reports, including one on the mining resources of the Lackawanna Basin and the lands of the Delaware, Lackawanna and Western Railroad Company.[2] The Helfensteins continued to be one of his major employers. Another one was Charles M. Wheatley. Rogers's relationship with Wheatley meant that some of Rogers's work became part of the Exhibition of the Industry of All Nations, the first international fair held in the United States.

In 1850 or 1851 Wheatley, a major developer of mineral lands, asked Rogers for an opinion on the lead and copper deposits of Montgomery and Chester County. The result was a report for Wheatley completed in February of 1852 that strongly supported the potential of the areas for mining copper and lead.[3] Further work was subsequently done in the same area, and Wheatley commissioned Rogers to make several maps of the area to be shown as part of an exhibit on the Wheatley Mines at the Exhibition when it opened in New York in 1853. The fair was held in a building especially constructed for the occasion and called, like an earlier British counterpart, the Crystal Palace. One exhibit was on the ores of lead from the Wheatley Mines but was not limited to samples of minerals. It was, rather, a comprehensive exhibit on the ore and its mining, complete with machinery as well as ore samples. According to a contemporary report, Rogers's maps of the area were "suspended in the Exhibition, and upon them the metallic veins are traced in gold."[4] Rogers insisted on being allowed to retain the right to publish the maps himself, and at least one of the maps, without the tracings in gold, is probably the same as the one of the area in the final report. In addition to his own fee for the preparation of the maps, Rogers, ever on the lookout for ways to improve the financial condition of the survey, made an additional $1,000 contribution to the survey part of his arrangement with Wheatley.[5]

Rogers came close to having a deeper involvement with the Crystal Palace exhibit. Sometime in late 1852 or early 1853 he was contacted, through his friend George Hillard, about becoming the head of the mineralogical and geological department of the exhibition.[6] The contact was made by navy captain Charles H. Davis, who, along with Captain Samuel Francis DuPont, was in charge of the whole exhibition. Rogers accepted but then heard nothing more. Hillard finally wrote to Davis to find out the cause of the delay only to learn that someone else had been hired for the job. He did not say who had been chosen in Rogers's stead, or why, and since there was never anyone formally described as head of mineralogy and geology, the answer remains unclear.[7] There were several people involved in the exhibit with backgrounds in these fields, any one of whom might have replaced Rogers. Benjamin Silliman, Jr., was called the manager of the mineralogy and chemistry departments, and he probably assumed some of the duties that would have been Rogers's. Others that might have taken over Rogers's tasks included D. T. Ansted, a British geologist, James Hall, and Charles Wetherill. Exhibits that were labeled "geology" but were mostly paleontological in nature, were under the direction of Ansted.[8] Hall contributed a major essay entitled "On the Geological Relations of American Rocks and Minerals" to an annotated catalog of the exhibition, in which he detailed the New York stratigraphy. Wetherill contributed his "Report on the Iron and Coal of Pennsylvania" to the same catalog.[9] Wetherill's paper expanded on exhibits of these products that were part of the fair and made several references to Rogers, using his formation numbers throughout in order to place the various deposits stratigraphically. To help the reader, he reproduced a table from Rogers's *Second Annual Report* showing the thirteen numbered formations.

Rogers felt quite sure that he had been hired and was surprised to learn he had not been. What led to the decision to bypass him is no more evident than is the name of the person who replaced him. Since Rogers's views on geology had met with so much opposition, however, the exhibit's directors might have been persuaded that they needed a geologist more in the mainstream of thought. In addition, Wetherill had been involved in the difficult and unsuccessful effort to identify all the specimens in the survey cabinet at the Franklin Institute and as a consequence might have communicated some misgivings about Rogers.

As a result of his surveys of the coal lands for private companies and the reopening of the survey, Rogers felt it was necessary to refine the classification and nomenclature as given in his 1844 presidential address to the AAGN. Sometime after 1844, in an unpublished and incomplete manuscript for a textbook, slight alterations were made in the boundaries of four of the divisions, but between 1850 and 1852 Rogers increased the number of divisions to fifteen and made substantial changes in the nomenclature.[10] (See figure 8.) Some of the names he now used to identify the formations were the same as those in his 1844 paper, but others were new, and still others were taken from earlier plans. Another major change was his placement of the beginning of the Devonian lower in this scheme than he had done in any of his earlier classifications. Until now, he had gone along with Hall and Conrad, who tended to place the beginning of the Devonian somewhere above the Oriskany formation of New York, or Rogers's Formation VII. Accordingly, Rogers had variously placed the beginning of the Devonian at the beginning of his Formation VIII or in the middle of VIII.[11] He had, nevertheless, suspected for some time that it might be lower in the scale, and now he made the Devonian coincidental with the beginning of the Oriskany, or his Formation VII. Later, New York would do the same.

The 1852 classification continued to reflect Rogers's conviction that there was a definite structural break (an unconformity) between Formations III and IV (the Matinal and Levant) and that Formations I, II, and III represented a distinct group that was the equivalent of Sedgwick's

Cambrian. Most Americans, led first by Timothy Conrad and then by James Dwight Dana, argued that there was no such break in the stratigraphic sequence in the United States. Although arguments in Europe usually focused on the lack of a significant paleontological break, in the United States, arguments in favor of a Cambrian were often discouraged because the admission of one system below the Silurian might lend credence to the notion of a Taconic sequence of rocks below the Silurian.[12] Undisturbed by this concern and more interested in the structural evidence than in what the fossils suggested, Rogers was one of the few geologists in the United States who supported Sedgwick's argument. Like Rogers, Sedgwick also tended to emphasize the importance of structure. His controversy with Murchison erupted with renewed vigor in the early 1850s, and in a published paper, Sedgwick turned for support to Rogers's discussion of the unconformity between Formations III and IV.[13]

Sedgwick sent Rogers a copy of his paper in the fall or early winter of 1852, along with a volume of his three-volume work on a classification of British paleozoic rocks. After reading the paper, Henry blamed his and William's failure to publish some of their work for Sedgwick's inability to convince geologists in England about the Cambrian: "I think our friend Sedgwick has all the philosophy and the justice on his side, yet, through our fault of procrastination in publishing, he has allowed Murchison to encroach on his whole ground, and to secure a sort of title by mere priority of occupancy of what is not his."[14] Thus convinced that his work would assist Sedgwick, Rogers responded to Sedgwick's gifts with a long letter detailing the support American geology offered the Cambrian concept. Sedgwick was delighted with the letter, and for a short time, he considered publishing an extract from it.[15] Although Rogers did not directly pursue the issue with Sedgwick at this time, he was preparing a map of North America that would, he believed, offer additional support to Sedgwick.

The map was being done for a forthcoming revised edition of Alexander Keith Johnston's *Physical Atlas of Natural Phenomena*. Sometime after Johnston and Rogers met in 1848, Johnston asked Rogers to provide some corrections on American geological formations for his atlas of the world.[16] This request evolved into an invitation to Rogers to do a complete new map of North American geology for the revised edition, scheduled for publication in 1856. Rogers, with Lesley's help, had been working on a map of the United States since the mid-1840s, the same map that had caused him to refuse to help Hall. He had been unsuccessful in earlier attempts to complete the map, but now that he had a specific goal, he was determined to finish it and expand it to all of North

America. Rogers reviewed the formations of the entire United States and particularly the central and western regions, with which he was least familiar, entreating his longtime friend Spencer Baird to send him any information that might come to his attention concerning these regions.[17] He was confident that his map, when finished, would supersede the work of others and particularly that it would replace an 1853 map by Jules Marcou. Marcou, another friend and associate of Agassiz from Switzerland, had come to the United Sates in 1848 with Agassiz's help, in part to make a collection of American fossils for the Geological Society of France.[18] Marcou went back to Europe in 1850, returning in 1853 to serve as geologist with the U.S. Army Topographical Corps surveying a Pacific railroad route along the thirty-fifth parallel. Rogers saw Marcou's map in February of 1853, possibly a copy in the possession of Marcou's friend, Charles T. Jackson, who was a frequent participant in the meetings of the Boston Society. Most American geologists, including Hall and Dana, found fault with Marcou's effort, and Rogers was no exception. He found the map seriously flawed and at least twenty years out of date, "which," he told Desor, "in the U.S. geology was before the Flood." Among other things, it did not acknowledge the possibility of a Cambrian.[19]

At the same time that Rogers was working on a map in a scale suitable for the *Atlas,* he also worked on a larger version that measured six feet by four feet eight inches. Rogers believed the larger map would also be published at some point by Johnston, perhaps as an independent map. It was to show all of Rogers's fifteen formations in detail, including the formations that he thought represented the Cambrian, while some of them had to be combined on the smaller map because of space limitations. Rogers never finished his larger map, but he completed the smaller one in 1855.[20] The formations as shown on this map are grouped in six major units. The legend accompanying them, however, lists all of the formations and acknowledges the Cambrian.[21]

Rogers welcomed new directions in his personal life in 1853. First, William wrote to say that he had decided to leave the University of Virginia and come to Boston.[22] William had taught at the University for eighteen years. It was a position he loved, and he enjoyed the total respect of his colleagues. Nearing the age of fifty and financially secure, however, he wanted to have time to participate more fully in the scientific community, and like Henry, he found Boston a stimulating center of activity. Furthermore, through his wife, who was the daughter of the Honorable James Savage of Boston, there were close family ties to the area. Henry was ecstatic at the thought that William would be once more with him, and he looked forward to the opportunity to pursue their plans for a school. In addition, with Robert in Philadelphia, where he had replaced James as professor of chemistry at the University of Pennsylvania (a job that James had held from 1847 until his death), the three brothers were to be geographically closer to each other than they had been for many years.

As happy an occasion as this was, it was little more than a prelude to Rogers's decision in the summer of 1853 to marry Elizabeth Stillman Lincoln, who was called Eliza by family and close friends. Rogers had remained single because he felt that as long as he was paying the debt incurred by the failure of his iron furnace, he could not afford to support a wife. His debts were paid now, his consulting work provided enough income to support a family, and he wanted to marry. He told his friend Desor that he was "happier than I can tell you at the prospect [because] I have long needed a wife and home, and now I shall have both to my heart's content."[23] Eliza was the daughter of James O. Lincoln of Hingham, Massachusetts, and half sister of William's wife. They were married in Boston on March 5, 1854.[24] The following year their first child, a daughter christened Edith Lincoln, was born.

At the time of his marriage, everything seemed to be going as well as it possibly could for Rogers and the survey. He had no reason to look forward to anything but a future that would see the fulfillment of his work and dreams. The survey, however, was destined for more trouble. In 1854, the prospects for publication of the report were suddenly mired in confusion. The precipitating events began in 1851, but by the time Rogers and members of the legislature were aware of impending doom the situation had gotten out of hand. Before the end of 1851, the firm of Hogan and Thompson was driven out of business by financial failures. This firm was succeeded by Hogan and Company, which assumed the publishing obligations of Hogan and Thompson. By law, the publisher was to have received all of the allocated money for the survey, which was to pay Rogers's salary, the survey field expenses, and the cost of publication. The legislature was ready to allocate the money in 1852 to Hogan and Company, but while this company was willing to publish the report, it refused to pay the other expenses. Unable to carry the financial burden of the whole operation, Rogers turned to the state for help, and the state assumed responsibility for paying the expenses, including those for 1851, which Rogers had paid from his own pocket.[25] At the same time, the state decided to withhold all payments to the publisher until the work was finished, a decision that would have serious consequences.

The publication headed for potential disaster when Hogan and Company went out of business on the death of Hogan in May 1852. This company was succeeded by Parrish, Dunning and Mears. Parrish, Dunning and Mears entered into an agreement with the state and with Thompson, as the surviving partner of the original company, on November 1, 1852, to carry out the provision of the original contract for publication. At that time, Rogers had several maps and sections ready to be engraved, which the company placed in the hands of Philadelphia engraver Wellington Williams, but the company refused to let Rogers supervise the engraving as had been specified in the original agreement. This might have become a critical issue except that, like its ill-fated predecessors, Parrish, Dunning and Mears went out of business, and their assets were liquidated in April 1853, leaving the fulfillment of the company's commitments unclear. It was still not immediately apparent to Rogers or the state how serious a predicament they were in, and not until a year later did the legislature suddenly recognize that there was a real problem. Rogers was immediately summoned to Harrisburg (in March 1854) to make a full statement on the progress of the final report.[26]

Recognizing that the report was nearly ready for publication, the state appointed a committee to look into a new avenue for its publication. During the winter of 1854–1855, the committee tried to determine just who was responsible for the publication. The partners of the defunct Parrish, Dunning and Mears told them that Thompson was responsible. Thompson argued that the state had abrogated its contract by amending the contract in order to pay Rogers directly and by passing a bill that prevented any further payments to the publisher until the work was completed. Thus, the committee had little choice but to find a new publisher. Although Rogers did not fall into a debilitating depression as the result of the confusion, he was despondent, telling Desor there were few "to understand or take an interest in my volumes."[27]

At first, Rogers and the committee discussed the situation with what seemed to be little hope of a solution, but then, on February 10, 1855, Rogers sent a formal letter to the committee proposing that he assume the contract for publication and asking for three more years "for the deliberate and careful execution of this responsible duty."[28] Although he had estimated earlier that the engraving, once in the hands of an appropriate firm, would take about two years to complete, the state had called for a geological map on a scale twice as large as Rogers had planned, a scale that was the same as one used on a map of the state by William E. Morris that had just been published. As a result, Rogers would have to redraw the map entirely, which was undoubtedly the reason he needed the additional time.[29]

Rogers stipulated that he was willing to forgo any salary. This would make his bid for the contract less than might be expected from any other publisher.[30] With some minor dickering, the committee members decided that "they can suggest no more prudent or economical plan than to confide the whole work, in its supervision and publication, to the State Geologist himself. With him it has been the work of a life time; and his professional reputation is involved in the manner in which the results of so much labor and research shall be spread before the world."[31] On the third of May (1855), Act 440 of the Pennsylvania legislature formally awarded the contract for the publication of the final report to Rogers under the same conditions as those originally awarded to Hogan and Thompson. Stipulations included the completion of the report in three years, the posting of a security bond by Rogers, and the provision of copies to specific individuals and to state governments. In return, Rogers was to hold the copyright as any other publisher would have held it, but this arrangement, of course, did not relieve him of the responsibilities outlined in the contract.[32]

When Rogers filed his request with the state, he indicated that nearly all the maps, sections, and illustrations had been completed and were ready for engraving. In addition at least five hundred pages of the text were finished.[33] While he supervised their engraving, he planned to complete the remaining text and the enlarged map. Rogers decided to pursue the publication of the final report abroad, but since he was afraid of public objection, he did not want this generally known. With his objective in mind, he sailed quietly for England on June 20, 1855, taking samples with him in order to get cost estimates.[34] Because of his established association with Johnston, he left with the intention of having the engravings done by his company and the report published by Johnston's publisher, Blackwood and Sons, who maintained offices in Edinburgh and London.

Rogers arrived in Liverpool in early July and spent a few days traveling in Wales and England before settling in Edinburgh. Once settled, he began to work on the survey materials.[35] Rogers asked his nephew Willie, who was the only assistant he now had, to continue the survey fieldwork and to provide a continuous stream of new and revised information. He also expected Willie to deal with the state on his behalf as necessary to get the money due him for the report. Before leaving, Rogers had accepted consulting jobs for coal studies in the Scranton and Carbondale areas, and he asked Willie to do the lion's share of this work as well, including the writing. Rogers demanded weekly reports from Willie, and although he often complimented and encouraged him, he pushed him relentlessly, expecting him to be everywhere and do everything at once. He was not always kind to his young nephew. He chided him for wasting time with some of his drawings, for not working quickly enough, for spending too much time with the Scranton project, and for chewing tobacco.[36] Rogers assured Willie, however, that he knew his "leading wish in all you do is to meet my instruction and my approval."[37] Apparently it was, for Willie seldom complained and only once asked for help, an assistant to help him with the Carbondale project. Willie was devoted to his uncle and several years later commented that "my anxious and harassed uncle trusted to my naturally ready memory to supply the defects of his own, my pen, my pencil, all my acquired information and my zeal, were at his service by day and by night."[38]

Rogers went back and forth between Edinburgh and London, but he spent most of his time in Edinburgh, both writing and supervising the engraving of the sections and maps for the final report. He took time to go to the British Association meeting at Glasgow in September, where he gave a further report on reptilian footprints, another entitled "On Some Geological Functions of the Winds, Illustrating the Origin of Salt, etc.," and a third on the geology of the United States exhibited by a map, apparently the same that was to be published by Johnston. Sedgwick later expressed regret that he had been unable to attend the meeting because Rogers's paper would have given him an opportunity to comment on the Cambrian.[39] Rogers sent a copy of the map to Sedgwick the following December. He noted that it was too small in scale to do justice to the subject, but still planning to complete the larger map, he assured Sedgwick that the larger map he was preparing for Johnston would be more useful.[40]

When he went to Europe, Rogers left his wife and infant daughter with William, who had settled with his wife at Lunenberg, near Boston. Edith was not well when Rogers left, and in frequent letters during the fall and winter of 1855, William tried to reassure his brother that her health was gradually improving.[41] At the same time, Henry reported that his own health had been good in the Scottish environment. Although Rogers was busy working on the final report, surrounded by friends, and confident that all was well at home, the Christmas season was a lonely one for him. The season was one filled with childhood memories of family gatherings, but now he was away from his own daughter at her first Christmas. The family affection that existed among the Rogers's brothers is shown in this touching letter from William, written on Christmas Day:

The ground is white with snow and sleet, and the icy shower is rattling against my windows as I sit down to speak a loving word to my dear brother across the sea. There is an influence coming from early association which fills this holiday season with tender recollections of the past, and with kind as well as wise resolves for the future. With what an earnest solicitude for your happiness does my heart now warm towards you, my dear brother, and with what true joy do I dwell on your improved health and the prospect of future cheerful labour and mutual helpfulness for us all. A thousand wishes crowd to be expressed, but I can only say, God bless you, my own dear brother! and beg you to take as the type of my present thoughts the happy affection of our boyhood which, ever dwelling on and around us, overflowed our breasts in this festive season, making our home, even shadowed by poverty, a place full of earth's truest, sweetest happiness. The long interval of years has not dimmed the images of parental goodness or of loving brotherhood. To-day we may open the casket in which they are kept within our heart of hearts, and have sweet pleasure in dwelling on the dear memory of those who have left them to us.[42]

After a long summer and winter of work on the report, Rogers was eager to return to his family in the spring, but he remained in Edinburgh until October. His decision to stay was prompted by the possibility of a professorship at the University of Glasgow.

To Leave a Land Sterile of Friendship, 1855–1857

15

Rogers was approached late in 1855 about a potential position at the University of Glasgow.[1] The professorship was in natural history and had a long history. The faculty of the university had voted in 1803 to appoint a lecturer in natural history, and in 1807, when William Hunter's anatomical museum was about to be moved from London to the university and expanded to cover natural history, a royal warrant was issued, creating a professorship of natural history. The holder of this appointment was to serve also as keeper of the museum. The professor was to be appointed by the reigning monarch and was to have all the rights and privileges of any other professor and an annual salary. The ensuing argument between faculty factions over the propriety of such an action resulted in the natural history professor's being recognized only as a member of the university and not of any single faculty within the university. Therefore, he was not entitled to vote in meetings of the faculty or to claim one of the houses allotted to professors that belonged to a faculty.[2] Since 1829, William Couper, who taught zoology, had held the position under these terms.

While Couper still occupied the position there could be no formal word about a potential opening, but Couper was growing old, and it was obvious to the rector of the university, who was George Douglas Campbell, duke of Argyll, that the position would soon be vacant. The duke of Argyll had far-ranging interests in geology and from time to time contributed to the London Geological Society, where he may have met Rogers for the first time. A general familiarity with Rogers's theories, with his work in Pennsylvania, and with Rogers's well-known skill in speaking were incentives for interest in Rogers as a potential candidate to replace Couper. Furthermore, the duke of Argyll was later called the Darwinian Duke because he favored the evolutionary theory of Charles Darwin, and if he harbored an interest in evolution at this early date, he found a kindred spirit in Rogers. Rogers was invited to Inveraray Castle in April, where the two men spent time studying geology and hunting grouse and surely discussing the professorship.[3]

After waiting so long for the time that he and William could be together to continue their plans for a school, thought of a permanent move to Scotland must have filled Rogers with mixed emotions. Several factors encouraged him to think seriously of the possibility, however. One was his concern about the direction of science in the United States, particularly as he saw it embodied in the American Association for the Advancement of Science. Had there been a place in the geological community of the United States for Rogers, he might not have cared much about this, but he enjoyed no such place and felt a strong sense of isolation. He was convinced by his long struggle with the survey that no one in the United States cared about his work and that he was, as he had told Desor three years earlier, a "lonely heart" in a land "sterile of friendship."[4]

Another factor that gave him added incentive to leave the country was the proslavery stance of the United States government. Although Rogers had given up his radical social leanings long before, he had remained sensitive to human injustice, particularly as embodied in slavery. His closest alliances and friendships through the years tended to be with others who opposed slavery, including his scientific colleagues Haldeman and Lesley and almost all of his Boston literary and political friends. He had not engaged in open antislavery activity, but the issue weighed heavily on his mind.

The potential of a Glasgow professorship explains a wholehearted effort by Rogers in the spring of 1856 to establish himself as an active participant in scientific activities in Scotland and England and to make sure his ideas were thoroughly known. He did this through the presentation of papers at various scientific societies, beginning with the weekly meeting of the Royal Institution of Great Britain on February 8, two months before visiting Inveraray, and continuing with papers at the Royal Society of Edinburgh and at the British Association for the Advancement of Science after the meeting.[5] His papers were, for the most part, a recapitulation of ideas expressed elsewhere. The one given at Edinburgh was one that he had prepared for the final survey report and a very complete statement of his theories. Although essentially a repetition of ideas expressed elsewhere, this paper did include a new discussion of cleavage.

Rogers had argued earlier that the cleavage dip was in one direction and parallel to the average dip of the axis plane of a fold, a point on which he had been challenged several times, particularly in regard to cleavage in the Alps. Daniel Sharpe, who had made himself known as an opponent of Sedgwick's Cambrian early in the controversy with Murchison, had turned his attention to a variety of other geological matters since then. Sharpe's attention had been called to Rogers's work on cleavage by Charles Darwin in 1846 and again in 1851 when Darwin referred to Rogers's belief that cleavage paralleled the axis planes of elevation as "Rogers's Law."[6] Sharpe was not impressed. He visited Rogers in the late summer of 1855 and at that time pointed out to Rogers that his [Sharpe's] observations indicated that cleavage dips alternated in anticlinal or synclinal belts.[7] Because of this, Rogers introduced a second law of cleavage in his paper at the Edinburgh meeting, suggesting that where cleavage was fully developed, the planes were not parallel to the axis planes but radiated from them. With this concept, he could, he believed, explain the observations presented by Sharpe and others. He also related foliation (the separation of different minerals into layers in the rock) and cleavage for the first time, arguing that both were the result of the "parallel transmission of planes or waves of heat, awakening the molecular force, and determining their direction."[8]

Rogers showed himself to be unafraid of controversy and any potential effect it might have on his quest for the Glasgow position by continuing his support of the Cambrian in his papers. His strongest statement on the Cambrian was in his paper to the British Association. While admitting, as Murchison argued, that there was no clear paleontological distinction between the Silurian and the Cambrian, he stressed that the structural break evident in the United States supported the validity of Sedgwick's argument. Rogers's support for the Cambrian got an immediate reaction from Murchison and his followers, although Rogers thought his arguments "were too strong for them."[9] Rogers also introduced some new thoughts on the Silurian and Devonian in the BAAS paper, suggesting that paleontological evidence showed that the Silurian-Devonian break might not be valid in the United States and pointing out that Hall had recognized dual Silurian-Devonian affinities in New York.[10]

At the very beginning of 1856, before Rogers first met with the duke of Argyll, he was appointed an editor for the *Edinburgh New Philosophical Journal,* a position he held until his death.[11] His appointment, which was specifically as editor for America, reflected the desire of the scientific community in Scotland to have reliable contact with America, but it provided Rogers with another opportunity to show himself as part of the scientific community of Scotland. Rogers's stance on the Cambrian issue may have affected his standing with the journal, however, for he was unable to accomplish one of the first tasks he set for himself. He had expected to publish an abstract of his BAAS paper in the journal and promised Sedgwick he would circulate it widely in America "for your sake more than for my own, for I am resolved the Truth of Nature shall be heard."[12] The article did not appear in the *Edinburgh Journal,* leading to the suspicion that the other editors favored Murchison's view and kept the article out of the journal in spite of Rogers's wishes.

In retrospect, one of the more interesting contributions that resulted from Rogers's role as editor was the publication in the summer of an article by William on the discovery of fossil trilobites in altered slates near Braintree and Quincy, Massachusetts.[13] William Rogers identified the trilobites as *Paradoxidis spinosus,* a species associated with some of the earliest Paleozoic formations and suggestive of a distinct set of rocks below the Silurian. *Paradoxidis* was characteristic of the so-called Primordial, or First, Fauna of Joachim Barrande, who in the mid-1840s, had arranged the lower Paleozoic rocks near Prague, Czechoslovakia, in several stages. The stages were called Primordial Fauna, Second Fauna, and Third Fauna, and Barrande demonstrated that there were characteristic trilobites by which each Fauna could be recognized. On a trip to England in 1850, Barrande had discovered the Primordial Fauna in rocks beneath the Silurian. William believed the Primordial Fauna was the equal of Henry's Primal Formation at the base of his Paleozoic sequence, thus substantiating a formation below the Silurian and one that Henry showed on his map in Johnston's *Atlas* as being the equivalent of the Cambrian. The effect of William's article on his brother is difficult to assess, but, like anything else that supported the Cambrian, it probably generated controversy.

Rogers's map in Johnston's *Atlas* (1856) should have drawn attention to Rogers, particularly since it was accompanied by other maps and essays by Rogers, including one highlighting the work of Henry, William, and Desor.[14] It resulted in more criticism, however. William J. Hamilton, president of the Geological Society of London, took advantage of the appearance of the map to sum up the British feeling about the Cambrian, adding the comment that the fact that most Americans opposed its existence was very telling of the inappropriateness of Rogers's notion.[15]

Rogers came home in October of 1856 after fifteen months in Great Britain. Happy to be reunited with family and friends, he spent a quiet winter in Boston, attending the meetings of the Boston Society of Natural History and sharing his work, and especially his map, with its members. While the members' reaction to the map is not known, the map met with harsh criticism from Logan and T. Sterry Hunt of the Canadian survey on grounds that went well beyond the Cambrian issue. Rogers had sent Logan a copy of the map, and Logan, in turn, showed it to Hunt, his assistant on the Canadian survey. Hunt wrote to Hall, criticizing Rogers's geology so severely that Hall told Lesley, "if it be as he [Hunt] says, Rogers knows less of American geology than I have given him credit for." Rogers's map differed substantially from a nearly contemporary one by the Canadian geologists, and it is evident that there was, indeed, a vast gulf separating the views of Rogers and the others.[16]

Before the spring was over Rogers's attention was turned to what was a potentially serious and damaging plagiarism of survey materials. In late April, he learned, from an advertisement in Philadelphia's *Daily Evening Bulletin,* that a railroad and county map of Pennsylvania and New Jersey, engraved by Wellington Williams with corrections by J. Peter Lesley, had been published by Charles DeSilver, printer and seller of maps in Philadelphia. The advertisement indicated that a geological map of Pennsylvania and New Jersey would soon be available also. Williams was the engraver commissioned by the firm of Parrish, Dunning and Mears in the early 1850s to engrave the maps and sections for the final report. It occurred to Rogers immediately that these maps could be the survey maps, and he lost no time in getting to DeSilver's shop to examine the published map. Although many features had been removed, Rogers recognized it immediately as taken from the plate of the survey map.

While angry at seeing the map, his immediate concern was the imminent availability of a detailed geological map. Having recognized the origin of the county map, Rogers was sure that the geological map was also a survey map, a document to which he now had a claim, since the state had given him the copyright privilege. A few of the maps had been printed at this time, and Rogers immediately sought a legal injunction against all the involved parties, including Williams, DeSilver, Lesley, Parrish, and Dunning to stop any further publication or distribution. Most of what is known about this incident comes from the court records pertaining to the request for this injunction.[17] The records include long depositions from Rogers, Williams, Lesley, and Willie Rogers and a statement from N. B. Brown about the findings of the committee appointed by the legislature in 1855 to investigate the survey. Brown's statement was accompanied by letters the committee solicited at the time from Dunning, Mears, Thompson, and Williams.

According to Williams, he had returned the original map to Rogers in 1854 at Rogers's request, albeit he was reluctant to do so, since he had never been paid for his work. Rogers accused him of having made a copy of the map before returning it and using that copy, with some changes by Lesley, as the source of the published map and the forthcoming geological map. Williams, who felt he deserved to use the map in lieu of payment for his services, did not deny having a copy of the map, but vehemently denied that he had used any information save what was already in the public domain, and he further claimed that the direct source of the maps was J. L. Lippincott, whom he identified as a geologist. Lesley was equally vehement in his denial of involvement. He said that, although his name had been used in connection with the published map, his involvement was limited to pointing out a few corrections on a copy of the map he saw in the possession of Williams. He denied having anything to do with the forthcoming geological map or having given any approval to use his name. The court, however, decided in favor of Rogers and issued the injunction.

Although Lesley publicly denied any involvement with the map, he was neither uninvolved nor unaware of the effect the map would have. Lesley had spent arduous months working on the survey maps, tying in geological and topographical features with railroads and other county features. He either did not know, or was unwilling to admit, that anyone else had worked on these maps after he left the survey, and through the years, he had developed a strong proprietary interest in the maps. Because he bitterly resented the fact that Rogers held the copyright to the materials, he had been a willing partner with Williams in the preparation of the maps and hoped that the trouble that he foresaw stemming from their publication would force some investigation of the survey. He had told James Hall the preceding summer:

I can give you also information of the publication of a map of Pa intended to be top. & geol on the basis of mine for the Survey, but better viva voce than by letter. If you hear of any such thing do not be expecting any thing great or good. Its publication involves trouble to all parties concerned and like a child born out of lawful wedlock will be kicked round the world considerably. How far its appearance will force on an exposé of the history of the survey I do not know.[18]

Furthermore, he told Hall, there was information on the map that could be useful, thus providing justification on a less personal ground for his involvement.

Following their problems in 1851 and 1852, the once strong feelings of mutual respect between Rogers and Lesley had deteriorated precipitously, and it is not an overstatement to say that by 1857, Lesley despised Rogers. Only a year earlier, he had told Hall, without a trace of humor, that the British Association for the Advancement of Science, where

Rogers gave one of his papers, was doing Americans a favor by keeping Rogers in England.[19] His utter contempt was expressed fully and publicly in his *Manual of Coal* published in 1856. The manual is relatively brief but long enough to give a good account of the Appalachian coalfields based on Lesley's broad knowledge of the area gained in the course of his work with Rogers and in the course of similar work for others. Much of the latter was done for the Pennsylvania Rail Road Company, for which he worked after parting from Rogers in early 1852. Lesley mentions Rogers in several places in the manual, even referring to him in what seem to be flattering terms and saying that Rogers's elevation theory, whether proved wrong or right, will "remain in the history of our science one of those glorious dreams which do more for the intellect of man than verities can do, and therefore justly confer an immortal fame upon the dreamer."[20]

While Lesley may have been sincere in his comment, for in later years he sometimes defended and sometimes praised Rogers, he was also bitter and angry because he believed that Rogers had not given enough credit to his assistants for the work that they had done. His feelings are evident throughout the manual. He implied that by ignoring their specific contributions, Rogers was guilty of gross plagiarism and had prevented the various assistants from attaining a position of their own in geology. To make the latter point, Lesley opened the manual with a dedication to James D. Whelpley and Andrew A. Henderson, two former assistants on the survey who were "men of infinite scope and love in science, poets by Divine right, pure-hearts, true to every duty, to whom, as master in youth and friend in middle age, the author has owed what it would be presumptuous to attempt in word, and American science what it has never yet had the opportunity to acknowledge."[21] Had Rogers but given them due credit, the achievements of Whelpley and Henderson would be known and appreciated. In praising Henderson's topographical studies in particular, Lesley said that under any other organization than the Pennsylvania survey, Henderson's genius would have been recognized fifteen years earlier, and his would be a familiar name in science.

Lesley went on in the manual to discuss his view of the ideal organization of a survey, emphasizing its need to be democratic in giving credit where credit was due, and in a damning appraisal of Rogers, his work with the survey, and his treatment of his assistants, he said of him and his methods:

imagining himself accumulating a fortune of facts, he is but rolling up before him work which will crush him in the end. It is a sadly common error in observers to gorge themselves with undigested and finally indigestible facts; to spend a life-time in becoming intellectual stomachs, armed about the mouth like octopods with long tentaculi, which grope on all sides everlastingly to take but never give. There are representative men whose names stand high in science as appropriators, not as benefactors; who form a cuttlefish class among their fellows. [22]

Lesley adopted an extreme position on the matter of crediting individuals with specific ideas, and even the contemporary scientific press took exception to his extremism. Although the author of a brief review of Lesley's manual in the *American Journal of Science* mistakenly thought that Lesley meant to indicate that he, Henderson, and Whelpley had originated Rogers's theory of elevation, he made the reasonable observation that Lesley offered no proof of theft and that the question of proper relations between assistants and superintendent with regard to information gathered on the survey had to be settled before any such claim could be entertained. [23]

Lesley's ceaseless emphasis on credit for the assistants' work suggests that all of Rogers's assistants had been very concerned about this issue. Although Lesley's comment in 1839 that Rogers played a "cruel and shameful" game with his assistants and Frazer's departure from the survey in 1837 support the idea of dissatisfaction, there is surprisingly little evidence to suggest that this had been as great an issue as Lesley tried to make it in the 1850s. Even Lesquereux's attitude toward Rogers on this account was based primarily on what he learned from Lesley. After his statement in 1839, Lesley himself never again alluded to any problem until after he left the survey in 1852. If George Whelpley was any indication, even the assistants singled out by Lesley were not as upset as his comments indicate. In 1852, Whelpley asked Rogers for a job on the survey, which Rogers would have given him gladly had enough money been available. [24] Furthermore, in 1865, Whelpley, who had found a place as editor of the *American Whig Review* and other Whig newspapers, wrote to William that he was personally obligated to Rogers for "intellectual awakening and guidance in science" and went on to say that no one could be with Rogers without being made a deeper and broader thinker. [25] These hardly seem like the comments of an angry man.

In short, one is led to think that Lesley's hostility had some other source than a need to defend the honor of past assistants. Perhaps in reality, his attacks were fostered by a fear that *he* would be cheated of credit for his work on the maps for the survey and especially for his work on the map of the United States. Lesley had once described the preparation of this map as "a most tedious affair" that nearly ruined his eyes, but one, nevertheless, in which he took great pride.[26] When he was writing the manual it was no secret that Rogers was working on a map of the United States for Johnston's atlas, although it was not yet published. Rogers had made a point of asking Lesley for all the materials relevant to the map at the time of his departure from the survey, and this request may have made Lesley suspicious that he would be denied credit.[27]

In the early summer of 1857, Rogers returned to Edinburgh, this time with his wife and daughter because he expected to become a permanent resident. He continued to work on the survey report, but all the while he kept an eye on the developments at Glasgow. On August 5, a formal announcement of an opening occasioned by the death of Couper was made. Soon after, Rogers went by special invitation to a breakfast with the duke and duchess of Argyll. The professorship was the main topic of conversation. Their discussion was described by Rogers as "frank." He, Rogers, talked about Scottish geology and "the importance of resuscitating the long dead chair and of permitting it to be largely and at least fairly a geological one," and not, as it had been under Couper, "little else than a teaching of zoology."[28] Within days of this meeting, Rogers told Sedgwick that he was a formal candidate for the chair, supported by several friends at Glasgow, and that he was deeply interested in the opportunity thus presented. "Having a sincere desire," he told Sedgwick, "to take up my residence in Great Britain where I can, I feel assured, possess a better share of health and make better progress in Science, I am disposed to resume Collegiate tasks on the comparatively easy conditions imposed in a Scotch university, where I may have the leisure and the field for research."[29] He also told Sedgwick that Sir George Grey, home secretary, on whose decision the appointment legally rested, had recommended him and that recommendations from Murchison, Lyell, Leonard Horner, and Phillips seemed likely. Obviously, his disagreement with Murchison and others on the issue of the Cambrian was a scientific disagreement that did not color their opinion about the general abilities of Rogers. He asked Sedgwick to send letters of recommendation also to the duke of Argyll and to Grey, among others, and Sedgwick readily agreed.[30]

While Rogers was waiting for word on his appointment, William, who had not been well, decided that a sea voyage and short visit with Henry would be good for him. Leaving his wife in Boston, he arrived in England in August and went immediately to Edinburgh, where Henry was living. Together they went to the BAAS meeting in Dublin later in the month. While he was there, Rogers received word of his appointment to the Glasgow professorship, although it was November before his formal appointment by the Queen was read to the University of Glasgow senate.[31] It had taken strong friends in England and in Scotland to promote the appointment of an American to the professorship, and William reported to his wife that "[Henry] has had even greater success than I imagined in making powerful friends" to whom he owed his appointment.[32] The friendship that had developed between Rogers and the duke of Argyll was critical, but the support of persons like Sedgwick, Murchison, and Phillips was also important.

Rogers had intended to give a paper on the Cambrian in America at the Dublin BAAS meeting, still believing that his emphasis on the Cambrian in America could help change the minds of some of Sedgwick's more difficult opponents, but Murchison, unswayed by his generally friendly feeling toward Rogers, let it be known that the paper "would not be allowed if it touched on 'debatable ground.'"[33] Rogers did not give his paper, but it is likely that his appointment at Glasgow and time spent with his brother interfered with the preparation and presentation of a paper far more than any threat from Murchison.[34]

On signifying his willingness to accept the Glasgow appointment, Rogers was asked to prepare a Latin dissertation on the various uses of natural history *(De veris usibus historiae naturalis)* and to present it on December 7. This was a mandatory trial, following which he would be formally accepted by the faculty. Detained in London by the publication arrangements for the final report, Rogers asked that the date be rescheduled for December 21.[35] The presentation duly completed as rescheduled, Rogers's career at the University of Glasgow began. Since his teaching season was November to May, beginning the following year, however, he had no immediate responsibilities and was not required to be in residence at the university.

The Facts Are Better than the Theory, 1857–1858   16

The period without teaching obligations gave Rogers time to finish the final report. In letters to Willie written in the late summer and throughout the fall and winter, Rogers indicated that he was making rapid progress with the report.[1] The engraving of the maps and sections was completed, and pages from the printer arrived regularly for proofreading. He finally tendered some much deserved compliments to Willie, but as usual, sent him scurrying from one task to another. Among other things, Rogers had Willie redo sections by Lesley on the bituminous coal region. By the first of the year, publication was imminent, and in late March (1858) Rogers left for New York bearing the completed volume 1 and two maps and several sections that would accompany the volumes as a boxed set.[2] Although the second volume was nearly ready, he had to get to Harrisburg before the legislature adjourned in order to demonstrate that the work had, indeed, been finished in the promised three years. Rogers had hoped to bring copies of the work into the United States duty free and had made application at Washington to do this, but his plea was rejected. "Thus," he said, "I am to sustain a tax of another $1000 in the production of my book," money that would come from his own pocket.[3] Rogers had already incurred several thousand dollars of personal expense for the fieldwork and the publication, and efforts to get some of the money back from the state failed.[4] But after all that had happened to the survey through the years, he could hardly have expected anything to go in his favor.

By the agreement between Rogers and the state of Pennsylvania, the state was to receive 1,000 copies of the report for distribution, the same number that the state had set in 1848. A final legislative act for the survey was passed on April 21, detailing the various persons and institutions that were to receive the copies, including all the assistants, former governors, and an assortment of colleges.[5] Members of the house of representatives and the senate debated the fate of the cabinets for a last time, deciding that one should go to the recently chartered Farmers High School in Centre County (later Pennsylvania State University).[6]

The first volume was distributed quickly and was probably in the hands of most of those designated to receive it by late summer of 1858. Rogers stayed in the United States until midsummer and told his friend Charles Sumner that he was "at my old game, collecting material for the elucidation of the geology, geography and nat[ural] history of this continent."[7] Part, if not all, of this work was aimed at providing more information on the coal deposits, which were the focus of volume 2. He also attended meetings of his beloved Boston Society of Natural History whenever he could, and he went to the meeting of the AAAS in Baltimore in April. He emerged from the AAAS meeting thinking no more highly of what was happening than he had when he first tried to keep the Bache clique from becoming all powerful.

During his absence in the United States, Rogers was elected to membership in the Royal Society of London. He was recommended by John Phillips, Charles Babbage, Charles Darwin, John Tyndall, William Thomson (Lord Kelvin), and Thomas Henry Huxley, among others, many of them longtime friends. He may not have met Darwin yet, but Darwin had become familiar with his work, and although he thought Rogers's theory of fluid undulations was "monstrous," he was interested in his recognition of inversion, his work on earthquakes, his delineation of the coalfields of Pennsylvania, and as noted above, his "law" of cleavage. As early as 1846, Darwin had written to his friend Joseph Hooker to say that Rogers's work on the coalfields of North America was "eminently instructive and suggestive." A few months later, he told Robert Mallet that the Rogers brothers were "excellent geologists."[8]

The formal proposal for his membership had been made before Rogers left for the United States, but before any further action occurred Rogers had gone. Rogers's name was presented in his absence to the general membership on June 3, and he was elected on the first ballot. He first heard the news of his election when word of it came to William.[9] Formal admission was delayed until his appearance at a meeting, and that was postponed twice, first because he was in the United States and then, after his return, because his teaching prevented him from making the trip to

London in the winter. Sometime in the spring of 1859, his admission was completed.[10]

Rogers's teaching did not begin until November 1858. This gave him time to complete nearly all of the second volume of the final report after returning to Glasgow in midsummer. It also gave him time to attend to what had become his usual busy schedule of scientific meetings. He often saw his role now as a transmitter of information to and from the United States, a role made possible by his visit to the United States and by the direct and continuing contact he had with events occurring in America through William. An example was his effort to inform the British scientific community in regard to a controversy over the presence of Permian rocks in the United States. At the BAAS meeting in late September, he discussed the pros and cons of the existence in the United States of this formation, which was part of what had long been known as the New Red Sandstone in Europe. The formation lies between the Carboniferous deposits and a younger period called the Triassic. Although Murchison had suggested the name "Permian" in 1841 for a formation that had been recognized and studied under several names in Europe for centuries, the formation was not seriously discussed in the United States for several years. In 1853, Marcou suggested that it was to be found in the area of Lake Superior, a notion with which Rogers disagreed.[11] In 1856 Emmons, who was at that time the state geologist of North Carolina, argued that a group of rocks called the Newark Group that stretches from Connecticut through the southern Appalachians was divisible into two parts, the lower belonging to the Permian.[12]

Just before Rogers came to the United States in the spring, William wrote to him about a new discovery of supposed Permian fossils by the Missouri state geologist George C. Swallow, an event that William referred to as "the most striking news in geology."[13] The fossils Swallow identified had been sent to him by a civil engineer working in Kansas, Major Frederick Hawn, who also sent a group of fossils from the same area to Fielding B. Meek in Albany. Meek was a paleontologist who had assisted Rogers's friend David Dale Owen in 1848–1849 and had worked with Hall in New York during the 1850s. He had also been part of the geological survey of Missouri and had worked in the South Dakota Badlands with another of Hall's assistants, Ferdinand V. Hayden.

Meek thought the fossils were either Triassic or Permian, and, excited by the possibility of finding certain evidence of the Permian in the United States, he asked Hawn to get Swallow to send him any fossils that were not Carboniferous. Swallow didn't send them but instead claimed to have discovered the Permian fossils himself in a paper given at the St. Louis Academy of Sciences, the announcement to which William referred. In March, Meek, with his friend and colleague, Hayden, countered Swallow's claim of discovery in a paper read to the Albany Institute.[14] The controversy heightened interest in the Permian, and by the time Rogers arrived in the United States it was a major topic of conversation. The existence of the Permian was questioned by many Americans, including Rogers, and when he got back to Scotland, he told Sedgwick that he was anxious to study a good series of Permian fossils in Europe for a better comparison.[15] He reported on the researches of Meek, Hayden, and others at the BAAS meeting and discussed the evidence for and against the Permian in the United States. He told William that the geologists at the meeting were unconvinced by Emmons's arguments and tended to think that the "Permian" of the United States was intermediate between the Carboniferous and the Permian of Europe. Rogers wrote to Hayden and Meek after the meeting, giving them the consensus of the British view and requesting that they forward a set of their supposed Permian fossils as soon as possible so that they could be studied by British geologists. There is no indication that they sent any, but with or without them, Rogers became personally convinced that fossil evidence did not support the Permian in America.[16]

It was nearly mid-December 1858 before the second volume of the final report was finished, but finally *The Geology of Pennsylvania: A Government Survey* was a fait accompli.[17] Although the engravings and printing were done in England, it was bound both by Blackwood in England and by Lippincott in the United States and was issued under their joint names in Great Britain as publishers but under Lippincott's name alone in the United States. A boxed set of sections and two maps, one of the entire state measuring six feet by a little over three feet in size and one of the coalfields measuring six feet by almost two and one-half feet accompanied the completed publication. The two quarto volumes contain just over 1,600 pages of text and more than seventy-five full-page illustrations, including sections, maps, and sketches, twenty plates of coal fossils and three of Vespertine and Umbral fossils, and three full-color illustrations of outstanding geological features in the state by George Lehman. The volumes include 778 smaller sketches and sections.

In the first volume, Rogers discussed the Appalachian rocks below the coal, following a division and nomenclature like the one developed in 1852. More than half of volume 2 is given over to the description of the coal formations, including a small map of the anthracite and bituminous coalfields. A brief section on the Mesozoic formations (those immediately above the Paleozoic) and thirty pages on the iron deposits of the state, including a paper by William Rogers entitled "On the Origin and Accumulation of the Proto Carbonate of Iron in Coal-Measures" are in the second volume, as are several general studies by Rogers. The latter include a long paper on the geology of the United States, one on the physical conditions that surrounded the production of the Paleozoic rocks of the United States (the most extensive statement on this matter since the 1837 paper), and his paper comparing the disturbed zones of America and Europe. Lesquereux's paper on fossil plants of the coal and one by T. Rupert Jones entitled "Notes on the Beyrichiae and Leparotlae of Pennsylvania" are also included. In the preface, Rogers indicated that he would include a discussion of the survey's methodology, but he did not do so.

The two large maps, both color lithographs, are topographical and geological. Mountains and valleys are portrayed by hachures without indication of specific elevations. Although Rogers felt that this procedure was adequate, it was not ideal, because he had hoped from the very beginning of the survey to be able to show a more accurate vertical scale. He felt that "no map can be said to meet the wants, either scientific or practical, of a geological survey, which does not picture, approximately at least, the vertical element as well as the horizontal."[18]

The map of the state has only Rogers's name on it as director of the survey. The map of the anthracite areas has Rogers's name as director and lists the topographical assistants as Lesley, Dalson, and William B. (Willie) Rogers, Jr. In the preface to the final report, however, Rogers credited Dalson exclusively with this map. A smaller map appearing in volume 2 and entitled "Map of the Anthracite and Bituminous Coalfields of Pennsylvania Exhibiting Their Relation to Various Markets" is credited in the preface to Lesley.[19]

Although he had been disillusioned by the long years of struggle to produce the report, Rogers could not help but hope that it would meet a favorable reception, particularly so that he could anticipate enough sales to offset a major part of the additional costs he had personally incurred in its production. Amid broad general praise for the detailed observations, however, there was extensive criticism of Rogers's theory of elevation and his nomenclature, and reaction tended to concentrate on these subjects at the expense of the more general report. Typical reaction came in a review in the *American Journal of Science.* The journal complimented the volumes on their excellent style and "fullness of illustration and description that meets so well the demands of science and the interests of the State," and said "The work is a great one . . . and will ever rank among the most important of the reports on the geology of the United States." The reviewer, however, was pointed in his comments on the theory of elevation and the nomenclature. "We like the facts," he said, "far better than the theory adopted to account for them." As for the nomenclature, he added that Rogers had no justification "for diverging in these respects from the New York survey, in which the subdivisions had been founded upon a thorough study of organic remains."[20]

In Europe, the report attracted very little attention—indeed, as late as 1860 Desor reported to Lesquereux that he had neither seen a copy nor had the volumes "been heard of in Europe"—but what comment it did receive reflected the same concerns. The British *Saturday Review*, for example, was critical of the nomenclature and was predictably critical of Rogers's use of the Cambrian. While sympathizing with Rogers's long struggle to finish the final report, the reviewer suggested that the volumes would remain in Europe little more than lovely ornaments for the bookshelf.[21]

Criticism also came from persons with specific points of view. Geologists who adhered to the Mosaic account saw nothing good in the report. Eli Bowen, who has been described as "Schuylkill County's self-styled resident Professor of Geology," dismissed the report categorically on the grounds that Rogers's account of earth history did not conform to the six biblical days of creation but like the others he could not refrain from a comment or two on the nomenclature. In his waggish way, Bowen supposed that, should a formation intermediate to those described be found, it would be called "snail-ent" or "crawling-day rock."[22]

A review of American geology by T. Sterry Hunt in 1861 all but ignored Rogers's final report and his work in general except for Rogers's ideas on the classification of pre-Paleozoic rocks, with which Hunt disagreed.[23] This silence can be attributed in part to Hunt's distrust of Rogers's geology in general, but it was also a graphic comment on catastrophism. Hunt was a uniformitarian and by ignoring Rogers's final report and his theory of elevation, he downplayed the opposing view. His interest in doing so is evident in that he did devote part of his review to a uniformitarian theory of elevation by James Hall, first published in 1857.[24]

Hall had been struck early in his work on the New York Survey with the immense thickness of the strata that make up the Appalachians, which Rogers had calculated in 1838 to be more than 40,000 feet. To account for this thickness, Hall envisioned deposition taking place in a huge basin, later called a geosyncline, filled with a relatively shallow sea.

The great thicknesses of sediment that marked the Appalachian chain had, Hall argued, been carried to this spot on strong currents from a paleocontinent lying to the east. Since the evidence showed that the sea itself had remained shallow during deposition, the accumulation of so many thousands of feet of sediment could be explained only by the continual subsidence of the floor of the basin as sediment was added. Hall's interpretation of this much of the process was similar to Rogers's concept, and it was a concept that remained a critical part of future attempts to understand the elevation of mountains.[25]

Rogers and Hall diverged in their views of what happened after the sediment began to accumulate. Hall looked at the depression of the sea-bed as an event that had created pressure on the sides of the basin, causing them to curve. As the top of the basin narrowed, smaller anticlines and synclines formed within the larger syncline. While this was a contributing factor, it was not the sole explanation for the origin of the mountains, for Hall believed a critical factor in the creation of mountains was continental elevation and the subsequent denudation of the area. This was linked to the depression of the basin. As the sediment accumulated and the bottom of the basin was depressed, an adjoining area was forced upward. This process of depression and associated elevation is called isostasy, a name applied to it in 1899 by Clarence E. Dutton.[26]

Although Hall's theory was not much more immediately successful than Rogers's theory, uniformitarianism grew more important as the century advanced, and it became an ever more important factor in discounting the work of Rogers, so much so that Gregory tried to argue that Rogers was a uniformitarian because he believed that paroxysmal upheavals, although in a less intense form, continued to affect the earth.[27]

The most severe criticism of Rogers's report came, not unexpectedly, from Lesley and was concerned primarily with his perception of Rogers's failure to credit his assistants with the work they did. Lesley, as a former assistant, was one of those designated by the state to receive a copy of the final report. As soon as he received volume 1, he began to prepare a circular that attacked the "awkward, fanciful, unscientific, and unusable" nomenclature of Rogers and argued that the final report was little more than "the rewritten, condensed and diluted, first reports of the geologists of the Survey." He stopped just short of calling Rogers a thief, saying instead that he had "suppressed" the assistants' work. He complained that too few copies of the report were printed and that they were too expensive. Consequently, he said, Rogers's purse benefited, while the people of Pennsylvania were denied their just due. He blamed this on Rogers's holding of the copyright to the survey materials and ignored the fact that the number of copies printed was the number mandated by the state long before Rogers was given the copyright.[28]

Lesley failed to distribute the circular and took a different opportunity to make his feelings about Rogers and the report public. At the time he wrote the circular, he was secretary of the American Iron Association. When, early in 1859, Lesley prepared a guide to the manufacture of iron in the United States for the association, he used it as a vehicle for his denunciation of Rogers. The fact that he had now seen both volumes did nothing to lessen his resentment as his attack waxed more grand than ever, this time openly calling Rogers a thief and noting that he (Lesley) was:

one of a few persons whose unfortunate connection with the subject gives them a right to say that the most brilliant imposture and the most extensive scientific theft of the present age has just appeared in the form of the Final Report of one of the State surveys. In this immense work of nearly two thousand pages, magnificently illustrated with maps, sections, and pictures of all kinds to the number of nearly a thousand, are the results of the toil of many men for many years, all appropriated by one man to himself,—a man who, apparently upon principle, gives credit to no one else, but practically asserts and compels the world to say, by the way he publishes this book, that he has done it all, has thought it all, and owns it all.[29]

In an effort to correct some of Rogers's supposed oversights, Lesley recited a long litany of accomplishments by the various assistants, noting that their contributions had all vanished with "the wave of one small hand, repressed forever into nothingness and night," a hand "almost never lifted to assist, almost always lifted to obstruct or to defeat through an imbecile selfishness, the work it was itself devoted to."[30]

Lesley's statements on the maps that accompanied the report were equally strong. Although the state map did not bear his name, he argued that he, and he alone, had constructed it. On the other hand, he said that he had had nothing to do with the map of the anthracite area. He did not make clear whether he addressed this disclaimer to the larger boxed map of the coalfields, where he was linked with Dalson and Willie Rogers, or the smaller map in volume 2 that Rogers attributed to him in the preface of the final report. In later comments, however, he said the larger boxed map was a reduction of his sheet maps to one map on a smaller scale, and his disclaimer must therefore have referred to the smaller map.[31]

Lesley's attack met with immediate responses from Willie Rogers and from Robert Rogers. In June, Willie prepared a pamphlet in which he hoped to expose Lesley's "misstatements and groundless pretensions" and argued that Lesley spoke only from the standpoint of wanting to discredit Rogers so that he could publish a work of his own on Pennsylvania's geology, a point that Lesley made clear in his unpublished circular. In fact, the circular was written for the purpose of soliciting subscriptions for the work.[32] Since so much of what Lesley said reflected his anger over the failure to give credit to assistants, the basic point of Willie's argument was simple: Rogers was hired by the state as the principal geologist who assembled and interpreted the data gathered by assistants, it was he who had to answer to the state, and it was, therefore, reasonable to expect the work to appear under his name. Observing that his uncle felt it was impossible to mention the many achievements of each assistant through the years, he emphasized that the assistants had been acknowledged by name in the annual reports and as a group in the preface to the final report. Furthermore, he said, he and others had learned "by the commendations which we have heard from the lips of the State Geologist himself, to admire the faithfulness and ability with which they [the assistants] performed the duties he assigned them."[33]

Willie methodically dissected every statement that Lesley made, pointing out at each place why Lesley was wrong in suggesting that Rogers used material without giving proper credit. He made it clear that the maps and the raw notes of the assistants had undergone extensive or complete revision in the early 1850s, and that it would have been unjust and incorrect for Rogers to have attributed them to earlier assistants when the work of others had superseded them.[34] With more tact than Lesley had shown, Willie was ready to believe, as unlikely as it was, that Lesley was simply unaware of the work that was done on the survey after his departure. Since Dalson and Willie Rogers himself had resurveyed so much of the area after Lesley's departure, since the state had mandated a change in size for the state map, on which Rogers alone worked in Edinburgh, and since a new nomenclature was used reflecting Rogers's new classification, Willie had ample reason to argue that the final form of the maps was substantially different from those on which others had worked in earlier years.

Willie relished the idea of turning Lesley's own argument against him and did not hesitate to point out that Lesley had, after all, been a draftsman who compiled the map partly on the basis of information from various assistants. If Lesley's concern about credit for the assistants was valid, Willie wondered, shouldn't they and not Lesley be given credit for the map? With regard to the anthracite map (presumably the smaller one in volume 2 for the reasons noted above), Willie was perplexed as to why

Lesley claimed it was not his, saying that, although the scale had been changed, it was based almost entirely on work that Lesley had done for the survey and while working for Rogers's private surveys in the Shamokin/Mount Carmel area. There is some evidence to support Willie's contention, because the map in volume 2 bears a similarity to an extant 1850 map of the area with Lesley's name on it.[35]

Since Lesley's report was written for the American Iron Association, Rogers's brother, Robert, wrote to the association demanding assurance that the views expressed by Lesley were not shared by the members. As a result, the association passed a simple resolution disclaiming any responsibility for the geological opinions and personal sentiments of Lesley.[36]

Lesley's attacks on Rogers continued into the 1870s. In 1876, he wrote a history of the survey in which he pointed out several instances in which he thought Rogers had harmed the survey by failing to listen to the assistants. For example, he said that Jackson, whom he described as a "man of singular but erratic genius," had correctly concluded in 1838 that free brown iron ores came from stratified limestone beds and were "set free by mechanical and chemical action." But Rogers's view, as published in his *Second Annual Report,* held that they were formed by water percolating through sediment laden with peroxide of iron.[37] Rogers's view prevailed, to the detriment of the survey in Lesley's opinion. Lesley also said that Whelpley had become convinced that mountains were wholly the result of erosion rather than upheaval (the very opposite of Rogers's view) and suggested that had Rogers paid attention to him, the survey results would have been more meaningful.[38]

It is only fair to note again that Lesley sometimes spoke well of Rogers when he was discussing matters that did not have to do with the contributions of the assistants. When Logan and Hall attacked Rogers in 1864 for what they considered his belief that the New Jersey Highlands were all metamorphosed Paleozoic rocks instead of older primary rocks, Lesley defended Rogers, arguing that this was a mistake that Rogers would never have made.[39] In his 1876 history (written ten years after Rogers's death), Lesley had some genuinely complimentary things to say, noting that the survey report was "one of the most remarkable books the world has ever seen" and that for anyone who had read Rogers's papers on the geology of the United States for the report there "can be no sentiment but one of admiration for the breadth of his [theoretical] views, and the clearness, force and elegance of his delineations." Furthermore, he said, "no geological paper has ever appeared, excelling, in every good scientific quality, his memoir on coal." Of Rogers's sketch of physical geography in the final report, Lesley thought that it was very sound and would make a fine textbook. Lesley also discussed Rogers's fossil collection, which was given to the Boston Society of Natural History after his death in 1866, excusing what Lesley called its disarray because Rogers had had a "miserably small annual appropriation" and "no proper headquarters; no museum; no available paid assistance for museum work; and his surplus funds were spent in freight charges which were then high."[40] In 1880, Lesley prepared a memoir of the life of Samuel Haldeman for the National Academy of Sciences in which he called Rogers a "keen intellect, that poet, that called teacher of men, eloquent above the common, and ambitious to attain the highest summits of observation."[41]

Although Lesley's attacks gave a distorted view of Rogers personally and Rogers's theory of elevation and his nomenclature were widely dismissed, the final report was, as the reviewer in the *American Journal of Science* had said, filled with important information on the structure and economic deposits of the state, and it remained a significant source of information on Pennsylvania's geological structure and economic importance for many years. Its information was significant enough for the work to be reissued in the United States in 1868, quite probably in response to renewed interest in the location of iron and coal deposits following the introduction of the Bessemer process, which promised to revolutionize the steel industry.[42] On this occasion, one reviewer emphasized again that "no other geological report upon territory previously explored by him can be expected to be altogether original, or in anywise an improvement upon his fulness [*sic*] and accuracy of detail."[43]

As the years passed, Rogers was repeatedly praised for his description of the folded structure of mountains, as a result of which mountains of this nature were henceforth called the Appalachian type. As late as 1888 the eminent Austrian geologist, Edward Suess, spoke highly of "the picture faithfully drawn by Henry Rogers many years ago" and went on to say that "each time we return to the work of this distinguished investigator we are filled anew with astonishment that he should have been able, at such an early period, to arrive at so correct a conception of the structure of mountain ranges of this kind."[44]

A later, interesting use of Rogers's ideas came from Bailey Willis. Many geologists in the last years of the century found the idea of isostasy appealing but doubted that it alone could account for elevation as Hall had suggested. Instead, they linked it with one form or another of contractualism, giving the whole a catastrophic dimension. Willis, of the United States Geological Survey, thought Rogers had been close to this same conclusion and believed that Rogers had anticipated the idea of isostasy. The vertical oscillation of the earth that took place as the folds were formed in Rogers's theory would, Willis claimed, "to-day [be] . . . attributed to slow adjustments of balance."[45] It stretches the mind unnecessarily to see Rogers's oscillations as isostatic movements, and Rogers's theory was not a contractual one, but Willis thought the contractual hypothesis and isostasy should be complementary, and Rogers's work seemed to him to be a good example of such complementarity.

A Greatly Respected Man, 1859–1866

For the final years of his life, Rogers experienced a leisurely academic career that allowed him ample time to participate to whatever extent he wanted in the scientific society of Great Britain. Earning a salary of about £300 per year, Rogers settled easily and quickly into life in Glasgow. He found the climate conducive to his good health and thought that with "reasonably good classes," he would prosper.[1] Just before Rogers commenced his tenure at the university, the Regius professors, of whom he was one, were made equal to other professors at the university. Rogers looked forward to an increase in his privileges and his salary as well.[2] Residences for a few of the professors were made available by the university in Professors Court, and although Rogers's position did not normally carry this privilege, he was able to secure a house in the court when the professor of medicine chose not to live there. Thus Henry, Eliza, and Edith took up their residency at Number 1, Professors Court. Number 2 was occupied by Rogers's friend William Thomson (Lord Kelvin), whose work on the dissipation of energy from a cooling earth tended, Rogers thought, to support the possibility of paroxysmal upheavals.[3]

Rogers expected to teach for about six months of the year, and he was free to spend the other months away from Glasgow if he wished. It was not long before he decided to spend most of his free months in Edinburgh. Although the climate in Glasgow was not disagreeable, he found the climate in Edinburgh even better for his health, and he gradually came to think of Edinburgh as his real home. Because he spent so much time in England, Rogers also maintained a home at Sudbrook Park, Petersham, after 1863.

Rogers's first class on natural history began in November 1858 and was offered over two sessions, each three months long. Although his class was on the general topic of natural history, Rogers emphasized geology through lectures and lengthy field trips. In the spring of 1859, the faculty senate received a recommendation from a university committee that zoology should be included in the curriculum of every medical student. On hearing the recommendation, Rogers moved to make a three-month course on zoology a requirement. His motion passed, and the second session of his natural history course henceforth became one devoted more directly to this subject. In preparation for this course, Rogers spent part of the summer in London attending lectures on comparative anatomy by Thomas Henry Huxley and spending every Saturday with Huxley in what Rogers called private demonstrations of points in zoology that he felt demanded a master's attention.[4] Huxley was an acknowledged expert in the study and a person with whom Rogers was acquainted through the Royal Society.

Soon after his arrival at the university, Rogers began an association with the Hunterian Museum that lasted until his death. The museum, described by a contemporary as "an elegant Doric temple . . . [that] contains a valuable collection of subjects in natural history and anatomy" was the creation of the anatomist, William Hunter, in London in the eighteenth century.[5] By the terms of his will, it was moved to the University of Glasgow in the early nineteenth century. The museum was developed as an anatomical and pathological museum to be used in teaching, but because its collection included natural history, it served dual purposes at the university. Rogers's predecessor, Couper, had been the keeper (curator) of the museum. Since Couper's death, the job had fallen informally to one of his assistants, but because of the lack of funds, a new keeper was not appointed immediately. When the assistant became ill, the museum closed.

Rogers wanted to use specimens from the museum in his classroom and to make the museum itself a place for instruction. In answer to his requests for this privilege, the trustees of the Hunterian Museum appointed him to head a committee whose task it was to prepare a report on the state of the museum. At the same time, they also gave notice that Rogers would be nominated as interim keeper at the next meeting and that an application would be made to the college for salary funds to cover his work in this capacity.[6] Rogers presented his report to the museum trustees and the university commissioners in March 1859, and in April his report was approved and he was appointed interim keeper.

Rogers hired John Young as his assistant. Young, a geologist, eventually succeeded Rogers both as professor of natural history and as keeper of the museum. Records of the meetings of the Hunterian trustees are sporadic for the next several years, but they do indicate that at some point the designation "interim" keeper was dropped. Henceforth, Rogers was the keeper.[7] The museum experienced growth and prosperity under Rogers's direction, and its use for classes increased. Early in 1861 a university committee on improved accommodations for the museum found that at least a temporary room for natural history was necessary and suggested that it should be set up in the back of the museum with the necessary equipment, chairs, and table for teaching. Later that same year, Rogers asked for £200 from surplus admission receipts to purchase specimens and diagrams with which to illustrate his lectures on zoology, a possibility that Rogers pointed out was entirely feasible, since the receipts were steadily increasing. His request was approved without question.[8]

Rogers's schedule outside the university was as full as he cared to make it, and he was a familiar and frequent presence at scientific meetings. The Philosophical Society of Glasgow was one of his favorites, and in 1859, he became a member of the society's council at a time when Kelvin was president and Kelvin's father-in-law, Walter Crum, was on the council. From 1860 to 1862 Rogers served as the society's vice president and was its president from 1862 to 1865. His affiliation with British societies continued with his election to membership in the Royal Geographical Society in mid-1860, for which he was recommended by Alexander Keith Johnston and Roderick Murchison, among others.[9] By now, he was a member of nearly every major scientific body in England and Scotland.

The great scientific event of the era was the 1859 publication of Charles Darwin's *On the Origin of Species by Means of Natural Selection.* Already convinced that species could change in accordance with natural law, Rogers found Darwin's work "full of ingenious arguments in favour of the Lamarckian hypothesis." He believed Darwin's principle of natural selection was indisputable, although he thought that it would be as impossible to find scientific proof of species rising from species as to find proof of supernatural creation.[10] His only criticism of Darwin was that "in his geology Darwin outdoes Lyell himself in ignoring paroxysmal actions."[11] Darwin's failure to acknowledge paroxysmal actions notwithstanding, Rogers was convinced that within a few years all scientific opinion would support Darwin. Darwin was pleased enough with Rogers's support, which he learned about from Huxley, to take note of it in a letter to Lyell.[12]

Although Rogers did not give any papers supporting Darwin directly, he did speak out on a subject that would be the topic of a later work by Darwin, the antiquity of man. The question was whether man had existed in Europe with animals now extinct. In the late 1830s, Jacques Boucher de Crevecoeur de Perthes, a customs official at Abbeville in northern France, found what appeared to be man-made tools in the Somme River Valley with the bones of these animals. His discovery garnered some attention, particularly as similar discoveries were made in other areas, but the scientific community remained unconvinced about the contemporaneity of the bones and tools even when, in the late 1850s, several British discoveries, particularly at Brixham, and a reexamination of the Somme Valley offered more evidence. Favorable support came from many quarters, including remarks by Charles Lyell in his presidential address to the British Association for the Advancement of Science held at Aberdeen.[13]

While studying with Huxley in London in the summer of 1859, Rogers heard considerable conversation on the topic and probably attended the session of the BAAS in which Lyell presented his paper. Henry and William talked about the general subject of the antiquity of man in their letters, but Rogers had little opportunity to pursue the matter until the summer of 1860, when he went to Abbeville to study the material firsthand. The trip resulted in a paper published in the *Edinburgh Magazine*. William thought the evidence for man's antiquity was strong, and Henry agreed with him, but in this paper he took a middle-of-the-road stance on the issue. He argued the impossibility of proving *or* disproving the contemporaneity of the bones and tools and the impossibility of arriving with certainty at a great age for either.[14] The question would be settled, he thought, only when a better understanding of the geology of the area was available, and he argued that the beliefs currently held depended to a great extent on whether the person was a "quietist" (uniformitarian) or a "paroxysmist." Surveying the Somme Valley from the point of view of each, he presented their respective views. The former, he argued, saw contemporaneity, while the paroxysmist argued that any mixture of non-contemporary pieces was possible as the result of violent geological changes in the area. He emphasized that paroxysmal forces did not rule out contemporaneity; they simply obscured it. Thus the geology needed to be studied in more detail in order to understand the true situation.

Rogers was paid "liberally" for the article, but the magazine's editor, John Blackwood, who also headed the firm that had published the final report of the survey of Pennsylvania, found fault with the paper, calling it too tedious for the general reader. Concerned, Henry wrote to William for an opinion, and William, in usual brotherly fashion, found it "an entirely fair and rational view of the question" and urged him to continue his researches and studies in the Somme Valley.[15] In November, as classes began again, Rogers was still pursuing the subject and went to London to gather more information for a paper on the subject that he planned to give, at the encouragement of Michael Faraday, to the Royal Institution the following March.[16] There is no indication that he delivered such a paper. He gave one instead on his earlier investigation of the parallel roads of Glen Roy.[17]

Throughout 1860 and 1861, Rogers's time was filled with lectures and teaching. He presented papers at the British Association meeting in Oxford on the metamorphism of coal and several at the Glasgow Philosophical Society, including one on the origin of petroleum, or rock oil, and one on the relations of salt and climate. All were reminiscent of his earlier papers.[18] In addition to his regular classes, his teaching responsibilities increased for a short time early in 1861 when he taught several classes for Thomson, who broke his leg in December and had to give up most of his activities for several months. Besides taking over some of Thomson's teaching responsibilities, Rogers read a paper for him at the Manchester meeting of the BAAS entitled "Physical Considerations Regarding the Possible Age of the Sun's Heat." This was a preliminary statement of a new theory, one of Thomson's most important, on solar energy, but in his absence "its reading apparently went unnoticed."[19]

As Rogers's activities with the university and with professional organizations indicate, he enjoyed a broad range of professional contacts. As in the United States, he also enjoyed the company of men whose liberal stances on politics, economics, and philosophy appealed to him, and his correspondence suggests that he did not want for social involvement. He especially cultivated a circle of friends who shared his concerns about slavery and with whom he could discuss the issues. They included John Stuart Mill, the philosopher; the economists Richard Bright and Richard Cobden of free-trade fame; Edward Chadwick, who was well known for his interests in social reform; David Livingstone, the African explorer; and George Combe, phrenologist and author of *The Constitution of Man* (1828).

His easy relationship with these friends was illustrated early in his residence by an incident involving Combe and Rogers's longtime friend, Charles Sumner. Rogers and Combe became acquainted through Charles Sumner when Sumner was in England in 1857, trying to recuperate from a serious injury. Since Rogers had first met him, Sumner had become a United States senator, a great orator, and one of the nation's most outspoken critics of slavery. When the Kansas-Nebraska bill was passed in

1854, effectively opening western lands to slavery, Sumner brought the full force of his oratorical power to bear in opposition. A particularly stinging speech delivered in May of 1856 resulted in a physical attack on Sumner by Preston S. Brooks, congressman from South Carolina and relative of one of the authors of the bill. Brooks beat Sumner severely, and for months thereafter the ailing senator suffered many physical problems. He finally went to Europe in the hope of speeding his recovery.[20] His friends became concerned about him and feared he would try to resume political activities too soon. Combe, one of Sumner's friends, requested an opinion on Sumner's condition from Sir James Clark, the Queen's physician. Since Rogers and Sumner were old friends, Rogers joined in Combe's concern for Sumner, and when Clark advised that Sumner should be very cautious in any effort to resume normal activities, Rogers felt comfortable in engaging in a small flurry of letter writing with Clark and Combe about the need for Sumner to follow Clark's advice.[21]

Rogers was content with his professional surroundings, and his family life was enhanced in 1860 with the birth of a second daughter, Mary Otis Rogers. On February 4, 1862, however, his first child, Edith, died at the age of seven. Rogers referred to her as his idol, and although little of this tragedy has been preserved in notes or correspondence, a single sentence written to Desor three years later says all that needs to be said of the effect of her death on his life: "the world has not seemed the same to me since her departure."[22] This tragedy was followed within the year by two others. Early in 1863, Rogers's beloved younger brother, Robert, was seriously injured. Robert was then serving as surgeon at the West Philadelphia Medical Hospital. His right hand was caught between the rollers of an ironing machine and crushed. Infection made it necessary to amputate, but Robert recovered well in both spirit and body. Scarcely a month later, however, Robert's wife died unexpectedly.[23]

Shattered by Edith's death and the subsequent tragedies, Rogers plunged into his work with even more tenacity than he had shown in the preceding years. His teaching for the winter session (February-March-April) had just gotten under way at the time of Edith's death. He lectured three times a week and included field trips and work at the Hunterian Museum whenever possible.[24] A spring session that began in May consisted of four lectures per week and again used field trips and the museum as integral parts of the instruction. In late March 1862, Rogers joined a delegation under Cyrus Field to urge the British government to renew its attempt at laying a transatlantic cable. An attempt in 1858 had failed, but Field, an American industrialist whose efforts were largely responsible for the cable, pursued the issue, and as a result of the deputation Rogers felt it "likely this government will allow another expedition to sound the Atlantic bottom as a preliminary."[25] Field did succeed in his efforts, and the cable was finally in use by mid-1866.

Rogers turned most of his energy to the antislavery cause. There was concern throughout Great Britain over the issue of slavery in general, and strong prosouthern sympathy among those in Scotland who relied on slave cotton for the textile mills directed the attention of abolitionist groups to the American Civil War. Furthermore, there was a high level of concern even among the North's supporters over how the nation could ever return to prosperity at the war's conclusion.[26] Rogers discovered that some of his friends, like Sedgwick, who had supported the North's position at first, had decided that the war would lead to military control by the North and had changed their allegiance.[27] Rogers was so affected by the war and the support for the South in Glasgow that for a time he isolated himself from others in the city and even shunned the Geological Society of Glasgow, and he admitted to William that "for a while Eliza and I could not refrain from engrossing ourselves overmuch with details of American war news," so much so that it became an unwholesome fascination.[28]

His isolation was short-lived. As an American in Glasgow who could speak forcefully and eloquently on the evils of slavery, and on the issue of whether the United States could survive economically without it, Rogers was an attractive speaker for the British abolitionist groups, and he was often asked to address them. He became well known for his position but was unable to do as much as he or others might have liked because his health was steadily declining. In 1862, for example, Rogers contemplated the preparation of four or six lectures on America's industrial and physical resources "as tending to show its capacity to recover from its present reverses."[29] He was encouraged in this endeavor by Alexander Keith Johnston, but it does not appear that he ever gave the lectures. In 1863, Rogers again contemplated, but never gave, another set of lectures for the 'Emancipation Society of London' (possibly the London Emancipation Committee, founded in 1859) "on the Benumbing Influences of African Slavery in the United States."[30] But, he told William that "I shrink from all display of the sort, partly from my growing repugnance to all excitements, the result, perhaps, of a certain lassitude of health."[31] He did write two papers that made a statement on slavery, both for *Good Words,* a kind of Sunday magazine filled with inspirational words, sermons, and the like as well as stories, natural history, and science. Rogers was in good company here, since men of scientific stature, like Thomson, often contributed. Although it was couched in scientific papers on coal, the point he clearly wanted to make was that God had given mankind a source of mechanical energy in coal, thus rendering unnecessary an institution that used human energy.[32]

In the early fall of 1863, Rogers attended a meeting of the Social Science Congress, at which he intended to make some comments on the relative statistics of the growth of the North and South in the United States and at which he distributed pamphlets that William had sent, hoping that, by "diffusing sound information about the North and South," the country's best interests would be served.[33] A little over a year later, in 1865, he was consulted about a project to establish an international weekly or semiweekly paper in New York. The paper, the idea of John H. Eastcourt, was to be liberal in politics and aimed at bringing greater understanding between the two countries. Because the concept fit in with his ideas on educating Britishers about American issues, pointing out that slavery was economically unnecessary, Rogers encouraged Eastcourt to proceed, but the paper was never established.[34]

That Rogers had to curtail his activities because of his failing health was apparent in many ways besides his failure to give lectures. In 1863, he declined an invitation to accompany Thomson and others to Teneriffe during the summer to conduct experiments in atmospheric electricity, fearing "a lack of the required strength and health."[35] When, shortly before the end of the summer session in 1863 Rogers went off with John Young and six members of his class on a geological tour to Arran, he found that he had to make many stops to rest during the ascent of "one of the loftiest [peaks] in Scotland." He could still recount the trip to William with something of his old excitement and enthusiasm when he described "a grand anticlinal folding in the *bedded* granite," that gave him a "clue to the structure of the whole central mass of the island," but although he wanted to call attention to this discovery in a paper, he never had the strength to do it.[36]

As usual, after Rogers completed his classes in July, the family went to Edinburgh. Shortly after arriving this summer, Rogers was invited to visit John Blackwood at his home in St. Andrews specifically to spend some time with Captain John Speke, explorer of the Nile. Speke had returned from Africa in 1859 and had written a few articles that appeared in Blackwood's *Edinburgh Magazine*. In 1863, Blackwood was about to publish Speke's journal of his Nile explorations.[37] Speke wanted to include some geology of the area but needed help with it. Blackwood thought immediately of Rogers.

Rogers found Speke's adventures "astonishingly strange, amusing, [and] perilous" but told William that Speke was "green in authorship, and untaught in science."[38] He agreed to help, and with a few specimens that Speke had brought back and Speke's recollections as his only source of information, Rogers set out on a task that he found "exceedingly interesting." As the summer waned, the family went to their residence in Sudbrook Park in England, where Rogers prepared his chapter for Speke quickly. Considering the evidence of Rogers's declining health, perhaps the chapter was insufficient, for Blackwood rejected it. Angered by Blackwood's refusal and the loss of the time he had spent on the manuscript, Rogers privately lashed out at Speke and the paucity of material that he had had to work with, which he blamed for any shortcomings in his chapter. He also criticized Blackwood, who, Rogers thought, for reasons he did not explain, may never have had any real intention of publishing his chapter.[39]

The family spent the early winter in England, returning in time for Rogers's class that was to begin in February 1864. The class was canceled, and Rogers decided to devote most of his time to the museum, overseeing its physical maintenance, the care of its collections, and additions to them. When his class was canceled, he requested permission from the university senate to combine the lectures that he would normally have given in this winter session with those he was scheduled to give in the summer. Therefore Rogers presented fifty lectures in the summer session of 1864.[40]

It had become the family custom to leave Professors Court during the warmer months when Rogers was teaching for a more comfortable spot in Glasgow. Rogers found this to be so beneficial to his family's health and well-being that he purchased a villa at Shawlands, just four miles from Glasgow. In spite of the favorable conditions at Shawlands, his heavy teaching burden aggravated his health problems, and he found himself nearly exhausted within a month. His energy was revived when William visited Europe in June, and they managed to spend a considerable amount of time together. In September, they went to Bath for the meeting of the BAAS where Henry, at William's request, called the members' attention to a cast of fossil bones from Connecticut, but this was the extent of activity that his health would permit.[41]

Rogers's health worsened rapidly as 1864 drew to a close, and early in 1865 his physician advised him to go to southern France, in the hope that the warmer and dryer climate would help him. Rogers and family hurried to Mentone, located a short distance from Nice. Mentone had long been a refuge for invalids because of its exceptional climate and tropical vegetation. It suited Rogers's condition, and by March, he was feeling well enough to write to William:

I am busy at my pen for some three or four hours daily and on some days for five hours at my text-book, or rather my course of lectures on Geology. I alternate this pleasant light sedentary work with a lounge in the olive groves and lemon orchards in the rear of our hotel, or sometimes spend the afternoon in the saddle on a donkey, with a little girl in attendance, searching into the geology of this fascinating coast region. Imagine the finest flexures and foldings of Jura limestone and cretaceous shells and nummulitic, Middle Eocene Tertiary Limestone, coming out in bold relief, open to the view and waiting my study for miles to the east and west of me along all this picturesque coast.[42]

Rogers had a chance to put his interest in the area to work when Dr. James Henry Bennet asked him if he would assist him with a book he was writing. In 1862, Bennet had written a book on the virtues of the area for those who sought the warmer areas of southern Europe for comfort and health in the winter. The book was so popular that Bennet was at work on a revision when he met Rogers, and Rogers agreed to aid Bennet "in making it more correct in its scientific portions, especially the geological."[43] Chapter 3 of Bennet's revised edition, entitled "Geology," is based on the work of many authors. It makes several references to Rogers, however, and includes a stratigraphic chart that Rogers prepared for the book. Among other things, Bennet discussed the glacial period and its effects on the area, noting at one point that his "learned friend Professor Rogers" believed the glacial period "to have been much exaggerated by recent writers." There is a suggestion in this comment that Rogers may have come to accept some small role for ice in shaping the landscape, and, indeed, Lesley said several years later that he thought Rogers had begun to yield on the issue of ice before his death.[44]

While Rogers found the area, the climate of Mentone, and his companions agreeable in many respects, he continued to be distressed by the war. He felt that the British were grossly misinformed about American affairs, and he grew "sick" of the falsehoods reported in the newspapers and "of the cantings and misstatements I habitually hear from people around me."[45] Constant news from William on the progress of the war encouraged him, and he was further cheered when, ready to return to Glasgow in April, he stopped at Avignon to visit with J. Stuart Mill, whose liberal attitude and antislavery position renewed his hope. When he reached London on April 26, he learned of Lincoln's assassination. "I only hope and trust," he wrote to William, "there will be no cruel retaliations, but that the soul of the North will rise to the demands of the occasion, a firm observance of justice, with a strict discrimination between the guilty and the not guilty in all the punishments to be visited."[46]

He was back in Glasgow in time for the beginning of his class in the late spring of 1865. Among his students was a young clergyman, identified only as the Reverend Mr. Steward, who had been with David Livingstone in some of his African explorations. Steward was taking the class to further his training in scientific studies in preparation for future trips with Livingstone.[47] Whether through this connection or some other one, Rogers met Livingstone in June and spent two days with him at the Crums' residence. Livingstone had returned from his travels convinced of the capability of the African to be self-sustaining and not, as proslavery advocates often argued, in need of constant care. Although this was the first and only meeting between Rogers and Livingstone, Livingstone was fully aware of Rogers's stand on the issue of slavery even before their meeting, and had mentioned to his sister-in-law that Rogers also believed that slaves were capable of self-support.[48] He and Rogers must have had much to discuss during their brief meeting.

After his classes ended in August, Rogers, his wife, and his daughter returned to Boston, by the "new commodious *screw* steamer" that was then in service between Glasgow and New York. Such ships, powered by a steam-driven screw propeller, were still relatively new in passenger service, and Rogers had thoroughly investigated their reliability. Since he had never overcome his tendency to seasickness, he was especially interested in their stability. He was convinced of their merit and noted that "Eliza, ever heroic when a crisis demanding courage arrives, is quite willing to take [it]."[49] Planning to stay in Boston until it was time to return to Glasgow for his class, which was once again to be only one three-month session starting in the late spring of 1866, Rogers kept busy seeing and corresponding with old friends, visiting the Boston Society of Natural History as often as possible, and even preparing a paper "On Petroleum," for the Glasgow Philosophical Society.[50] The sale of the final report had not yielded enough to pay costs not covered by the state, and he made one last unsuccessful effort to recoup some of his losses from the state.

Rogers left Boston in early April in order to be ready for his class, which was to begin in May. Eliza and Mary remained in Boston, expecting to meet him in the south of France at the close of the summer session. His stay in Mentone the year before had been so beneficial that the family expected to spend as many months as possible there. Rogers appeared to be in good health when he left Boston, but the return voyage was stormy, and when he reached Glasgow on April 22, he was gravely ill. He was scheduled to give two public lectures on coal and petroleum in Norwich on April 27 and May 1, but he was so weak that he could hardly talk. Indeed, he had to sit throughout most of his lectures. Not surprisingly, Rogers was unable to begin his class at Glasgow. A few weeks later, on the morning of May 29, 1866, two months and three days before his fifty-eighth birthday, Henry Rogers died at his villa in Shawlands. Eliza received the first word of her husband's illness on June 6, and on June 9 she, Mary, and his brothers sailed from New York. They were greeted in Queenstown with the news of his death. Rogers was buried in Dean Cemetery in Edinburgh beside his daughter, Edith.[51]

Rogers's death was announced with deep regret by the University of Glasgow on June 7, and obituaries appeared in many places soon after that date. They included the *Glasgow Herald*, the *American and Gazette* (Philadelphia), the *Proceedings of the Royal Society of Edinburgh*, the *Proceedings of the Royal Society of London*, the *Quarterly Journal of the Geological Society of London*, and the *American Journal of Science*.[52] They were a standard lot. None are lengthy and none, save the one in the *American and Gazette*, which was written by G. W. Biddle, a friend, and the one in the *American Journal of Science*, seems to have much precise knowledge of Rogers. Almost uniformly, however, they praised his work in geology. As Warrington Smith, president of the Geological Society of London put it, "it must be conceded that Professor Rogers contributed a noble quota to the unraveling of some of the grandest phenomena which geologists have been called upon to investigate."[53]

Consistently the eulogizers applauded Rogers for his manner and skill as speaker and teacher, and some of them waxed eloquently on this point. For example, the author of the notice in the *Proceedings of the Royal Society of London* noted: "In society, and as a lecturer, his great and varied knowledge gave him an advantage which he exercised with graceful facility, and on favourite topics he would at times surprise and charm his hearers by bursts of eloquence." The writer for the *American Journal of Science* added:

His great knowledge on many subjects he was able to impart in a style equally clear and graceful, whether in public speaking or as a writer. Few teachers of sciences have excelled him in power of illustration of difficult subjects, or in commanding the attention of large audiences to themes not commonly discussed in public lectures. . . . His amiable manners and remarkable powers as a conversationalist had won for him the same social distinction in Great Britain which he long enjoyed in America, and a numerous body of personal friends deplore his loss on both sides of the Atlantic.

The journal also noted that the 1842 paper on Appalachian structure and dynamics had been "communicated with an eloquence and fascination of style never surpassed at a meeting of the American Association of Geologists and Naturalists."

Through a lifetime of work, Rogers had amassed a large library and large collections of specimens and maps. His widow gave a "large suite of rocks and fossils," his maps, and his library to the Massachusetts Institute of Technology. The maps have disappeared. His library remains there today in the archives, intermixed with that of William. William moved the rocks and fossils to the Boston Society of Natural History, and in 1875 the collection was mentioned as part of the society's "American Collection," which also included William's collection. William thought that this was the best place for the two collections, since he had made arrangements to use rooms in the Society for lectures and laboratory work for MIT students. According to a report in 1875, the collection had been packed in haste in Glasgow after Rogers's death, and many of the labels had been lost at that time. William, however, was reported to be working on organizing them. The Boston Museum of Science, the society's descendant, still has a few of his specimens; others once in their possession were given, for reasons unknown, to the Museum of Comparative Zoology at Harvard and to the University of Maine in Orono. Rogers himself had donated a few fossils to the Hunterian Museum in Glasgow, where they remain today.[54]

The cabinets of specimens prepared as part of the survey of Pennsylvania and the cause of such constant irritation for Rogers have disappeared. One cabinet in Pittsburgh was destroyed in a fire, while the one in Harrisburg was put in the Pennsylvania State Lunatic Hospital, was subsequently given to the American Philosophical Society, and later disappeared. In 1879, a collection was placed in the Academy of Natural Sciences of Philadelphia, which may have been the Harrisburg cabinet. A note at the academy written about 1935, says that "the collections of the First Geological Survey of Pennsylvania under Rogers, are in boxes and have perhaps not been unpacked since 1858!"[55] They have not been seen since. Whether the fourth cabinet was ever finished or whether it went to the Farmers High School, as mentioned in the legislative proceedings of 1857, is uncertain, but it is not at Pennsylvania State University today.

Soon after Rogers died, Eliza returned to Boston with their daughter, Mary, who, in 1898, became the second wife of the Reverend Charles F Russell of Weston, Massachusetts. Eliza Rogers spent her last years with her daughter in Weston where she died on February 23, 1906. Although the Reverend Russell and his first wife had children, he and Mary had none, and thus there are no direct descendants of Henry Rogers.

Rogers had not been happy or contented during his last years of residency in the United States. He had made enemies; some of his work that was most important to him had not been given, in his eyes, due consideration; he had grown weary of his struggle to complete the Pennsylvania survey; and he felt that American science had taken a direction that left him no room. Things had been different when he moved to Glasgow. Always more at ease with liberal literary and political figures in the United States, he found a similar group in Great Britain. More important to Rogers was a sense of place in the scientific community, where he participated comfortably in the major scientific organizations and continued to enjoy the friendship of Murchison, Phillips, Sedgwick, and others whom he had come to know during his first visit to England as a young, would-be social reformer. The personal animosities that Rogers experienced in America were largely absent in Great Britain. Although they often disagreed with him, his British friends were able to regard his views as representing simple difference of scientific opinion, without the personal overtones that were present in the United States. In this environment, Rogers displayed a personal warmth that was often missing in the United States, and his friends in Scotland and England described him as quiet, lovable, shy, retiring, modest, serious-minded, and fatherly. He was, as Thomas Henry Huxley told William Rogers many years after Henry's death, "greatly respected and liked in this country."[56] The comment would have pleased Rogers.

Abbreviations

AN	Archives de l'Etat, Neuchâtel, Switzerland
ANSP	Academy of Natural Sciences of Philadelphia, Philadelphia, Pennsylvania
APS	American Philosophical Society, Philadelphia, Pennsylvania
CUL	Cambridge University Libraries, Cambridge, England
HDR	Henry Darwin Rogers
HSP	Historical Society of Pennsylvania, Philadelphia, Pennsylvania
JBR	James Blythe Rogers
LL1	*Life and Letters of William Barton Rogers,* volume 1
LL2	*Life and Letters of William Barton Rogers,* volume 2
MIT	Massachusetts Institute of Technology, Cambridge, Massachusetts
NYSL	New York State Library, Albany, New York
PKR	Patrick Kerr Rogers
RER	Robert Empie Rogers
WBR	William Barton Rogers

Notes

I. To Have Some Certain and Definite Object in View, 1808–1829

1. The principal published sources of biographical and historical information on Henry Darwin Rogers and his family are: J. W. Gregory, *Henry Darwin Rogers, An Address to the Glasgow University Geological Society, 20th January, 1916* (Glasgow: James MacLehose & Sons, 1916); W. S. W. Ruschenberger, "A Sketch of the Life of Robert E. Rogers, M.D., LL.D., with Biographical Notices of His Father and Brothers," *Proceedings of the American Philosophical Society* 23 (1885): 104–146; and Emma Rogers, *Life and Letters of William Barton Rogers*, 2 vols. (Boston: Houghton Mifflin, 1896). Emma Rogers made use of information found in Ruschenberger for some of her commentary on Patrick Rogers. Other sources of help include W. H. Ruffner, "The Brothers Rogers," in *The Scotch-Irish in America: Proceedings and Addresses of the Seventh Congress of the Scotch Irish at Lexington, Virginia, June 20–23, 1895* (Nashville, Tenn.: Barber & Smith, Agents, 1895), pp. 123–139, also published in the University of Virginia *Alumni Bulletin* 5 (1898): 1–13; an autobiographical fragment by Patrick Rogers published in *LL1*, pp. 8–10; Edgar Fahs Smith, *Biographical Memoir of Robert Empie Rogers, 1813–1884* (Washington, D.C., 1904); Joseph Carson, *A Memoir of the Life and Character of James B. Rogers, M.D.* (Philadelphia: T. K. & P. G. Collins, 1851). There were several obituaries of Henry Rogers that contributed little to his life's story, and there is a short sketch of him by John Rodgers in C. C. Gillespie, ed., *Dictionary of Scientific Biography*, vol. 11 (New York: Scribners, 1975).
2. Ruschenberger, "A Sketch of the Life," p. 105.
3. Carson, *A Memoir of the Life*, p. 8.
4. Ruschenberger, *A Sketch of the Life*, pp. 106–109.
5. Patrick Rogers, *An Investigation of the Properties of the "Liriodendron tulipfera," or Poplar Tree* (Philadelphia: Benjamin Johnson, 1802).
6. Joseph Carson, *A History of the Medical Department of the University of Pennsylvania* (Philadelphia: Lindsay & Blakiston, 1869), pp. 107–108.

7. Ruschenberger, *A Sketch of the Life,* p. 109.
8. The date of James's birth is variously given as February 11 and February 22.
9. Patrick Rogers, Autobiographical Fragment, *LL1,* p. 8.
10. Ibid., pp. 8–9.
11. Several years later Patrick mentioned these lectures in a letter to Thomas Jefferson, saying that they began in 1808 (*LL1,* p. 10). He published *A Syllabus of a Course of Lectures on Natural Philosophy and Chemistry with the Application of the Latter to Several of the Arts* in 1810 to accompany his course (*LL1,* p. 10). Circular dated Philadelphia March 12, 1811, Rogers Papers (MC1), MIT.
12. PKR to Thomas Jefferson, May 21, 1819, *LL1,* p. 10.
13. Patrick Rogers, Autobiographical Fragment, *LL1,* p. 9.
14. Ruschenberger, *A Sketch of the Life,* p. 111; Autobiographical Fragment, *LL1,* p. 9; John Arnest to PKR, Rogers Papers (MC1), MIT; Eugene Fauntleroy Cordell, *The Medical Annals of Maryland, 1799–1899* (Baltimore, 1903), pp. 21–22; John B. Caldwell to PKR, December 1817, Rogers Papers (MC1), MIT.
15. John R. Quinan, M.D., *Medical Annals of Baltimore* (Baltimore: Press of Isaac Friedenwald, 1884), p. 152.
16. From notes by WBR, quoted by Emma Rogers in *LL1,* pp. 14–15.
17. WBR to George Hillard, January 27, 1846, Incoming Letters of the University Treasurer (UAI.50.8VT), Harvard University Archives, Cambridge, Massachusetts.
18. On his unsuccessful attempt at Virginia, see Jefferson to PKR, June 23, 1819, *LL1,* pp. 11–12.
19. Patrick Rogers, *An Introduction to the Mathematical Principles of Natural Philosophy* (Richmond, 1822). It is probable that this was an extension of the work he wrote in 1810. (See n. 11.)
20. PKR to Thomas Jefferson, March 14, 1824, *LL1,* pp. 29–30.
21. A report on the standing of William and James at William and Mary appears between pp. 18 and 19, *LL1.*
22. Information on Rogers's attendance at the College of William and Mary was provided to me by archives assistant Sharon Garrison. Records of students at the college before 1827 have been compiled from many sources, including bursar's books, professors' class books, and letters.
23. WBR to PKR, November 22, 1825 (typescript), Rogers Papers (MC1), MIT.
24. Ibid.
25. PKR to WBR, December 1, 1825, Rogers Papers (MC1), MIT.
26. WBR to PKR, November 3, 1826, *LL1,* pp. 34–36; WBR to PKR, December 5, 1826, Rogers Papers (MC1), MIT. Ruschenberger, *A Sketch of the Life,* refers to the school as having opened in 1821, but the date is incorrect. There is a letter from Patrick to WBR published in *LL1,* p. 31, congratulating the brothers on their new enterprise. It is dated October 17, 1825, but should be October 17, 1826.
27. [John N. B. Latrobe], *Picture of Baltimore, Containing a Description of All Objects of Interest in the City and Embellished with Views of the Principal Public Buildings* (Baltimore: F. Lucas, [1832]), p. 196.
28. WBR to PKR, January 25 and January 27, 1827, *LL1,* p. 37.
29. HDR to PKR, April 13, 1827, *LL1,* p. 42.

30. HDR to PKR, January 8, 1827, *LL1*, p. 40.
31. JBR to PKR, April 20, 1827, *LL1*, pp. 43–44.
32. HDR to PKR, April 12, 1828, *LL1*, p. 48.
33. "William to the Governors of the Maryland Institute, April 13, 1828," *LL1*, p. 49.
34. Bruce Sinclair, *Philadelphia's Philosopher Mechanics: A History of the Franklin Institute, 1824–1865* (Baltimore: Johns Hopkins University Press, 1974), pp. 124–125.
35. WBR to PKR, May 19, 1828, *LL1*, p. 51.
36. WBR to PKR, June 26, 1828, *LL1*, p. 53.
37. HDR to PKR, June 7, 1828, *LL1*, p. 52.
38. Quoted in an editorial comment, *LL1*, p. 65.
39. HDR to WBR, November 14, 1828, *LL1*, p. 67.
40. HDR to Michael Keyser, April 15, 1829, Misc. Mss., New York Historical Society, New York City. Information on the china warehouse appears in an editorial comment, *LL1*, p. 15.
41. Editorial comment, *LL1*, p. 69; Quinan, *Medical Annals of Baltimore*, pp. 33–34.
42. For a summary of Maclure's role at New Harmony, see John B. Patton, Anne Millbrooke, and Clifford M. Nelson, "The New Harmony Geologic Legacy," in *Field Trips in Midwestern Geology* (Boulder, Colo.: Geological Society of America, 1983), especially 1:225–243, or the introduction to *The European Journals of William Maclure,* ed. with notes and introduction by John S. Doskey (Philadelphia: American Philosophical Society, 1988), pp. xv–xlviii.
43. Alice Felt Tyler, *Freedom's Ferment: Phases of American Social History from the Colonial Period to the Outbreak of the Civil War* (1944; reprint, New York: Harper Torchbooks, 1962), chapter 9.
44. HDR to WBR, December 6, 1828, *LL1*, pp. 68–71.
45. Ibid.
46. HDR to WBR, January 6, 1829, *LL1*, p. 73.
47. JBR to HDR, March 13, 1829, Rogers Papers (MC1), MIT.
48. HDR to Michael Keyser, March 10, 1829, Misc. Mss., New York Historical Society; JBR to WBR, August 8, 1829, Rogers Papers (MC1), MIT.

2. Acquiring an Intimacy with Geology, 1829–1833

1. Unidentified correspondent to James Rogers, September 30, 1829, Rogers Papers (MC1), MIT.
2. JBR to WBR, October 5, 1829, Rogers Papers (MC1), MIT.
3. Ibid.; Charles Coleman Sellers, *Dickinson College: A History* (Middletown, Conn.: Wesleyan University Press, 1973), p. 175.
4. For details of the college's problems, see Sellers, *Dickinson College,* especially chapter 8, "The Duffield Years."
5. Broadus Mitchell, "Henry Vethake," in *Dictionary of American Biography* (New York: Charles Scribner's Sons, 1928).
6. Sellers, *Dickinson College,* p. 185; *Carlisle Republican,* June 9, 1831.
7. Minutes of the Board of Trustees, October 17, 1829, Dickinson College Archives, Carlisle, Pennsylvania.

8. Stanley M. Guralnick, *Science and the Ante-Bellum American College* (Philadelphia: American Philosophical Society, 1975), especially chapters 3 and 4.

9. In an editorial comment (*LL1*, p. 75), Emma Rogers mistakenly says he began teaching in January 1830, as does Sellers, *Dickinson College*, p. 186.

10. HDR to WBR, February 17, 1830, and JBR to WBR, Christmas, 1829, Rogers Papers (MC1), MIT.

11. HDR to WBR, January 15 and February 17, 1830, Rogers Papers (MC2), MIT.

12. HDR to WBR, December 23, 1829, *LL1*, p. 82.

13. Ibid., pp. 82–83.

14. HDR to WBR, February 2, 1830, *LL1*, pp. 83–84.

15. HDR to Uncle James, March 15, 1830, Rogers Papers (MC2), MIT.

16. Minutes of the Board of Trustees, March 30, 1830, Dickinson College Archives. See Sellers, *Dickinson College*, p. 186, for information on what McFarlane and Rogers taught at this time.

17. Dickinson College, 1834 Catalog, Dickinson College Archives.

18. Sellers, *Dickinson College*, p. 188. George Fleming was the printer.

19. "Editorial Notices," *Messenger of Useful Knowledge* 1 (November 1830): 62.

20. HDR to James Hamilton, October 5, 1830, James Hamilton Papers, HSP.

21. HDR to WBR, October 24, November 3, 1830, and December 15, 1830, Rogers Papers (MC2), MIT. The state of medicine in the first half of the century has been discussed by many authors. For a general sense of the fringe cures and the recommendation of diet as a pathway to health, see William G. Rothstein, *American Physicians in the Nineteenth Century: From Sects to Science* (Baltimore: Johns Hopkins University Press, 1972); or Norman Gevitz, ed., *Other Healers: Unorthodox Medicine in America* (Baltimore: Johns Hopkins University Press, 1988).

22. HDR to WBR, November 3, 1830, Rogers Papers (MC2), MIT. This letter is faded and fire-damaged to the point of being almost impossible to read. Most of the letters that comprise the MC2 file were involved at some time in a fire. Rogers's daughter married a minister in Weston, Massachusetts, and Rogers's widow lived with the couple. It is known that the parsonage burned, and the letters were possibly damaged at that time.

23. Henry Darwin Rogers, "Education," *Messenger of Useful Knowledge* 1 (December 1830): 65–70. The quoted passage is on pp. 69–70.

24. HDR to Uncle James, February 7, 1831, Rogers Papers (MC2), MIT; JBR to WBR, December 13, 1829, *LL1*, p. 80.

25. Rogers's exchanges are mentioned in a letter from his Uncle James, December 8, 1830, Rogers Papers (MC1), MIT.

26. Minutes of the Board of Trustees, February 10, 1831, Dickinson College Archives.

27. HDR to James Hamilton, March 13, 1831, Hamilton Papers, HSP. For information on Featherstonhaugh, see Edmund Berkeley and Dorothy Smith Berkeley, *George William Featherstonhaugh, the First U.S. Government Geologist* (Tuscaloosa: University of Alabama Press, 1988).

28. HDR to Uncle James, May 15, 1831, Rogers Papers (MC2), MIT.

29. HDR to Hamilton, July 22 and August 25, 1831, Hamilton Papers, HSP.

30. Editorial comment, *LL1*, p. 88.

31. For information on McNeill and the growing importance of railroad engi-

neers at this time, see Darwin Stapleton, "The Origin of American Railroad Technology, 1825–1840," *Railroad History* 139 (Autumn 1978): 65–77. Rogers's acquaintance with McNeill is evident in HDR to WBR, October 3, 1828, *LL1*, p. 60.

32. Typescript of a letter, probably from WBR to JBR, August 12, 1831, Rogers Papers (MC1), MIT.

33. HDR to Hamilton, August 25, [1831], Hamilton Papers, HSP; RER to WBR, September 18, 1831, Rogers Papers (MC2), MIT.

34. HDR to WBR, September 25, 1831, *LL1*, pp. 89–90.

35. RER to WBR, September 18, 1831 (typescript), Rogers Papers (MC1), MIT.

36. HDR to WBR September 25, 1831, *LL1*, p. 90. The publication is in *Report of the Board of Directors to the Stockholders of the Boston and Providence Rail-Road Company, Submitting the Report of Their Engineer, with Plans and Profiles, Illustrative of the Surveys, and Estimates of the Cost of a Rail-Road from Boston to Providence to Which Are Annexed the Acts of Incorporation* (Boston: J. E. Hinckley, 1832), pp. 52–68.

37. HDR to WBR, September 10, 1831, Rogers Papers (MC2), MIT, mentions that Henry and Robert saw "our" friend Robert Dale Owen and his father.

38. Rogers, "Education—Essay," *Free Enquirer*, May 5 and 12, 1832.

39. Ibid., June 2, 1832.

40. Ibid., April 28, 1832.

41. Robert D. Owen, "Scientific Lectures at Concert Hall," *Free Enquirer*, May 12, 1832.

42. Walter B. Hendrickson, *David Dale Owen, Pioneer Geologist of the Middle West*, Indiana Historical Collections, vol. 27 (Indianapolis: Indiana Historical Bureau, 1943), p. 21, states that Rogers helped Owen at an institute on Gray's End Road, as does Richard William Leopold, *Robert Dale Owen: A Biography* (Cambridge, Mass.: Harvard University Press, 1940), p. 115.

43. William's support is reported in HDR to Uncle James, May 12, 1832, *LL1*, p. 92.

44. Ibid.

45. An editorial comment in *LL1*, p. 92, gives the sailing date as May 19, but a note in the *Free Enquirer* says they sailed on May 16. Chronicles of the trip by Robert Dale Owen published in the *Enquirer* do not mention David Dale Owen, and my contention that he was with this group is based on the fact that Hendrickson says David Dale Owen went to England in 1831.

46. Hendrickson, *David Dale Owen*, passim.

47. A survey of shipboard life at the time appears in Robert E. Spiller, *The American in England During the First Half Century of Independence* (New York: Henry Holt, 1926), pp. 8–13.

48. HDR to the *Free Enquirer*, July 13, 1832, *Free Enquirer*, September 8, 1831.

49. RER to WBR, November 28, 1832, Rogers Papers (MC1), MIT; Editorial comment, *LL1*, p. 95.

50. WBR to Uncle James, November 26, 1832, Rogers Papers (MC1), MIT.

51. Uncle James to WBR, December 19, 1832, Rogers Papers (MC1), MIT.

52. HDR to Uncle James and WBR, November 14, 1832, *LL1*, pp. 95–96.

53. On the Cambridge tradition, see James A. Secord, *Controversy in Victorian Geology: The Cambrian-Silurian Dispute* (Princeton, N.J.: Princeton University Press, 1986), pp. 63–64.

54. HDR to Uncle James, December 14, 1832, *LL1*, p. 97.
55. HDR probably to WBR, 1832, Rogers Papers (MC1), MIT.
56. HDR to Uncle James, December 14, 1832, *LL1*, p. 98.
57. HDR to Uncle James, December 21, 1832, Rogers Papers (MC2), MIT.
58. HDR to Uncle James, December 14, 1832, *LL1*, pp. 98–99.
59. HDR to WBR, January 5, 1833, *LL1*, pp. 99–100.
60. Quoted from a letter of Henry to Robert in RER to WBR, March 20, 1833, Rogers Papers (MC1), MIT.
61. HDR to WBR, March 6, 1833, *LL1*, p. 105.
62. HDR to WBR, May 22, 1833, *LL1*, pp. 107–108.
63. Gideon Mantell to Benjamin Silliman, October 3, 1833, Silliman Family Papers, Yale University Library, Manuscripts and Archives, New Haven, Connecticut.
64. HDR to WBR, March 30, 1833, *LL1*, p. 106, mentions both his nomination and his work with de la Beche. The nomination is also recorded in the Minute Book of the Geological Society of London, March 27, 1833, London Geological Society, London, England.
65. HDR to WBR, March 30, 1833, *LL1*, p. 106; HDR to de la Beche, March 18, 1834, National Museum of Wales, Cardiff, Wales. The book was Henry de la Beche, *A Geological Manual* (London: Treuttel & Wurtz, 1831). Rogers gave as a reason for his failure the recent reprint of the first edition of the *Manual* by Philadelphia publishers Carey and Lea.
66. Robert Rogers referred to William's encouragement in a letter he wrote to William, RER to WBR, May 6, 1833, Rogers Papers (MC1), MIT.
67. The Faraday paper is mentioned in a letter from HDR to Benjamin Silliman, May 20, [1834], Simon Gratz Collection, HSP.

3. Promoting an Interesting Branch of Science, 1833–1835

1. HDR to Silliman, May 20 [1834], Simon Gratz Collection, HSP. The article to which he referred is "On the Proposed Method of Analysing Mineral Waters by Alcohol," *Journal of the Philadelphia College of Pharmacy* 5 (1833): 279–284. It was written in response to an article in the same journal by C. C. C. Cohen, "Essay on the Analysis of Mineral Waters, Together with a New Analysis of Saratoga Water, etc.," *Journal of the Philadelphia College of Pharmacy* 5 (1833): 186–194.
2. For his general acceptance in the Philadelphia scientific community, see RER to WBR, December 4, 1833, Rogers Papers (MC1), MIT. On a national organization, see Silliman to HDR, December 22, 1834, Rogers Papers (MC1), MIT.
3. I quote Bache's eulogy of James P. Espy, in the *Annual Report of the Board of Regents of the Smithsonian Institution, Showing the Operations, Expenditures, and Condition of the Institution for the Year 1859* (Washington, D.C.: Thomas Ford, 1860), p. 109, quoted in Sinclair, *Philadelphia's Philosopher Mechanics*, p. 152.
4. Records of membership, Franklin Institute Archives, Philadelphia, Pennsylvania.
5. Alexander D. Bache, Diary, March 12, 1837, A. D. Bache Papers (RU7053), Smithsonian Institution Archives, Washington, D.C.

6. Patsy Gerstner, "The Influence of Samuel George Morton on American Geology," in *Beyond the History of Science: Essays in Honor of Robert E. Schofield,* ed. Elizabeth Garber (Bethlehem: Lehigh University Press; London: Associated University Presses, 1990), pp. 126–136. For a summary of the development of American geology in general at this time, see Leonard G. Wilson, "The Emergence of Geology as a Science in the United States," *Journal of World History* 10 (1967): 416–437. Also helpful is Cecil Schneer, "Ebenezer Emmons and the Foundations of American Geology," *Isis* 60 (1969): 439–450.

7. Berkeley and Berkeley, *George William Featherstonhaugh,* pp. 107–114.

8. Patsy A. Gerstner, "Vertebrate Paleontology, an Early Nineteenth-Century Transatlantic Science," *Journal of the History of Biology* 3 (1970): 137–148. Harlan gave a paper at the British meeting on fossil reptiles found in America: *Report of the Third Meeting of the British Association for the Advancement of Science Held at Cambridge in 1833* (London: John Murray, 1834), p. 440.

9. "Annual Report of the Board of Managers," *Journal of the Franklin Institute* 13 (1834): 227. The lectures are also mentioned in Sinclair, *Philadelphia's Philosopher Mechanics,* p. 116, and RER to WBR, December 4, 1833, Rogers Papers (MC1), MIT.

10. Bache and Rogers, "Analysis of Some Coals in Pennsylvania," *Journal of the Academy of Natural Sciences of Philadelphia* 7, pt. 1 (1834): 158–177.

11. *Geology at Penn* (Philadelphia: University of Pennsylvania, 1973). See also Carol Faul, "A History of Geology at the University of Pennsylvania: Benjamin Franklin and the Rest," in *Geologists and Ideas: A History of North American Geology,* ed. Ellen T. Drake and William M. Jordan (Boulder, Colo.: Geological Society of America, 1985), pp. 377–383.

12. Minutes of the Trustees of the University of Pennsylvania, March 10, 1834, volume 8, University of Pennsylvania Archives, Philadelphia, Pennsylvania. All sources that mention Rogers and his appointment to the university indicate that he first taught in 1835. This is incorrect. He was named professor of geology in 1835, during his second year at the university.

13. Richard Harlan, *Remarks on Prof. Rogers' Geological Report to the British Association for the Advancement of Science During Their Recent Meeting Held at Edinburgh* (Philadelphia: Printed for the Publisher, 1835), p. 3.

14. For information on the "club," see Sinclair, *Philadelphia's Philosopher Mechanics,* pp. 153–154, and Nathan Reingold, ed., *The Papers of Joseph Henry* (Washington, D.C.: Smithsonian Institution Press, 1975), 2:290–291n.

15. "Forty-fourth Quarterly Report of the Board of Managers," January 1835, Franklin Institute Archives.

16. An account of his paper appears in the *Proceedings of the Geological Society of London* 2 (1833–1838): 103–106. The paper was read on November 19 and December 3.

17. HDR to de la Beche, March 18, 1834, National Museum of Wales.

18. Samuel George Morton to Gideon Mantell, January 31, 1835, Mantell Collection, Alexander Turnbull Library, Wellington, New Zealand. Morton's paper was published in 1834 (Philadelphia: Key & Biddle, 1834).

19. Rogers, "Report on the Geology of North America," in *Report of the Fourth Meeting of the British Association for the Advancement of Science Held at Edinburgh in 1834* (London: John Murray, 1835), pp. 1–66.

20. HDR to Roderick Murchison, July 13, 1834, Murchison Letters, Geological Society of London, London, England, on microfilm at the APS.
21. Murchison to HDR, September 14, 1834. Phillips reported his reading in a letter to Rogers, October 1, 1834, Rogers Papers (MC1), MIT.
22. Martin J. S. Rudwick, *The Great Devonian Controversy: The Shaping of Scientific Knowledge Among Gentlemanly Specialists* (Chicago: University of Chicago Press, 1985), pp. 122–123.
23. Rogers, "Report on the Geology of North America," pp. 30–32.
24. Ibid., pp. 63–64.
25. Turner to HDR, September 28, 1834, Phillips to HDR, October 1, 1834, and Murchison to HDR, September 14, 1834, Rogers Papers (MC1), MIT.
26. Lyell's comments are reported in a summary of the paper in the *Edinburgh New Philosophical Journal* 17 (1834): 426–427.
27. Murchison to HDR, September 14, 1834, and Phillips to HDR, October 1, 1834, Rogers Papers (MC1), MIT.
28. Minutes of the Trustees of the University of Pennsylvania, December 2, 1834, and January 6, 1835, volume 8, University of Pennsylvania Archives. In his *History of the University of Pennsylvania*, p. 225, Edward Potts Cheyney suggests that Rogers was especially desirable because he was willing to work without formal pay.
29. Rogers, *A Guide to a Course of Lectures* (Philadelphia: Printed by W. P. Gibbons, 1835), pp. 11, 22–32. Lyell's glossary is from *Principles of Geology, Being an Attempt to Explain the Former Changes of the Earth's Surface, by Reference to Causes Now in Operation* (London: John Murray, 1833), 3:61.
30. Rogers, "On the Falls of the Niagara and the Reasonings of Some Authors Respecting Them," *American Journal of Science* 27 (1835): 326–335.
31. George Fairholme, "On the Falls of the Niagara," *Philosophical Magazine*, 3d ser. 5 (1834): 11–25. Fairholme and Rogers are discussed in Richard J. Chorley, Antony J. Dunn, and Robert P. Beckinsale, *The History of the Study of Landforms; or, The Development of Geomorphology*, Vol. 1: *Geomorphology Before Davis* (London: Methuen, New York: John Wiley, 1964), pp. 253–255. George P. Merrill, *One Hundred Years of American Geology* (New York: Hafner, 1964), pp. 165–167, also mentions this point.
32. HDR to WBR, February 2, 1835, Rogers Papers (MC2), MIT. For a discussion of the journal and its change, see Sinclair, *Philadelphia's Philosopher Mechanics*, chapter 8.
33. "Report on the Proceedings of the British Association for the Advancement of Science Held at Edinburgh in September 1834," *Edinburgh New Philosophical Journal, April–October, 1834* 17 (1835): 426–427; *American Journal of Science* 28 (1835): 74–75.
34. Phillips to HDR, September 28, 1836, *LL1*, p. 136.
35. Harlan is identified as Amphibole in a letter from Morton to Mantell, October 20, 1835, Mantell Collection, Alexander Turnbull Library.
36. Morton to Mantell, January 31, 1835, Mantell Collection, Alexander Turnbull Library; Harlan, "Critical Notices of Various Organic Remains Hitherto Discovered in North America," *Transactions of the Geological Society of Pennsylvania* 1, pt. 1 (1834): 46–112. The article actually dealt almost exclusively with paleontology.
37. Harlan, *Remarks on Prof. Rogers's Geological Report*.

38. Morton to Mantell, October 10, 1835, Mantell Collection, Alexander Turnbull Library.
39. Morton to Mantell, April 28, 1836, ibid.
40. "Introduction," *Report of the Fourth Meeting of the British Association for the Advancement of Science Held at Edinburgh in 1834* (London: John Murray, 1835), p. xx.
41. Mantell to Silliman, April 14, 1835, Silliman Family Papers, Yale University Library.
42. HDR to WBR, February 2, 1835, Rogers Papers (MC2), MIT.
43. Phillips to HDR, September 28, 1836, *LL1*, p. 136; Bache Diary, March 12, 1837, Bache Papers (RU7053), Smithsonian Institution Archives.
44. George Daniels, *Science in the Age of Jackson* (New York: Columbia University Press, 1968), remains one of the best studies of the changes taking place in American science at this time.
45. Richard Harlan, "Tour to the Caves in Virginia," *Monthly American Journal of Geology and Natural Science* 1 (1831): 58–67; Richard C. Taylor, "Memoir of a Section Passing Through the Bituminous Coal Field Near Richmond, in Virginia," *Transactions of the Geological Society of Pennsylvania* 1, pt. 2 (1835): 275–297; "Report of the Committee Appointed by the Geological Society of Pennsylvania to Investigate the Rappahannock Gold Mines, in Virginia," *Transactions of the Geological Society of Pennsylvania* 1, pt. 1 (1834): 147–166. For Featherstonhaugh's work, see Berkeley and Berkeley, *George William Featherstonhaugh*, pp. 106–112.
46. William B. Rogers, "Magnesian Marl of Hanover," *Farmers Register* 1 (1834): 462–463; "On the Discovery of Green Sand in the Calcareous Deposit of Eastern Virginia," and "On the Probable Existence of This Substance in Extensive Beds Near the Western Limits of Our Ordinary Marl," *Farmers Register* 2 (1835): 129–131. On Ruffin's approach to Rogers, see Robert C. Milici and C. R. Bruce Hobbs, Jr., "William Barton Rogers and the First Geological Survey of Virginia, 1835–1841," *Earth Sciences History* 6 (1987): 3.
47. ANSP Minutes, vol. 6, October 21, 1834.
48. Ibid. Featherstonhaugh, like Harlan, had been miffed by Rogers's appointment at the university. Judging from a letter to him from Girard Troost, a longtime friend from Philadelphia who was the state geologist of Tennessee, Featherstonhaugh had written to him about the appointment of Rogers. Troost responded only that the university usually made good appointments and that in fact he had been a candidate once to replace Thomas Cooper, who had taught mineralogy and chemistry until 1822. Troost to Featherstonhaugh, April 1, 1835, Featherstonhaugh Papers, originals at the Minnesota Historical Society, St. Paul, Minnesota, on microfilm at the APS.
49. HDR to WBR, March 31, 1835, Rogers Papers (MC2), MIT.
50. William Rogers and Henry Rogers, "Contributions to the Geology of the Tertiary Formations of Virginia," *Transactions of the American Philosophical Society* 6 (1839): 329–341. A continuation was read on March 3, 1839, and was printed in the same volume as the first paper, pp. 347–370.
51. That Featherstonhaugh wanted the job is implied in HDR to WBR, March 31, 1835, Rogers Papers (MC2), MIT.
52. "Mr. Featherstonhaugh's Geological Report," *Transactions of the Geological Society of Pennsylvania* 1, pt. 2 (1835): 413; "Review of a 'Geological Report of

An Examination, made in 1834, of the Elevated Country between the Missouri and Red Rivers. By G. W. Featherstonhaugh, U.S. Geologist, Published by order of both Houses of Congress, Washington: Printed by Gales & Seaton, 1835,'" *Journal of the Franklin Institute* 17 (1836): 109–117, and 184–190. Featherstonhaugh's report is: *Geological Report of the Examination Made in 1834 of the Elevated Country Between the Missouri and Red Rivers by G. W. Featherstonhaugh, U.S. Geologist,* House Document, 23d Cong., 2d Sess., vol. 4, no. 51 (Washington, D.C., 1835).

53. "Review of a 'Geological Report,'" p. 190.
54. HDR to WBR, November 1, 1835, Rogers Papers (MC2), MIT.

4. Field Research of a Scientific Kind, 1835–1836

1. HDR to WBR, March 30, 1833, *LL1*, p. 106.
2. Editorial comment, *LL1*, p. 110.
3. HDR to John F. Frazer, April 28, 1834, Frazer correspondence, APS.
4. HDR to Uncle James, July 27, 1834, Rogers Papers (MC2), MIT.
5. Milici and Hobbs, "William Barton Rogers and the First Geological Survey of Virginia," p. 4. For a brief summary of the Virginia survey, see George P. Merrill, *Contributions to a History of American State Geological and Natural History Surveys,* U.S. National Museum Bulletin 109 (Washington, D.C., U.S. Government Printing Office, 1920), pp. 428–34. The Virginia survey and other state surveys were summarized recently in Arthur A. Socolow, ed., *The State Geological Surveys: A History* (Champaign, Ill.: Association of American State Geologists, 1988).
6. WBR to HDR, November 30, 1834, *LL1*, pp. 112–114; "Report of the Select Committee," *Farmers Register* 2 (1835): 688–692.
7. Bache liked William so much that he had tried to get him to accept a position in Philadelphia in 1834. WBR to HDR, November 30, 1834, *LL1*, pp. 112–113.
8. HDR to WBR, February 2, 1835, Rogers Papers (MC2), MIT. The proposal has not survived. Vroom's plea is in [Message to the Legislative Council], *Journal of the Proceedings of the Legislative Council of the State of New Jersey, Convened at Trenton, on the Twenty-eighth Day of October, A.D. 1834* (Somerville, N.J., 1835), pp. 18–19, and in *Votes and Proceedings of the Fifty-ninth General Assembly of the State of New Jersey, at a Session Begun at Trenton, on the Twenty-third Day of October, 1834* (Freehold, N.J., 1835), pp. 20–21.
9. *Votes and Proceedings of the Fifty-ninth,* p. 329.
10. Bache to Vroom, March 11, 1835, A. D. Bache Papers, APS.
11. Joseph Henry to Green, March 16, 1835, in Reingold, *The Papers of Joseph Henry,* 2:366–367.
12. Joseph Henry to HDR, March 9, 1835, Rogers Papers (MC1), MIT. Also printed in Reingold, *The Papers of Joseph Henry,* 2:364–366.
13. Joseph Henry to Jacob Green, March 16, 1835, in Reingold, *The Papers of Joseph Henry,* 2:366–367.
14. Rogers mentions Clemson and Pierce in an April 11 letter to Joseph Henry in the Rogers Papers (MC1), MIT and printed in Reingold, *The Papers of Joseph Henry,* 2:374–375. Clemson married the daughter of John C. Calhoun and subsequently donated Calhoun's homestead to South Carolina for the college

that now bears Clemson's name. Conrad was recommended for the job by Samuel George Morton. Morton to Vroom, March 1835, Morton Papers, HSP.

15. HDR to Joseph Henry, April 11, 1835, Rogers Papers (MC1), MIT, and printed in Reingold, *The Papers of Joseph Henry*, 2:373–375. The New Jersey survey under Rogers is discussed briefly by Jean W. Sidar, "New Jersey Geological Surveys in the Nineteenth Century," *Northeast Geology* 3 (1981): 52–53, and in her *George Hammell Cook: A Life in Agriculture and Geology, 1818–1889* (New Brunswick, N.J.: Rutgers University Press, 1976), pp. 58–66.

16. HDR to WBR, February 8, 1835, Rogers Papers (MC2), MIT.

17. Peter Browne, *An Address Intended to Promote a Geological and Mineralogical Survey of Pennsylvania* (Philadelphia: P.M. Lafourcade, 1826). His lecture to the Franklin Institute is also mentioned in Sinclair, *Philadelphia's Philosopher Mechanics,* p. 255, n. 39. For a study of Browne's efforts in behalf of a survey, see Anne Millbrooke, "The Geological Society of Pennsylvania, 1832–1836, Pt. 1: Founding the Society," *Pennsylvania Geology* 7/6 (1976): 7–11, and her Ph.D. dissertation, "State Geological Surveys of the Nineteenth Century: Pennsylvania, a Case Study" (University of Pennsylvania, 1981), pp. 74–133.

18. "Appendix, Note A," in William H. Dillingham, *A Discourse on the Advantages of the Study of Natural Science, Delivered by Request of the Chester County Cabinet, Introductory to Their Course of Lectures: December 5th, 1835* (Philadelphia: William Brown, Printer, 1835).

19. Sinclair, *Philadelphia's Philosopher Mechanics,* chapter 3.

20. Ibid., p. 255n.

21. Questions were raised about the society's motivations. S. G. Morton wrote to a friend, saying that "something else than science is at the bottom of it [Geological Society]." Morton to Benjamin Silliman, February 2, 1832, Silliman Family Papers, Yale University Library. George Featherstonhaugh noted that the society had many enemies, including the American Philosophical Society. Berkeley and Berkeley, *George William Featherstonhaugh,* pp. 111–112.

22. Circular, *Monthly American Journal,* no. 9, March 1, 1832.

23. A plea, for example, was presented to the house from Montgomery County, which it rejected on February 11, 1831. *Journal of the Forty-first House of Representatives of the Commonwealth of Pennsylvania* (Harrisburg, 1830–1831), 2:761.

24. *Journal of the Forty-third House of Representatives of the Commonwealth of Pennsylvania* (Harrisburg, 1832–1833), 1:24.

25. Ibid., pp. 149, 217, 258, and 307; J. Peter Lesley, *Historical Sketch of Geological Exploration in Pennsylvania and Other States* (Harrisburg: Published by the Board of Commissioners for the Second Geological Survey, 1876), p. 31.

26. "Document 214, Report upon the Geological Survey of the State," *Journal of the Forty-third House,* 2:710–717. The report and act were reproduced in *Hazard's Register of Pennsylvania* 40, no. 15 (April 13, 1833).

27. Millbrooke's dissertation summarizes the financial situation of Pennsylvania at the time. For a more complete picture, see Philip Klein and Ari Hoogenboom, *A History of Pennsylvania* (New York: McGraw-Hill, 1973), chapter 11.

28. *Journal of the Forty-fourth House of Representatives of the Commonwealth of Pennsylvania* (Harrisburg, 1833–1834), 1:456 and 742, and Report 194, 2:847–854.
29. HDR to J. Vanderkemp, December 27, 1835, APS.
30. Ibid.
31. "Governors Message," *Journal of the Senate of the Commonwealth of Pennsylvania* (Harrisburg, 1835–1836), 1:16–17, read December 2, 1835.
32. HDR to WBR, January 23, 1836, Rogers Papers (MC2), MIT.
33. Ibid.
34. *Journal of the Forty-sixth House of Representatives of the Commonwealth of Pennsylvania* (Harrisburg, 1835–1836), 1:435.
35. Charles B. Trego, *Report of the Committee Appointed on So Much of the Governor's Message As Related to a Geological and Mineralogical Survey of the State of Pennsylvania, Read in the House of Representatives, Feb. 3, 1836* (Harrisburg, 1836).
36. Ibid., p. 11.
37. Henry Rogers, *Report of the Geological Survey of the State of New Jersey* (Philadelphia: Desilver, Thomas, 1836), p. 6. This volume was also printed for the state by Bernard Connolly in Freehold, New Jersey. References are to the Philadelphia edition. Rogers's recommended lines for New Jersey were used for some of the subsequent investigations of the state. George H. Cook, "Historical Notes on the Geological Surveys of New Jersey," *Annual Report of the State Geologist, for the Year 1885* (Trenton, N.J.: John L. Murphy, 1885), pp. 152–203.
38. Allan Nevins, "Thaddeus Stevens," *Dictionary of American Biography*, ed. Dumas Malone (New York: Charles Scribner's Sons, 1935). Stevens went on to be a powerful foe of slavery in the United States Congress but earned a reputation for a vindictive temperament. Fraley presented bill 288 to the senate, February 28, 1838. *Journal of the Senate of the Commonwealth of Pennsylvania* (Harrisburg, 1837–1838), 1:428. Geological surveying was often tied directly to transportation in the minds of the legislators, but although Pennsylvania was in the midst of massive canal construction, it seems never to have occurred to anyone to make a specific link between the geological survey and canals or even railroad construction. On the frequent link with transportation, see Michele L. Aldrich, "American State Geological Surveys, 1820–1845," in *Two Hundred Years of Geology in America* (Hanover, N.H.: University Press of New England for the University of New Hampshire, 1979), p. 135.
39. "To Provide for a Geological and Mineralogical Survey of the State," *Laws of the General Assembly of the Commonwealth of Pennsylvania Passed at the Session of 1835–1836* (Harrisburg, 1836), pp. 225–227.
40. The final memorial of 1834 is recorded in the *Journal of the Forty-fifth House of Representatives of the Commonwealth of Pennsylvania* (Harrisburg, 1834–1835), 1:60.
41. See, for example, Anne Millbrooke, "The Geological Society of Pennsylvania, 1832–1836, Pt. 2: Promoting a State Survey," *Pennsylvania Geology* 8 (1977): 12–16, and her Ph.D. dissertation: "State Geological Surveys of the Nineteenth Century: Pennsylvania, a Case Study." Similar claims are repeated in William M. Jordan, "Geology and the Industrial Transportation

Revolution in Early to Mid Nineteenth-Century Pennsylvania," in *Two Hundred Years of Geology in America*, ed. Cecil J. Schneer (Hanover, N.H.: University Press of New England for the University of New Hampshire, 1979), pp. 91–103, and by Donald Hoskins, "The First Geological Survey of Pennsylvania: The Discovery Years," *Pennsylvania Geology* 18 (February 1987): 1–2, which was condensed from an article entitled "Celebrating a Century and a Half: The Geologic Survey," *Pennsylvania Heritage* 12 (Summer 1986): 26–31. Michele Aldrich has pointed out that political affiliation had little bearing on the outcome of survey legislation in the various states but that strong governors were often key factors in the approval of surveys. Aldrich, "American State Geological Surveys," p. 134.

42. Sinclair, *Philadelphia's Philosopher Mechanics*, p. 151.
43. S. W. Roberts, "An Obituary Notice of Charles B. Trego," *Proceedings of the American Philosophical Society* 14 (1875): 356.
44. HDR to WBR, January 23, 1836, Rogers Papers (MC2), MIT.
45. *Journal of the Forty-seventh House of Representatives of the Commonwealth of Pennsylvania* (Harrisburg, 1837–1838), 2:30, also quoted in Millbrooke, *State Geological Surveys*, p. 92.
46. "To Provide for a Geological Survey," *Laws of the General Assembly . . . 1835–36*, p. 226.
47. Information on allocation of salaries in Ohio, New York, and Virginia is from George Merrill, *Contributions to a History of American State Geological and Natural History Surveys*.
48. RER to WBR, December 16, 1835, *LL1*, p. 127.
49. Rogers, *Report of the Geological Survey . . . New Jersey*. The letter asking for two weeks is HDR to Vroom, February 1, 1836, Southard Hay Collection, Yale University Library.
50. *Journal of the Proceedings of the Legislative Council of the State of New Jersey, Convened at Trenton, on the Twenty-seventh Day of October, A.D., 1835* (Somerville, N.J., 1836), p. 15.
51. RER to WBR, December 16, 1835, *LL1*, p. 128. Michele Aldrich has noted that Rogers and Booth both reported on the importance of potassium in the greensand in their respective reports on New Jersey and Delaware. Aldrich, "American State Geological Surveys," p. 135. On the state of agricultural science, see Margaret Rossiter, *The Emergence of Agricultural Science: Justus Liebig and the Americans, 1840–1880* (New Haven: Yale University Press, 1975).
52. Rogers, *Report of the Geological Survey . . . New Jersey*, p. 42. He repeated this statement, essentially unchanged, in the *Description of the Geology of the State of New Jersey, Being a Final Report* (Philadelphia: Sherman, Printers, 1840), pp. 226–228.
53. HDR to WBR, [Spring] 1836, Rogers Papers (MC2), MIT.
54. For the New Jersey decision to continue, see the [Report of committee on the subject chaired by Henry Hillard], in *Votes and Proceedings of the Sixtieth General Assembly of the State of New Jersey*, pp. 552–554 especially. The increase in funds met some opposition, but an effort to hold the allocation at $1,000 failed by a vote of 23 to 18. *Votes and Proceedings of the Sixtieth General Assembly of the State of New Jersey, at a Session Begun at Trenton, on the Twenty-seventh Day of October, 1835* (Freehold, N.J., 1836), p. 596.

55. WBR to HDR, January 21, 1835, and February 11, 1835, *LL1*, pp. 115, 117. William publicly acknowledged Henry's help with the stratigraphy of the more western portions of the state (the part bordering Pennsylvania) and in locating the shoreline of the ancient sea that had once covered the area in his *Second Report of the Progress of the Geological Survey of the State of Virginia for the Year 1837*, pp. 49–57, copy in the Rogers Papers (MC1), MIT.
56. HDR to WBR, [Spring] 1836, Rogers Papers (MC2), MIT.
57. Editorial comment, *LL1*, p. 141; WBR to HDR, April 30, 1837, *LL1*, p. 141. When he announced Henry's resignation to the Virginia government, William recommended James to take his place.
58. HDR to WBR, April 10, 1836, *LL1*, p. 131.
59. HDR to WBR, 1836, Rogers Papers (MC2), MIT; RER to WBR, December 16, 1835, *LL1*, pp. 127–128.

5. Questions of the Highest Importance, 1836–1837

1. HDR to WBR, April 10, 1836, *LL1*, pp. 130–132. On official notification regarding the job, see HDR to John Frazer, April 13, 1836, J. F. Frazer Papers, APS. That Rogers was especially glad to have Frazer is clear in a letter from Uncle James to WBR, June 14, 1836, Rogers Papers (MC1), MIT.
2. For information on Booth, see Thomas E. Pickett, "James C. Booth and the First Delaware Geological Survey, 1837–1841," in *Two Hundred Years of Geology in America*, ed. Cecil J. Schneer, pp. 167–174. On Booth's background in Germany, see letters to his mother (Ann Booth) and brother (C. W. Faber), March 22, July 14, and December 10, 1833, Booth Papers, University of Delaware Archives, Newark, Delaware.
3. Walker may have stayed on in Virginia, because a later letter in *LL1* from William to Henry mentions that Walker was working on maps. Although the letter as published is dated June 26, 1838 (*LL1*, p. 160), I suspect that the date was really 1836.
4. Robert had completed his medical education by now and had been offered a position as physician to the Philadelphia Almshouse, but because he preferred chemistry to medical practice, he chose to remain with the geological survey.
5. Assistants for William are discussed in two letters from Henry to William: HDR to WBR, April 10, 1836, *LL1*, pp. 131–132, and April 3, 1836, Rogers Papers (MC2), MIT. William found his own assistants later in the persons of his brother, James, who joined the Virginia Survey in 1837, and C. B. Hayden, who joined in 1838.
6. HDR to WBR, June 1, 1836, Rogers Papers (MC2), MIT.
7. Booth to C. W. Faber, June 5, 1836, Booth Papers, University of Delaware Archives.
8. On some of their experiences during the reconnaissance, see Booth to Ann Booth, June 16 and June 30, 1836, University of Delaware Archives; on Rogers's report to the legislature, see HDR to WBR, June 7, 1836, Rogers Papers (MC2), MIT; on the path followed by the reconnaissance, see Rogers, *First Annual Report of the State Geologist*, (Harrisburg: Printed by Emanuel Guyer, 1836), p. 6.
9. Booth to Ann Booth, June 30, 1836, Booth Papers, University of Delaware Archives.

10. See, for example, HDR to Booth and Frazer, [summer] 1836, Frazer Papers, APS; HDR to WBR, June 7, 1836, Rogers Papers (MC2), MIT.
11. Rogers, *First Annual Report*, p. 8.
12. Anthony F. C. Wallace, *St. Clair: A Nineteenth-Century Coal Town's Experience with a Disaster Prone Industry* (New York: Alfred A. Knopf, 1987), p. 202; Rogers, *First Annual Report*, p. 10.
13. The study for the state is Samuel J. Packer, *Report of the Committee of the Senate of Pennsylvania, upon the Subject of the Coal Trade* (Harrisburg: H. Welsh, 1834); Richard C. Taylor, "On the Mineral Basin or Coal Field of Blossburg, on the Tioga River, Tioga County, Pennsylvania," *Transactions of the Geological Society of Pennsylvania* 1, pt. 2 (1835): 204–219; Benjamin Silliman, "Notice of the Anthracite Region in the Valley of the Lackawanna and Wyoming on the Susquehanna," *American Journal of Science* 18 (1830): 308–328, and "Notes on a Journey from New Haven, Conn., to Mauch Chunk and Other Anthracite Regions of Pennsylvania," *American Journal of Science* 19 (1830): 1–21. In addition to the work he did with Bache, cited in Chapter 3, Rogers published "Anthracite Coal Region of Pennsylvania," *Messenger of Useful Knowledge* 1, no. 8 (March 1831): 113–115.
14. Rogers, *First Annual Report*, p. 11.
15. Ibid., pp. 8–9.
16. Rogers, *First Annual Report*. All annual reports were read in the house of representatives soon after submission and were printed in full in the *Journal of the Senate*.
17. The formations are discussed on pp. 12–17 of the *First Annual Report*. Lesley discussed Frazer's work in his *Historical Sketch*, pp. 53–55.
18. Rogers, *First Annual Report*, p. 12.
19. Ibid., p. 20.
20. Rogers, *The Geology of Pennsylvania: A Government Survey*, 2 vols. (Philadelphia: J. B. Lippincott, 1858), 1:v.
21. That Rogers was influenced in this by de la Beche was observed by Mott T. Green, *Geology in the Nineteenth Century: Changing Views of a Changing World* (Ithaca: Cornell University Press, 1982), p. 124.
22. "A Sketch of What Has Been Achieved Towards a Geological Survey of New Jersey During the Past Year," included in the legislative message of the governor, Philemon Dickerson, *Journal of the Proceedings of the Legislative Council of the State of New-Jersey Convened at Trenton on the Twenty-fifth Day of October, A.D. 1836* (Woodbury, N.J., 1837), pp. 67–68.
23. Rogers, "Geology [U.S.]," in Hugh Murray, *The Encyclopedia of Geography Comprising a Complete Description of the Earth, Physical, Statistical, Civil, and Political; Exhibiting Its Relation to the Heavenly Bodies, Its Physical Structure, the Natural History of Each Country, and the Industry, Commerce, Political Institutions, and Civil and Social State* (Philadelphia: Carey, Lea, & Blanchard, 1837), 3:373–403.
24. Henry Rogers, "On the Geology of the Ancient Secondary Basin of the United States," Brown University Library, Providence, Rhode Island. For a summary of past and current understanding of Appalachian dynamics, see Robert T. Faill, "Evolving Tectonic Concepts of the Central and Southern Appalachians," *Geologists and Ideas: A History of North American Geology*, ed. Ellen T. Drake and William M. Jordan (Boulder, Colo.: Geological Society of America, 1985), pp. 19–46.

25. For a detailed and enlightening study of the elevation theories of the nineteenth century, see Greene, *Geology in the Nineteenth Century.* Chapters 3 and 4 are especially important with regard to Elie de Beaumont.
26. Martin J. S. Rudwick, "Uniformity and Progression: Reflections on the Structure of Geological Theory in the Age of Lyell," *Perspectives in the History of Science and Technology,* ed. Duane H. D. Roller (Norman: University of Oklahoma Press, 1971), pp. 209–227. For other studies that relate to the question of uniformitarianism and catastrophism, see R. Hooykaas, *Natural Law and Divine Miracle: A Historical-Critical Study of the Principle of Uniformity in Geology, Biology, and Theology* (Leiden: E. J. Brill, 1959); Leonard G. Wilson, "Geology on the Eve of Charles Lyell's First Visit to America, 1841," *Proceedings of the American Philosophical Society* 124 (1980): 168–202; Greene, *Geology in the Nineteenth Century,* especially chapters 3 and 4.
27. Charles Lyell, *Principles of Geology,* especially volume 3.
28. HDR to McIlvaine and Biddle, May 20, 1837, Brown University Library.
29. Information on their first meeting appears in McIlvaine to John A. Lowell, February 17, 1846, Incoming Letters of the University Treasurer (UAI.50.8VT), Harvard University Archives.
30. The testimonials are with the manuscript in the Brown University Library.
31. HDR to Biddle, May 20, 1837, Brown University Library.
32. HDR to Frazer, April 12, 1837, Frazer Papers, APS.
33. Details of the legislation allocating the additional funds, including $1,000 for the Franklin Institute for meteorological studies, appear in the *Journal of the Forty-seventh House of Representatives of the Commonwealth of Pennsylvania* (Harrisburg, 1837), 1:329, 656, 725, 814, and 855, and in "A Supplement to the Act Entitled 'An Act to Provide for a Geological and Mineralogical Survey of the State,'" *Laws of the General Assembly of the Commonwealth of Pennsylvania Passed at the Session of 1836–1837* (Harrisburg, 1837), pp. 72–73.
34. HDR to Frazer, April 12, 1837, Frazer Correspondence, APS. I assume that each assistant received the same letter because, several years later, J. Peter Lesley quoted such a letter to Haldeman. Lesley, "Memoir of Samuel Stedman Haldeman, 1812–1880," *National Academy of Sciences Biographical Memoirs* (Washington, D.C.: Judd & Detweiler, Printers, 1886), 2:152–153.
35. RER to HDR, April 28, 1837, Rogers Papers (MC1), MIT.
36. Booth to Ann Booth, March 4, 1837, Booth Papers, University of Delaware.
37. HDR to Frazer, May 17, 1837, Frazer Papers, APS.
38. For a brief history of the breakwater, see Frank E. Snyder and Brian H. Guss, *The District: A History of the Philadelphia District U.S. Army Corps of Engineers, 1866–1971* (Philadelphia: U.S. Army Engineers, District Philadelphia, January 1974).
39. "Report on the Stone of Leiper's Quarries," *Journal of the Franklin Institute* 19 (1837): 367–369.
40. HDR to WBR, June 14, 1837, *LL1,* pp. 144–145.
41. JBR to WBR, March 21, 1837, Rogers Papers (MC1), MIT. When Henry refused the job, he nominated James, who also refused it.
42. HDR to WBR, June 14, 1837, *LL1,* pp. 144–145.
43. Ibid. In the *Life and Letters,* Emma Rogers used only the initial "P" to refer to Park and Patterson, which makes the letter by itself confusing. Patterson, however, is identified as being in opposition to Park in a letter from Robert

to William, June 17, 1837, Rogers Papers (MC2), MIT. Robert suggested that Park had made it possible for Rogers's ideas to get across. An unsigned report dated June 12, 1837, on file at the Franklin Institute recommends trappean rocks. The reasons given for preferring the trappean rock include its higher specific gravity, desirable regularity in its joints, which permitted it to break into pieces well suited to packing together, and its high level of chemical indestructibility. Another report in the archives of the institute (Case 148), signed by Joseph A. Totten, S. H. Thayer, and John S. Smith and dated September 1, 1837, says that either the gneiss or trappean rock is all right.

44. HDR to WBR, June 14, 1837, *LL1*, pp. 144–145. Rogers was not the only one kept from the field. Booth was also unable to begin his new job on June 1 as scheduled. The breakwater debate forced him to delay until mid-June. Booth to Ann Booth, July 1, 1837, Booth Papers, University of Delaware Archives.

45. Lesley, *Historical Sketch*, p. 56.

46. Rogers, *Second Annual Report of the Geological Exploration of the State of Pennsylvania* (Harrisburg: Printed by Packer, Barrett & Parke, 1838), p. 89. Rogers stayed true to this definition in coming years.

47. Ibid., p. 82.

48. Ibid., p. 37.

49. Henry and William Rogers, "On the Geological Age of the White Mountains," *American Journal of Science*, n.s. 1 (1846): 8–11. The brothers argued in this paper that the strike of the White Mountains was different from that found commonly in the Appalachians because of the nature and location of the earlier disturbance. They also said that the area was difficult to understand geologically unless one took into account the effects of both the earlier disturbance that lifted the White Mountains above the sea and the later one that affected the whole of the Appalachian area.

50. Among the state geologists, Rogers's ideas on metamorphism met an early acceptance by W. W. Mather who quoted extensively on metamorphism from Rogers's final New Jersey report in his New York report. W. W. Mather, *Geology of New York*, Pt 1: *Comprising the Geology of the First Geological District* (Albany: Printed by Carroll & Cook, 1843), pp. 468–475.

51. Rogers, *Second Annual Report*, p. 5.

52. *Journal of the Forty-eighth House of Representatives of the Commonwealth of Pennsylvania* (Harrisburg, 1837–1838), 1:348.

53. HDR to WBR, March 6, 1838, Rogers Papers (MC2), MIT.

54. Ibid.

55. Emma Rogers, editorial comment, *LL1*, p. 157; WBR to HDR, March 14, 1838, *LL1*, p. 153.

56. Amos Lawrence to Biddle, 1839, Biddle Correspondence, Library of Congress, Washington, D.C.

57. Robert Walter Smith, *History of Armstrong County, Pennsylvania* (Chicago: Waterman, Walker, 1914), 1:286, says the furnace opened in 1839. The date 1838 for Rogers's furnace is based on Rogers's statement in his report of the 1838 field season that iron ore for Rogers's furnace was taken from the west side of Laurel Hill and from the fact that John McKinney, who was the furnace's manager of operations, was hired by Rogers as a survey assistant in

1838. Rogers, *Third Annual Report of the Geological Survey of the State of Pennsylvania* (Harrisburg: Boas & Coplan, Printers, 1839), p. 68.
58. Smith, *History of Armstrong County*, p. 286.
59. Roswell Colt to John Cadwalader, July 21, 1842, Cadwalader Collection, HSP.
60. HDR to Biddle, December 14, 1838, Biddle Papers, Library of Congress.

6. Cautious and Laborious Research, 1838–1840

1. An act "Supplementary to an Act Entitled Act Authorizing the Governor to Incorporate the Meadville and Titusville Turnpike Road Company, and for other Purposes," *Laws of the General Assembly of the Commonwealth of Pennsylvania, Passed at the Session of 1837–38* (Harrisburg, 1838), p. 380.
2. The information on the chemical laboratory comes from rental receipts in the "Records of the Department of the Auditor General: Geological Survey: Financial Records, 1837–1875," Microfilm Roll RG2, Pennsylvania State Archives, Harrisburg, Pennsylvania. About this time Boye and Rogers coauthored a paper, "Upon a New Compound of Deuto-Chloride of Platinum Nitric Oxide, and Chlorohydric Acid," *Transactions of the American Philosophical Society*, n.s. 7 (1841): 59–66.
3. Lesley, *Historical Sketch*, p. 70.
4. Henry Rogers, *Third Annual Report*, p. 9.
5. *Laws of the General Assembly . . . 1837–38*, p. 380.
6. Wallace, *St. Clair*, pp. 85–93, summarizes the development of the use of anthracite and mentions the prize. See also Alfred D. Chandler, Jr., "Anthracite Coal and the Beginnings of the Industrial Revolution in the United States," *Business History Review* 46 (1972): 141–181, and James M. Swank, *History of the Manufacture of Iron in All Ages, and Particularly in the United States from Colonial Times to 1891* (Philadelphia: American Iron & Steel Association, 1892), p. 359.
7. HDR to Thomas H. Burrows, November 5, 1838, Physicians of Pennsylvania Collection, HSP.
8. J. P. Lesley refers to hiring coal surveyors in a letter to his father, April 22, 1839, Lesley Papers, APS.
9. Rogers, *Third Annual Report*, p. 26.
10. Lesley, *Historical Sketch*, p. 79. Lesley also mentions in his history (p. 80) that McKinley completed a topographical map of the mountainous country west of the Susquehanna in 1838. Neither map is extant.
11. Timothy Conrad, "Observations on Characteristic Fossils and upon a Fall of Temperature in Different Geological Epochs," *American Journal of Science* 35 (1839): 244.
12. Ibid., p. 244.
13. Leonard Wilson, "The Emergence of Geology," pp. 417–423; Amos Eaton, *A Geological and Agricultural Survey of the District Adjoining the Erie Canal in the State of New York* (New York: Packard & Van Benthuysen, 1824).
14. Rogers, *Second Annual Report*, p. 90; Lyell's glossary is in his *Principles*, 3:79.
15. Rogers, *Second Annual Report*, p. 82.
16. On the new red sandstone, see Rogers, *Description of the Geology of the State of*

New Jersey, pp. 114–117; *Reports of the First, Second, and Third Meetings of the Association of American Geologists and Naturalists at Philadelphia, in 1840 and 1841, and at Boston in 1842, Embracing Its Proceedings and Transactions* (1842) (Boston: Gould, Kendall, & Lincoln, 1843), pp. 63–64; HDR to WBR, December 11, 1841, *LL1*, p. 202. In 1853, or shortly before, Rogers decided that the red sandstone was slightly older than he first thought but still well above the coal. HDR to Desor, August 16, 1853, Desor Papers, AN.

17. *American Journal of Science* 35 (1839): pp. 250–251.
18. Amos Eaton, "Cherty Lime-rock, or Corniferous Lime-rock, Proposed as the Line of Reference, for State Geologists of New York and Pennsylvania," *American Journal of Science* 36 (1839): 51–71.
19. "Miscellaneous," *American Journal of Science* 37 (1839): 380.
20. Donald Hoskins, "Henry Rogers and James Hall of the Pennsylvania and New York Geological Surveys, 1836–1842," *Earth Science History* 6 (1987): 18.
21. James Hall, *Geology of New York*, Pt. 4: *Comprising the Survey of the Fourth Geological District* (Albany: Carroll & Cook, 1843), p. 24.
22. *Journal of the Forty-ninth House of Representatives of the Commonwealth of Pennsylvania* (Harrisburg, 1838–1839), 1:724; WBR to HDR, April 22, 1837, *LL1*, p. 169.
23. Rogers, *Second Annual Report*, p. 74.
24. In his study of the coal town St. Clair, Anthony Wallace argued that Rogers encountered a problem with the state legislature because of his statements on the area. Wallace, *St. Clair*, p. 207.
25. Lesley, *Historical Sketch*, p. 111. Lesley wrote that residents were suspicious of the geologists in a letter to his father, May 15, 1839, Lesley Papers, APS.
26. C. K. Yearley, Jr., *Enterprise and Anthracite: Economics and Democracy in Schuylkill County, 1820–1875* (Baltimore: Johns Hopkins University Press, 1961), p. 100. Yearley also suggests that Rogers's failure to use local formation names further confused the situation for miners (p. 103).
27. Bache and Aldrich, "Report of the Commission Relative to the Topographical Survey of the State of Pennsylvania," *Journal of the Forty-ninth House*, 2:1068–1071.
28. *For the Establishment of a School of Arts: Memorial of the Franklin Institute, of the State of Pennsylvania, for the Promotion of the Mechanic Arts, to the Legislature of Pennsylvania* (Philadelphia: Printed by J. Crissy, 1837); HDR to WBR, March 8, 1846, *LL1*, p. 258. Part of the handwritten manuscript for the memorial is in the Rogers Papers (MC1) at MIT, lending credence to Henry's claim.
29. Sinclair, *Philadelphia's Philosopher Mechanics*, pp. 230–233.
30. Rogers, *Fourth Annual Report of the Geological Survey of the State of Pennsylvania* (Harrisburg: Holbrook, Henlock, & Bratton, Printers, 1840), pp. 176–177. Rogers does not list Lehman by name in the annual report for this year, but his presence is documented in letters that Lesley wrote, for example, to his stepmother, Mrs. S. E. Lesley, May 31, 1839, Lesley Papers, APS.
31. Lesley to his father, June 4, 1839, and to Mrs. S. E. Lesley, June 10, 1839, Lesley Papers, APS.
32. Rogers, *Fourth Annual Report*, pp. 176–177 and p. 50.
33. Ibid., p. 11.

34. Lesley, *Historical Sketch*, p. 10.
35. A few notebooks of assistants survive. A field notebook of Alexander McKinley, who wrote humorous poems at the beginning of his notebook "to relieve the horrible dread of writing my report" is at the APS in the Lesley Papers. R. M. S. Jackson's notebooks are at the Pennsylvania State University Archives (State College, Pennsylvania) in the Jackson Papers. An 1837 field notebook of Samuel Haldeman's work in Pennsylvania is in the Haldeman Papers at the ANSP.
36. *Journal of the Fiftieth House of Representatives of the Commonwealth of Pennsylvania* (Harrisburg, 1840), 1:289–90, 299, and 307, February 8, 10, and 12. The number was reduced from a record 5,000 copies in English to 2,000 and from 2,000 in German to 500.
37. Ibid., p. 1031.
38. Ibid., p. 271.
39. Ibid., p. 301.
40. "Records of the Department of the Auditor General: Geological Survey, Financial Records, 1837–1875," Microfilm Roll RG2, Pennsylvania State Archives.
41. See Aldrich, pp. 190–191, on the issue in New York. Aldrich points out that many state geologists objected to the practice and that in later years clauses were written into state survey legislation prohibiting profit by the geologists before the state profited.
42. Rogers, *Description of the Geology of the State of New Jersey, Being a Final Report* (Philadelphia: Sherman & Company, Printers, 1840). The report, a 174-page document, was also reprinted in 1865 by J. R. Freese in Trenton, New Jersey, and by John W. Lyon in Jersey City.
43. *Votes and Proceedings of the Sixty-first General Assembly of the State of New Jersey, at a Session Begun at Trenton, on the Twenty-fifth Day of October, 1836* (Belvidere, N.J., 1837), p. 122.
44. The assistant is mentioned in a letter from Rogers to Governor Dickerson, in *Journal of the Proceedings of the Legislative Council of the State of New Jersey, Convened at Trenton, on the Twenty-fifth Day of October, A.D., 1836* (Woodbury, N.J., 1837), p. 68.
45. "Governor's Message," *Journal of the Proceedings of the Legislative Council of the State of New Jersey, Convened at Trenton, on the Twenty-fourth Day of October, A.D. 1837* (Somerville, N.J., 1838), p. 15.
46. For a summary of the distribution plans, see *Votes and Proceedings of the Sixty-fifth General Assembly of the State of New Jersey, at a Session Begun at Trenton, on the Twenty-seventh Day of October 1840* (Freehold, N.J., 1841), pp. 78–79.
47. Rogers's response to Pennington, dated October 17, 1839, is in *Votes and Proceedings of the Sixty-fourth General Assembly of the State of New Jersey, at a Session Begun at Trenton, on the Twenty-second Day of October 1839* (Belvidere, N.J., 1840), p. 21.
48. Ibid., p. 404.

7. A Capricious Master, 1840–1842

1. Benjamin Silliman to HDR, December 22, 1834, Rogers Papers (MC1), MIT.
2. Hall to HDR, October 22, 1839, and HDR to Hall, November 6, 1839, Hall

Papers, NYSL. Also printed (with his comments) in John M. Clarke,
James Hall of Albany, Geologist and Paleontologist, 1811–1898 (Albany, 1923),
pp. 76–78.

3. Hall, "Association of American Geologists" (typescript), James Hall Papers,
NYSL.

4. Sally Gregory Kohlstedt, *The Formation of the American Scientific Community:
The American Association for the Advancement of Science, 1848–1860* (Urbana:
University of Illinois Press, 1976), pp. 62–65.

5. HDR to WBR, September 26, 1838, *LL1*, pp. 155–156.

6. See Kohlstedt, *The Formation of the American Scientific Community*, especially
p. 49.

7. HDR to WBR, April 9, 1839, *LL1*, p. 164.

8. See, for example, letter to Booth, November 6, 1839, Booth Papers, University of Delaware Archives.

9. A list, signed by those present, is at the Academy of Natural Sciences of
Philadelphia in AAGN Collection 305. There are at least three versions of the
list of attendees. This one, one published in the association's proceedings,
and a typescript by Hall. See Kohlstedt, *The Formation of the American Scientific Community*, p. 184n., for a discussion.

10. *Reports of the First, Second, and Third Meetings* (1840), pp. 9–11.

11. Ebenezer Emmons, *Geology of New York*, Pt. 2: *Comprising the Survey of the
Second Geological District* (Albany: W. & A. White & J. Visscher, Printers,
1842), p. 99.

12. Rogers, *Fifth Annual Report of the Geological Survey of Pennsylvania* (Harrisburg: James S. Wallace, Printer, 1841), pp. 10–11.

13. Lesley to his father, June 13, 1839, Lesley Papers, APS.

14. Lesley to Mrs. S. E. Lesley, May 31, 1839, Lesley Papers, APS. Lesley at first
thought Lehman's name was Sessman. He corrected himself in a letter to his
father on June 25, Lesley Papers, APS.

15. Lesley to his father, September 26, 1839, Lesley Papers, APS.

16. Lesley to his father, June 4, 1839, Lesley Papers, APS.

17. HDR to WBR, November 4, 1836, Rogers Papers (MC1), MIT. This volume
was never completed.

18. Rogers, [Summary of a paper on the Berkshires], *Proceedings of the American
Philosophical Society* 2 (January 1841–June 1843): 3. Hall and James Dwight
Dana were some of the outspoken critics. For a history of the Taconic controversy, see Cecil Schneer, "The Great Taconic Controversy," *Isis* 69 (1978):
173–191.

19. On being idled by Rogers's absences, see Lesley to his father, May 31, 1840,
and to Mrs. S. E. Lesley, August 29, 1840, Lesley Papers, APS.

20. Lesley to his grandmother, Mrs. Sarah Allen, August 31, 1840.

21. Lesley to Mrs. S. E. Lesley, October 9, 1840, Lesley Papers, APS.

22. Rogers's plea is in a letter to the house printed in his *Fourth Annual Report*
before the text of the report itself, n.p.

23. JBR to WBR, February 8 and March 3, 1841, Rogers Papers (MC1), MIT.

24. "Annual Message of the Governor," *Journal of the Senate of the Commonwealth
of Pennsylvania, 1841* (Harrisburg, 1841), 1:21.

25. *Journal of the Fifty-first House of Representatives of the Commonwealth of Pennsylvania* (Harrisburg, 1841), 1:47, January 11, 1841.

26. Report 101, "Communication from the State Treasurer, Accompanied with a Statement Exhibiting the Amount Paid to Each Individual Engaged in the Geological Survey," *Journal of the Fifty-first House*, 2:354–355, February 19, 1841.
27. JBR to WBR, March 3, 1841, Rogers Papers (MC1), MIT; "Report 277," *Journal of the Fifty-first House*, 1:514, March 13, 1841.
28. Report 144, "Report of Mr. Brodhead of Northampton, a Member of the Select Committee to Whom Was Referred That Part of the Governor's Message, Relative to the Geological Survey of the State," *Journal of the Fifty-first House*, 2:439–44.
29. HDR to WBR, April 25, 1841, *LL1*, p. 190.
30. HDR to WBR, May 4, 1841, *LL1*, p. 190.
31. H. Pritikin to Lesley, April 14, 1841, Lesley Papers, APS.
32. Lesley to his father, July 24, 1841, Lesley Papers, APS.
33. William Rogers and Henry Rogers, "Observations on the Geology of the Western Peninsula of Upper Canada, and the Western Part of Ohio," *Transactions of the American Philosophical Society*, n.s. 8 (1843): 274, read December 3, 1841. Very little information exists about the summer trips, but WBR refers to the 1841 "rambles" in a letter to his Uncle James Rogers, September 11, 1841, *LL1*, pp. 191–192. William noted at an AAGN meeting in 1842 that Hitchcock had joined them. *Reports of the First, Second, and Third Meetings* (1842), p. 173.
34. On the meeting between Lyell and Rogers, see WBR to his Uncle James Rogers, August 10, 1841, *LL1*, p. 191. For Rogers's and Lyell's meeting, see, for example, Lyell to HDR, September 19 and October 13, 1841, *LL1*, pp. 193–194; Lesley to Elizabeth Lesley, October 7, 1841, Lesley Papers, APS.
35. Rogers, *Sixth Annual Report on the Geological Survey of Pennsylvania* (Harrisburg: Henlock & Bratton, Printers, 1842).
36. "Annual Message of the Governor," *Journal of the Senate of the Commonwealth of Pennsylvania, 1842* (Harrisburg, 1842), 1:36; [Resolution asking the secretary of the Commonwealth to let the house know what efforts had been made toward the cabinet], *Journal of the Fifty-second House of Representatives of the Commonwealth of Pennsylvania* (Harrisburg, 1842), p. 1175, July 12.
37. "Act 176 to Provide for the Ordinary Expenses of Government, and for Other Purposes," *Laws of the General Assembly of the Commonwealth of Pennsylvania, Passed at the Session of 1844, Including Two Acts Passed by Both Branches of the Legislature at the Session of 1843* (Harrisburg, 1844), n.p.
38. Lesley, *Historical Sketch*, p. 115.
39. Ibid., p. 110.
40. Some information on the Appalachians would certainly have been included in William's final report of the Virginia survey, but the act creating the survey of Virginia was repealed in 1841 with no provision for a final report, and William was never able to persuade the legislature to fund a final report. A volume of the annual reports published during the survey was published in 1884: J. Hotchkiss, ed., *A Reprint of Annual Reports and Other Papers on the Geology of the Virginias by the Late William Barton Rogers* (New York: D. Appleton, 1884).

8. A Theory So Much More Satisfactory, 1842–1843

1. *Reports of the First, Second, and Third Meetings of the Association of American Geologists* (1842), pp. 474–531. Rogers's published report was the basis for my earlier study, "A Dynamic Theory of Mountain Building: Henry Darwin Rogers, 1842," *Isis* 66 (1975): 26–37.

2. HDR to WBR, September 26, 1838, *LL1*, p. 156. A note with the manuscript indicates that it was returned on May 11, 1838.

3. Rogers summarized the events of the 1840 meeting and Hitchcock's views in his 1842 paper, "On the Physical Structure," pp. 481–483.

4. Ibid.

5. Rogers, [Summary of a paper on the Berkshires], pp. 3–4.

6. Edward Hitchcock, *First Anniversary Address Before the Association of American Geologists at Their Second Annual Meeting in Philadelphia, Ap. 5, 1841* (New Haven: Printed by B. L. Hamlen, 1841), p. 13.

7. Charles Lyell, "A Letter Addressed to Dr. Fitton, by Mr. Lyell and Dated Boston the 15th of October, 1841," *Proceedings of the Geological Society of London* 3 (1838–1842): 554–558.

8. HDR to WBR, February 13, 1842, *LL1*, p. 207.

9. Lyell, "A Letter Addressed to Dr. Fitton," pp. 555–556.

10. HDR to WBR, January 14, 1842, *LL1*, p. 206.

11. Hall to Silliman, March 28, 1842, Silliman Family Papers, Yale University Library, also printed in *Science in Nineteenth Century America: A Documentary History*, ed. Nathan Reingold (New York: Hill & Wang, 1964), p. 168. Hall later learned that he had been misinformed about Lyell's intention and wrote to Lyell saying that he had jumped to the wrong conclusions. Hall to Lyell, April 3, 1842, James Hall Papers, NYSL.

12. HDR to WBR, May 11, 1842, Rogers Papers (MC2), MIT. William Buckland, "Address Delivered on the Anniversary, February 19th [1841], by the Rev. Professor Buckland, D.D.," *Proceedings of the Geological Society of London* 3 (1841): 476–479. Hopkins's ideas are in a series of works: "Researches in Physical Geology," *Transactions of the Cambridge Philosophical Society* 6 (1838): 1–84, which is mainly about crustal elevation and fracturing; "Researches in Physical Geology," *Philosophical Transactions of the Royal Society* 129 (1839): 381–423; 130 (1840): 193–208; and 132 (1842): 43–55 are on precession and nutation and their probable effects on the crust and interior of the globe; "On the Geological Structure of the Wealden District and of Bas Boulonnais," *Proceedings of the Geological Society of London* 3 (1841): 363–366.

13. Rogers, "On the Physical Structure," p. 512.

14. Ibid., p. 508; *Fourth Annual Report*, p. 106.

15. Rogers, "On the Physical Structure," pp. 517–522; John Phillips, *A Treatise on Geology* (London: Longman, Orme, Brown, Green & Longmans, 1839), 2:209; Charles Darwin, "On the Connexion of Certain Volcanic Phenomena in South America and on the Formation of Mountain Chains and Volcanos, as the Effect of the Same Power by Which Continents Are Elevated," *Transactions of the Geological Society of London* 5 (1840): 620–621. In discussing Phillips's ideas, Darwin mentioned that the idea that an earthquake resulted from an undulating fluid had occurred to him in 1835. Henry told his

brother, William, that Darwin's ideas confirmed his own in a letter dated October 22, 1841, *LL1*, p. 199.

16. Henry Rogers, "An Inquiry into the Origin of the Appalachian Coal Strata, Bituminous and Anthracitic," *Reports of the First, Second, and Third Meetings* (1842), pp. 433–474.

17. Lyell made use of Rogers's observations on gradation in "On the Probable Age and Origin of a Bed of Plumbago and Anthracite Occurring in the Mica Schist Near Worcester, Massachusetts," *Proceedings of the Geological Society of London* 4 (1842–1845): 416–419, and in the "Appendix: Analysis of Specimens of Bituminous and Anthracite Coal of the United States, and of the Plumbaginous Anthracite Alluded to in the Foregoing Paper," pp. 419–420. He did not, however, accept Rogers's explanation. Others who followed Rogers were no less charitable. See, for example, John J. Stevenson, "Origin of the Pennsylvania Anthracite," *Bulletin of the Geological Society of America* 5 (1894): 39–70. Gradation is understood today to be primarily the result of varying pressure.

18. For a summary of thoughts on coal formation among Rogers's contemporaries, see his paper, especially pp. 460–463, or William Buckland, "Address Delivered on the Anniversary, Feb. 19th [1841]," *Proceedings of the Geological Society of London* 3 (1838–1842): 487–492.

19. Rogers, "An Inquiry into the Origin," pp. 454–455n; Edward Mammat, *A Collection of Geological Facts and Practical Observations, Intended to Elucidate the Formation of the Ashby Coal Fields* (Ashby-de-la-Zouch: Published by George Lawford, London, 1834), passim.

20. Rogers, *An Inquiry*, pp. 466–470.

21. Louis Agassiz, *Etudes sur les Glaciers* (Neuchâtel: Jent & Gassmann, 1840), translated by V. Carozzi as *Studies on Glaciers* (New York, London: Hafner, 1967).

22. Proceedings, *Reports of the First, Second, and Third Meetings* (1842), p. 59.

23. Ibid., p. 48.

24. Ibid., pp. 72–74.

25. Rogers, *The Geology of Pennsylvania: A Government Survey*, 2:773–775.

26. On the nature of the presentation, see extract of a letter from John F. Hayes, Esq., to Emma Rogers, June 4, 1882, *LL1*, p. 210, and "Professor Rogers's Lectures," *Boston Daily Advertiser*, November 29, 1843. On the reaction, see *American Journal of Science* 42 (July 1842): 146–184.

27. Charles Lyell, *Travels in North America, in the Years 1841–2, with Geological Observations on the United States, Canada, and Nova Scotia* (New York: Wiley & Putnam, 1845), pp. 77–79. The argument between Rogers and Lyell over the idea of gradual rising is in Proceedings, *Reports of the First, Second, and Third Meetings* (1842), pp. 47–48.

28. Timothy Conrad, "Observations on the Characteristic Fossils," p. 239. Throughout their careers, Conrad and Rogers were at odds on almost every issue. In addition to their differences on those points already discussed, they disagreed on the age of the Appalachians, which Conrad thought were Upper Tertiary, and on organic evolution, which Rogers would accept but Conrad would not.

29. William Mather, "On the Origin of the Sedimentary Rocks of the United States, and on the Causes That Have Led to Their Elevation Above the Level

of the Sea," in Henry D. Rogers, *Address Delivered at the Meeting of the Association of American Geologists and Naturalists, Held in Washington, May 1844, with an Abstract of the Proceedings at Their Meeting* (New Haven: B. L. Hamlen, 1844), pp. 2–5 (abstract of proceedings).

30. *Reports of the First, Second, and Third Meetings* (1842), p. 75. *American Journal of Science* 43 (July 1842): 182; HDR to Lyell, July 12, 1842, University of Edinburgh Archives, Edinburgh, Scotland.

31. Jackson to HDR, August 3, 1842, Rogers Papers (MC2), MIT.

32. Dana, "On the Temperature Limiting the Distribution of Corals," *American Journal of Science* 45 (1843): 130–131.

33. "Reply of J. P. Couthouy, to the Accusation of J. D. Dana," in "Proceedings of the Association of American Geologists and Naturalists," *American Journal of Science* 45 (1843): 388–389.

34. *Athenaeum: London Literary and Critical Journal* (April 16, 1842): 345–346.

35. Logan, "On the Coal Fields," *Proceedings of the Geological Society of London* 3 (November 1838–June 1842): 707–712.

36. Lyell to Rogers, May 19, 1842, *LL1*, pp. 214–215; HDR to Lyell, [June], Lyell to HDR, June 10, 1842, Rogers Papers (MC2), MIT; HDR to Lyell, July 4, 1842, University of Edinburgh Archives.

37. Rogers, "On the Physical Structure," pp. 525–531.

38. *Athenaeum: London Literary and Critical Journal*, July 2, 1842, pp. 591–592. The paper and reaction to it were abstracted for the *American Journal of Science* 44 (1843): 359–365.

39. Phillips to HDR, November 15, 1842, *LL1*, p. 218.

40. R. I. Murchison, "Anniversary Address of the President," *Proceedings of the Geological Society of London* 4 (1842–1845): 119.

41. HDR to WBR, May 27, 1843, *LL1*, pp. 223–225.

42. HDR to WBR, April 8, 1843, *LL1*, p. 220.

43. HDR to WBR, April 30, 1843, *LL1*, p. 223; "Abstract of the Proceedings of the Fourth Session of the Association of American Geologists and Naturalists," *American Journal of Science* 45 (1843): 341–347. He cited the Reverend John Michell as the basis for many of his ideas. For Michell's ideas, see his "Conjectures Concerning the Cause, and Observations upon the Phenomena of, Earthquakes, Particularly of that Great Earthquake of the First of November, 1755, Which Proved Fatal to the City of Lisbon, and Whose Effects Were Felt as Far as Africa, and More or Less Throughout Almost All of Europe," *Philosophical Transactions* 51, pt. 2 (1760): 566–634.

44. JBR to WBR, June 1, 1843, Rogers Papers (MC1), MIT. Rogers, [Report on his paper on earthquakes], in "Celebration of the Hundredth Anniversary," *Proceedings of the American Philosophical Society* 3 (May 25–30, 1843): 64–67. See also a letter acknowledging the invitation, HDR to John Ludlow and A. D. Bache, May 13, 1843, APS.

45. Rogers, *Address Delivered at the Meeting of the Association of American Geologists . . . , 1844 . . . , With an Abstract of the Proceedings* (New Haven: B. L. Hamlen), pp. 3–58. This is the "official" AAGN publication. An identical copy was also published by Wiley and Putnam in New York, and the address appeared in the *American Journal of Science* 47 (1844): 137–160, 247–278. Because Rogers found several errors in the official version, he made some corrections and had it republished in Philadelphia by C. Sherman

under the title *Address on the Recent Progress of Geological Research in the United States, Delivered at the Annual Meeting of the Association of American Geologists and Naturalists Held at Washington City, May 1844.* The corrections are minor and include such things as the substitution of "Prof. W. B. Rogers" for "my brother," the addition or deletion of italics, the capitalization of some words, and the combining of two paragraphs in one. An occasional word is corrected; for example, "attenuate" is replaced by "alternate." He mentions the republication in a letter to Nathan Appleton, October 14, 1844, Massachusetts Historical Society, Boston, Massachusetts. All references are to the New Haven edition.

46. Smith is generally considered to be a member of the Hudson River School of artists. He did geological illustrations for other geologists, including Charles Lyell, and also portraits of William and Robert Rogers and Robert's wife. Virginia E. Lewis, *Russell Smith, Romantic Realist* (Pittsburgh: University of Pittsburgh Press, 1956).

47. Rogers, *Address Delivered at the Meeting . . . 1844,* pp. 43–55 on drift. In 1845 in a paper read to the Boston Society of Natural History, the Rogers brothers applied the theory to the Richmond Boulder Train, a well-known train of glacial boulders first described by Hitchcock in 1842. Henry and William Rogers, "An Account of Two Remarkable Trains of Angular Erratic Blocks in Berkshire, Mass., with an Attempt at an Explanation of the Phenomena," *Boston Journal of Natural History* 5 (1845–1847): 310–330. Rogers's theory of drift is discussed briefly in Chorley, Dunn, and Beckinsale, *The History of the Study of Landforms,* 1:276–279. The authors feel that Rogers "exemplified well" the "intelligent reactionary approach [against the ice age] in the United States" (p. 276).

48. Rogers, *Address Delivered at the Meeting . . . 1844,* p. 54.

9. Names Make General Propositions Possible, 1843–1844

1. The twelve formations recognized in the first year are discussed on pp. 12–18 of that year's report.

2. A table of the thirteen formations appears between pp. 18 and 19 of the *Second Annual Report.* An earlier study of Rogers's work on nomenclature is my "Henry Darwin Rogers and William Barton Rogers on the Nomenclature of the American Paleozoic Rocks," in *Two Hundred Years of Geology in America,* ed. Cecil J. Schneer (Hanover, N.H.: University Press of New England for the University of New Hampshire, 1979), pp. 175–186.

3. WBR to HDR, April 22, 1839, *LL1,* p. 165.

4. Rogers, *Second Annual Report,* p. 51.

5. WBR to HDR, April 22, 1839, *LL1,* p. 165.

6. For the Cambrian-Silurian debate, see Secord, *Controversy in Victorian Geology.*

7. Martin J. S. Rudwick, *The Great Devonian Controversy: The Shaping of Scientific Knowledge Among Gentlemanly Specialists* (Chicago: University of Chicago Press, 1985).

8. WBR to HDR, April 22, 1839, *LL1,* p. 165. R. I. Murchison, *The Silurian System, Founded on Geological Researches in the Counties of Salop, Hereford, Radnor, Montgomery, Caermarthen, Brecon, Pembroke, Monmouth, Gloucester,*

Worcester and Stafford; with Descriptions of the Coal-Fields and Overlying Forma-tions (London: John Murray, 1839).

9. Rogers, *Final Report of New Jersey,* p. 171.
10. HDR to WBR, October 16, 1841, *LL1,* p. 196; John Phillips, *Figures and Descriptions of the Paleozoic Fossils of Cornwall, Devon, and West Somerset, Observed in the Course of the Ordnance Survey of the District* (London: Longman, 1841).
11. *Second Annual Report of T. A. Conrad on the Paleontological Department of the Survey,* [New York] Assembly Document 275, February 27, 1839, pp. 57–63; [New York] Assembly Doc. 50, January 24, 1840, p. 200; "Geological Reports of the State of New York for 1840, Communicated by the Governor to the Assembly, February 17, 1841," *American Journal of Science* 42 (1842): 231.
12. Rogers, *Address Delivered at the Meeting . . . 1844,* pp. 20, 22–23.
13. HDR to WBR, October 16, 1841, *LL1,* p. 196.
14. For example, see Schneer, "Ebenezer Emmons and the Foundations of American Geology," and James A. Secord, *Controversy in Victorian Geology.*
15. HDR to WBR, December 11, 1837, *LL1,* p. 150.
16. William Whewell, *The History of the Inductive Sciences* (London: John W. Parker, 1837), 3:188–189.
17. Although Rogers abandoned numbers, the idea of numbers was acknowledged as a good one by many later geologists, including Lesley, but the only places where they were used extensively were in Pennsylvania, where they continued to appear as late as 1931, and Virginia. Vanuxem, who surveyed the third district of New York, went along with the New York scheme but had at one time been "intending also to content myself with a numerical distinction of the parts." Lardner Vanuxem, *Geology of New York, Pt. 3: Comprising the Survey of the Third Geological District* (Albany: W. & A. White & J. Visscher, Printers, 1842), p. 12.
18. HDR to WBR, October 29, 1841, *LL1,* p. 199. Rogers noted that these names were suggested by McIlvaine and were better than those suggested by Parke. What Parke suggested is not known.
19. Ibid., p. 200.
20. HDR to WBR, February 22, 1842, *LL1,* pp. 206 and 208. Conrad's paper was his "Observations on the Silurian and Devonian Systems of the United States, with Descriptions of New Organic Remains," *Journal of the Academy of Natural Sciences of Philadelphia* 8, pt. 2 (1842): 228–280.
21. Henry Rogers, "A System of Classification and Nomenclature of the Paleozoic Rocks of the United States," Rogers Papers (MC1), MIT. A statement introducing the "System" (p. 45) says that they (Henry and William) had been gathering information for eight years in the field. Even if this remark referred only to survey fieldwork, the date of this manuscript could be no later than 1843. This manuscript, entitled "System," is one part of four distinct parts that at some time were assembled into one document. One part is a section of about a dozen pages that uses formation numbers and New York names and refers to several New York reports, the latest being Hall's final report, which was published in 1843. A second part is this one, titled "A System of Classification." A third part reflects a division used in a published paper in 1844, and the fourth part is titled "Division of the Subjects in the American Text-Book of Geology" and is of an uncertain but later date. The pages are numbered consecutively, but whether the numbering was done by

Rogers or someone else is not known. The term "Paleozoic" was suggested by John Phillips in 1840 and again in 1841. It incorporated the Cambrian, Silurian, Devonian, Carboniferous, and a formation above the latter that was later identified as Permian.

22. "A System of Classification," p. 45.

23. HDR to WBR, March 30 and April 8, 1843, *LL1*, pp. 219–220.

24. Rogers, *Address Delivered at the Meeting . . . 1844*, pp. 14–26. It is likely that this new nomenclature in this paper may have been developed soon after the "System" was written. Robert commented to William in 1842 that Henry was traveling in the late summer of that year to "settle every point essential to your proposed new system," a comment that may be taken to refer to the new nomenclature. RER to WBR, appended to a letter JBR to WBR, September 2, 1842, Rogers Papers (MC1), MIT. When I wrote my article on the Rogers's nomenclature cited above, I thought that this letter referred to a system that William was developing. I now believe that "your" was plural and referred to a system that both were working on.

25. Rogers, *Address Delivered . . . 1844*, pp. 22–23.

26. Rogers's identification of the blue limestone may be traced in William Rogers and Henry Rogers, "Observations on the Geology of the Western Peninsula of Upper Canada, and the Western Part of Ohio," *Transactions of the American Philosophical Society*, n.s. 8 (1843): 273–284; HDR to WBR, October 22, 1841, *LL1*, p. 198; James Hall, "Notes upon the Geology of the Western States," *American Journal of Science* 42 (1841): 51–62. See also Hall's address to the AAGN, "Notes Explanatory of a Section from Cleveland, Ohio, to the Mississippi River, in a Southwest Direction, With Remarks upon the Identity of the Western Formations with Those of New York," *Reports of the First, Second, and Third Meetings* (1843), pp. 267–293. Rogers's change in the position of Ohio and New York units relative to each other is recorded on p. 120 of the unpublished manuscript in the Rogers Papers (MC1) at MIT, cited in n. 21 above.

10. A Mind and a Heart with Scope to Unfold, 1843–1845

1. Rogers became angry with many of his antagonists at the Franklin Institute, and his anger was long-lived. In 1851, when Richard S. McCulloh, melter and refiner at the United States Mint until 1849, wrote a scathing memorial to the Congress accusing Booth, who had succeeded him, and Patterson, who was the mint's director, of refusing his work on a new method of refining California gold because of personal issues, Rogers was delighted that at last "Patterson, Peale and Booth will get their desserts." HDR to WBR, April 5, 1851, Rogers Papers (MC2), MIT; R. S. McCulloh, *Memorial to the Congress of the United States, Requesting an Investigation and Legislation in Relation to the New Method for Refining Gold* (Princeton, N.J.: John J. Robinson, Printer, 1851).

2. Minutes of the Committee on Minerals, 1835–1843, Franklin Institute Archives.

3. Trego, *A Geography of Pennsylvania* (Philadelphia: Edward C. Biddle, 1843), pp. 6, 42.

4. For a discussion of the National Institute and reaction to it, see Sally

Kohlstedt, "A Step Toward Scientific Self-Identity in the United States: The Failure of the National Institute, 1844," *Isis* 62 (1971): 339–362.

5. Roberts, "An Obituary Notice of Charles B. Trego," mentions that he, Trego, Charles Ellet, Jr., and Ellwood Morris, all members of the Franklin Institute, had become members in January 1843.

6. JBR to WBR, June 1, 1843, Rogers Papers (MC1), MIT.

7. HDR to Hall, September 25 and October 13, 1843, James Hall Papers, NYSL.

8. Donald Hoskins believes that Rogers failed to supply information for the New York map because he was unhappy with the way Hall used the information he sent for the first map. Hoskins, "Henry Rogers and James Hall of the Pennsylvania and New York Geological Surveys, 1836–1842," *Earth Science History* 6 (1987), pp. 19–20.

9. Frazer to Hall, September 22, 1844, James Hall Papers. Unable to get information from Rogers, Hall prepared the section on Pennsylvania himself, sending his map to Trego for comment. Hall asked both Trego and Frazer to provide the information on Pennsylvania for his U.S. map. Trego refused because "it might appear indelicate or improper" to associate himself with a map not authorized by the survey. Draft of a letter Trego to Hall, June 26, 1844, Trego Papers, APS. Frazer did help. Hall to Trego, June 4, 1844, APS and Hall to Frazer, August 29, 1844, Frazer Papers, APS.

10. For a study of the iron industry in America during this period, see Peter Temin, *Iron and Steel in Nineteenth-Century America: An Economic Inquiry* (Cambridge, Mass.: MIT Press, 1964), pt. 1.

11. McKinney to HDR, April 24, 1842, Biddle Papers, Library of Congress, Washington, D.C.

12. HDR to Roswell Colt, May 3, 1842, McKinney to HDR, May 23, 1842, and June 23, 1842, Heiskill and Hoskins to Biddle, June 23, [1842], and Hoge and Austin to Biddle, n.d., Biddle Papers, Library of Congress.

13. Colt to Biddle, June 6 and June 13, 1842, Biddle Papers, Library of Congress.

14. McKinney to HDR, June 23, 1842, and Hoskins to Biddle, n.d., Biddle Papers, Library of Congress; Colt to John Cadwalader, July 21, 1842, Cadwalader Collection, HSP.

15. Biddle to Colt, September 8, 1842, and October 11, 1842, Colt Collection, HSP.

16. Biddle to Colt, November 29, 1842, Colt Collection, HSP.

17. McKinney to Colt, October 14, 1843, Biddle Papers, Library of Congress.

18. HDR to Colt, November 20, 1843, Colt Collection, HSP.

19. Ibid.

20. Emerson to HDR, November 5, 1843, *LL1*, p. 229.

21. WBR to HDR, November 5, 1843, *LL1*, p. 229.

22. *Daily Advertiser*, November 28, 29, 30, and December 2, 1843; JBR to WBR, October 15, 1843, Rogers Papers (MC1), MIT.

23. *Daily Advertiser*, January 6, 1844.

24. JBR to WBR, December 30, 1844, Rogers Papers (MC1), MIT.

25. See *LL1*, p. 232, for mention of Lowell's invitation to Henry. The invitation was sent to Rogers in care of his brother at the University of Virginia.

26. JBR to WBR, January 10, 1845, Rogers Papers (MC1), MIT.

27. Joseph Lovering to WBR, February 9, 1845, *LL1*, pp. 236–237.

28. HDR to WBR, January 24, 1845, *LL1*, p. 238; *Portsmouth Journal of Literature and Politics*, Portsmouth, N.H., January 11 and 18; C. S. Gurney, *Portsmouth Historic and Picturesque*, A Strawberry Bank Publication (Hampton, N.H.: Peter E. Randall, 1981), p. 25. Reprint of the 1902 edition.

29. *Portsmouth Journal of Literature and Politics*, February 1, 1845.

30. HDR to RER, February 11, 1845, *LL1*, p. 239.

31. Rogers, [Abstract of his paper] "On the Direction of Slaty Cleavage in the Strata of the Southeastern Belts of the Appalachian Chain, and the Parallelism of the Cleavage Dip with the Planes of Maximum Temperature," *Abstract of the Proceedings of the Sixth Annual Meeting of the Association of American Geologists and Naturalists Held in New Haven, Conn., April 1845* (New Haven: Printed by B. L. Hamlen, 1845), pp. 49–50.

32. Rogers, *Second Annual Report*, p. 32; *Description of the Geology of the State of New Jersey*, p. 97; HDR to WBR, October 16, 1841, *LL1*, p. 197.

33. For a summary of the argument, see George L. Vose, *Orographic Geology; or, The Origin and Structure of Mountains: A Review* (Boston: Lee & Shepard, 1866), pp. 70–101. As understood today, the alignment of crystals occurs during metamorphism when shales are changed into slates by pressure and/or heat.

34. Adam Sedgwick, "Remarks on the Structure of Large Mineral Masses, and Especially on the Chemical Changes Produced in the Aggregation of Stratified Rocks During Different Periods after Their Deposition," *Transactions of the Geological Society of London* 3 (1835): 461–486.

35. Rogers, "On the Direction of Slaty Cleavage," p. 50.

36. Editorial comment, *LL1*, p. 251.

37. Henry D. Rogers, "On the Geological Age of the White Mountains," pp. 1–11.

38. HDR to Lucius Thayer, May 14, 1847, Lucius Lyon Papers, William L. Clements Library, University of Michigan, Ann Arbor, Michigan.

39. HDR to WBR, January 2, 1845, *LL1*, p. 237; HDR to WBR, January 16 [1845], Rogers Papers (MC2), MIT.

40. Hillard to WBR, October 18, 1845, *LL1*, p. 253.

41. For a discussion of science education in the United States at this period, see Stanley M. Guralnick, *Science and the Ante-Bellum American College* (Philadelphia: American Philosophical Society, 1975).

11. Faithful Labours Cruelly Repaid, 1846–1848

1. HDR to WBR, March 8, 1846, *LL1*, pp. 257–258.

2. The plan in its entirety is given in appendix C, *LL1*, pp. 420–427. The passage quoted is on p. 420.

3. WBR to HDR, March 13, 1846, *LL1*, p. 259.

4. For a history of the Rumford Professorship, see Mary Ann James, "Engineering and Environment for Change, Bigelow, Peirce, and Early Nineteenth-Century Practical Education at Harvard," in *Science at Harvard University, Historical Perspectives*, ed. Clark A. Elliott and Margaret W. Rossiter (Bethlehem: Lehigh University Press; London: Associated University Presses, 1992), pp. 55–75 and especially pp. 59–63.

5. McIlvaine to Lowell, February 17, 1846; C. C. Biddle to Lowell, February

17, 1846; Binney to Charles Loring, February 13 and 21, 1846; Bailey to Whom It May Concern, February 15, 1846; Maury to Whom It May Concern, February 12, 1846; Morton to JBR, February 10, 1846; Vethake to Biddle, February 10, 1846; Smith to McIlvaine, February 10, 1846; Hillard to B. B. Curtis, n.d.; WBR to Hillard, January 27, 1846; Edward Hitchcock to Binney, February 25, 1846; Incoming Letters of the Treasurer (UAI.50.8VT), Harvard University Archives. Margaret Rossiter, *The Emergence of Agricultural Science: Justus Liebig and the Americans, 1840–1880* (New Haven: Yale University Press, 1975), p. 71, says that J. D. Dana and Charles T. Jackson also supported Rogers. John White Webster referred to Amos Eaton's support for a Rumford candidate, whom I presume to have been Rogers. Webster to Lewis Norton, February 27, 1846, Horsford Papers, Rensselaer Polytechnic Institute, Troy, New York. Joseph Henry's support is mentioned in a letter from John White Webster to Hall, December 7, 1846, James Hall Papers, NYSL, and in Webster to Lewis Norton, February 29, 1846, Horsford Papers, Rensselaer Polytechnic Institute.

6. Jeffries Wyman to Morrill Wyman, December 21, 1845, Harvard Medical Library in the Francis A. Countway Library of Medicine, Harvard University, Cambridge, Massachusetts. Although Wyman does not mention James by name, it is likely that he was referring to him rather than William. William was well established in Virginia and was highly respected in the scientific community.

7. Jeffries Wyman to Elizabeth Wyman, January 31, 1846, Harvard Medical Library in the Francis A. Countway Library of Medicine.

8. HDR to Edward Desor, August 16, 1853, Desor Papers, AN.

9. For a discussion of Wyman, see Toby A. Appel, "A Scientific Career in the Age of Character: Jeffries Wyman and Natural History at Harvard," in *Science at Harvard University: Historical Perspectives,* ed. Margaret Rossiter and Clark Elliott, pp. 95–96.

10. Hillard to Curtis, n.d., and McIlvaine to Lowell, February 17, 1846, Incoming Letters of the Treasurer (UAI.50.8VT), Harvard University Archives.

11. For example, see Rogers's comments in *Proceedings of the Boston Society of Natural History* 4 (1851–1854): 169 (December 17, 1851); Rogers, "On the Position and Character of the Reptilian Foot-Prints in the Carboniferous Red Shale Formations of Eastern Pennsylvania," *Proceedings of the American Association for the Advancement of Science, Fourth Meeting, Held at New Haven, Connecticut, Aug. 1850* (Washington, D.C.: Published by S. F. Baird; New York: G. R. Putnam, 1851), pp. 250–251; and HDR to Sedgwick, February 15, 1863, Sedgwick Papers (ADD 7652 B73), CUL.

12. For a history of the theory of evolution, see Peter J. Bowler, *Evolution: The History of an Idea* (Berkeley: University of California Press, 1983).

13. Robert Chambers, *Vestiges of the Natural History of Creation* (London: J. Churchill, 1844). For helpful studies of the American conflict between science and religion on the evolution question, see Herbert Hovenkamp, *Science and Religion in America, 1800–1860* (Philadelphia: University of Pennsylvania Press, 1978), and John A. DeJong, "American Attitudes Toward Evolution Before Darwin" (Ph.D. diss., Iowa State University, 1962).

14. HDR to WBR, January 24, 1845, *LL1*, p. 238.

15. Haldeman, "Enumeration of the Recent Freshwater Mollusca Which are

Common to North America and Europe, with Observations on Species and Their Distribution," *Journal of the Boston Society of Natural History* 4 (1843–1844): 468–484. On Haldeman, see Conway Zirkle, "The Early History of the Idea of the Inheritance of Acquired Characters and of Pangenesis," *Transactions of the American Philosophical Society* n.s. 35, pt. 2 (January 1946): 118.

16. McIlvaine to Lowell, February 17, 1846, and Emerson to Samuel Eliot, February 27, 1846, Incoming Letters of the Treasurer (UAI.50.8VT), Harvard University Archives. Quoted by permission of the Harvard University Archives.

17. Gray to Torrey, February 12 and March 8, 1845, Jane Loring Gray, *Letters of Asa Gray* (London: Macmillan, 1893), 1:330–331; A. Hunter Dupree, *Asa Gray, 1810–1888* (Cambridge, Mass.: Belknap Press, 1959), p. 144. Torrey had solid connections with the Franklin Institute group and was a colleague of Joseph Henry, whom he had helped get a job at Princeton in 1832. Torrey had little reason to like the Rogers brothers, since he had lost out on a Philadelphia professorship to Robert in 1842. RER to HDR, August 28, 1842, Rogers Papers (MC2), MIT. He would also lose out on a job to James Rogers in 1847. This job was the professorship of chemistry at the University of Pennsylvania left vacant by the resignation of Robert Hare. Torrey to Gray, July 14, 1847, Gray Papers, Gray Herbarium, Harvard University, and Torrey to Frazer, October 24, 1848, Frazer Correspondence, APS. That Torrey's dislike of Rogers extended to the evolution question is doubtful, since he is said to have taken no stand on the evolution question. A. Hunter Dupree, "John Torrey," *Dictionary of Scientific Biography*, vol. 13 (New York: Scribners, 1976).

18. *New York Herald*, May 9, 1845. The reading of Webber's paper is mentioned in the *Abstract of the . . . Sixth Annual Meeting . . . AAGN*, p. 83.

19. Gray to Torrey, January 26, 1846, Torrey Correspondence, New York Botanical Garden, Bronx, New York. Also cited in Dupree, *Asa Gray*, p. 149.

20. HDR to WBR, October 7, 1846, *LL1*, p. 267.

21. Peirce to Bache, January 29, 1846, Peirce Mss., Houghton Library, Harvard University.

22. Hillard to Curtis, n.d., Incoming Letters of the Treasurer (UAI.50.8VT), Harvard University Archives.

23. Bache's comments for the National Institute are in a draft of his speech in the Bache Papers (RU 7053), Smithsonian Institution, and are cited in Robert V. Bruce, *The Launching of American Science, 1846–1876* (New York: Alfred A. Knopf, 1987), p. 99.

24. Bache to Peirce, February 2, 1846, Incoming Letters of the Treasurer (UAI.50.8VT), Harvard University Archives.

25. In *The Emergence of Agricultural Science* (p. 71), Rossiter suggests that a reason for Webster's strong feeling against Rogers was an attempt in 1844 by Rogers and Emerson to get Webster ousted from his position at Harvard. Although the correspondence is not clear on this point, I believe the supposed 1844 attempt was really an attempt that was made in 1846 by unidentified persons. Horsford to Hall, May 1, 1846, James Hall Papers, NYSL. In 1844, Rogers had few connections with Boston and no reason to concern himself with Webster's position.

26. Webster to Horsford, February 27, 1846, Horsford Papers, Rensselaer Polytechnic Institute Archives.
27. WBR to Hall, May 28, 1838, James Hall Papers, NYSL; Samuel Rezneck, "Eben Norton Horsford," *Dictionary of Scientific Biography,* vol. 6 (New York: Scribners, 1972); Webster to Edward Everett, May 15, 1846, (quoted), Harvard University College Papers, 1845–1846 (vol. 13, 2d ser., UAI.5.125), Harvard University Archives.
28. Webster to Horsford, February 27, 1846, Horsford Papers, Rensselaer Polytechnic Institute Archives.
29. Hall to an unknown correspondent, February 28, 1846, James Hall Papers, NYSL. Hall also wrote to Webster, telling him that he had misinterpreted his (Hall's) attitude toward Rogers. Hall to Webster, February 23, 1846, James Hall Papers, NYSL.
30. Samuel A. Eliot to Everett, February 26, 1846, Harvard Corporation Papers (UAI.5.130), Harvard University Archives.
31. Webster to Norton, February 27, 1846, and Norton to Webster, April 6, 1846, and May 13, 1846, Horsford Papers, Rensselaer Polytechnic Institute Archives; Webster to Everett, May 15, 1846, Harvard University College Papers, 1845–1846 (vol. 13, 2d ser., UAI.5.125). Among the letters recommending Horsford, the letter from Rogers is missing.
32. Webster to Hall, August 25, 1846, James Hall Papers, NYSL.
33. Horsford to Hall, July 5, 1846, James Hall Papers, NYSL.
34. Ibid.
35. Webster to Warren, August 20, 1846, Warren Papers, Massachusetts Historical Society, Boston, Massachusetts.
36. Webster to Hall, December 7, 1846, James Hall Papers, NYSL.
37. J. E. Teschemacher to Horsford, September 1, 1846, Horsford Papers, Rensselaer Polytechnic Institute Archives. These probably include efforts mentioned in a letter from Horsford to Hall, May 1, 1846, James Hall Papers, NYSL, discussed in n. 25 above.
38. For a review of the circumstances and case and a modern appraisal of the judicial proceedings involved, see Robert Sullivan, *The Disappearance of Dr. Parkman* (Boston: Little, Brown, 1971). The author, a judge, believes that the trial was unfair and that there was scant evidence on which to convict Webster of the crime.
39. Bruce, *The Launching of Modern American Science,* p. 24; Gray to Torrey, January 26, 1846, Torrey Correspondence, New York Botanical Garden. Several months later Gray said that Rogers disclaimed the evolutionary idea as a last effort to succeed at Harvard. Gray to Torrey, February 20, 1847, Torrey Correspondence, New York Botanical Garden. If he did so, it was uncharacteristic of his later statements on the theory. William Rogers was then, or soon became, a staunch supporter of evolution and is remembered for a later debate with Agassiz on the question. William Smallwood, "The Agassiz-Rogers Debate on Evolution," *Quarterly Review of Biology* 16 (1941): 1–12.
40. WBR to HDR, November 11, 1846, *LL1,* p. 267. His lectures were followed by a series given by Agassiz. *Boston Daily Advertiser,* October 10, 1846. Lesley also referred to these lectures, noting that Rogers, Potter, Agassiz, and

Hillard were the four lecturers for the winter. Lesley to Mrs. S. E. Lesley, December 16, 1846, in Mary Ellen Ames, ed., *Life and Letters of Peter and Susan Lesley* (New York: G. P. Putnam's Sons, 1909), 1:143.

41. WBR to HDR, October 3, 1847, *LL1*, p. 274. For a concise history of the development of the Lawrence School at Harvard, founded as a "School of Instruction in Theoretical and Practical Science" on February 13, 1847, see Edward Lurie, *Louis Agassiz: A Life in Science* (Chicago: University of Chicago Press, 1960), pp. 136–137.
42. WBR to HDR, October 20, November 3, and December 27, 1847, *LL1*, pp. 276–278.
43. Letters throughout the period refer to various consulting jobs. He identified himself as a practical geologist on his marriage certificate. The certificate, dated March 5, 1854, is in the Massachusetts Archives at Columbia Point, Boston, Massachusetts.
44. HDR to WBR, April 23 and June 6, 1848, Rogers Papers (MC2), MIT.
45. Kohlstedt, *The Formation of the American Scientific Community*, p. 74.
46. Ibid., pp. 76–78.
47. Circular, American Association for the Advancement of Science Archives, Washington, D.C. This is the same as the constitution published in the first volume of the new association's proceedings: *Proceedings of the American Association for the Advancement of Science, First Meeting, Philadelphia, 1848* (Philadelphia: Printed by John C. Clark, 1849), pp. xix–xxii. Henry had sent the circular to William for comments before publication, but it is not known whether William suggested any changes. HDR to WBR, May 30, 1848, *LL1*, p. 288.
48. Circular, AAAS.
49. HDR to WBR, May 16, 1848, *LL1*, p. 288.
50. HDR to WBR, June 3, 1848, Rogers Papers (MC2), MIT.
51. JBR to HDR, July 31, 1848, Rogers Papers (MC1); WBR to HDR, April 15, 1848, *LL1*, pp. 285–286.
52. HDR to WBR, June 3, 1848, Rogers Papers (MC2), MIT.

12. A Spirit Oppressed, 1848–1851

1. HDR to WBR, August 17, 1848, *LL1*, p. 290; Rogers, "On the Origin of the Parallel Roads of Lochaber (Glen Roy) Scotland," *Proceedings of the Royal Institution* 3 (1858–1862): 341–345, March 22, 1861.
2. Richard Owen, "Notes on Remains of Fossil Reptiles Discovered by Professor Henry Rogers of Pennsylvania, U.S., in Greensand Formations of New Jersey," *Quarterly Journal of the Geological Society of London* 5 (1849): 380–383. Owen said he had written a longer paper on these fossils but had lost it.
3. Proceedings, *Report of the Eighteenth Meeting of the British Association for the Advancement of Science, Held at Swansea in August, 1848* (London: John Murray, 1849), pp. 74–75.
4. Rogers gave a lengthy account of his travel in HDR to WBR, September 26, 1848, Rogers Papers (MC2), MIT.
5. Edouard de Verneuil, "Note sur le parallélisme des dépots paléozoiques de

l'Amérique Septentrionale avec ceux de l'Europe, suivie d'un tableau des espèces fossiles communes aux deux continents, avec l'indication des étages où elles se recontrent, et terminée par un examen critique de chacune de ces espèces," *Bulletin de la Société Géologique de France*, 2d ser., 4 (1847): 668 n. 1.

6. HDR to RER, November 5, 1848, *LL1*, p. 293.
7. HDR to WBR, August 17, 1848, *LL1*, p. 290. He mentions the French scientists in HDR to RER, November 5, 1848, *LL1*, p. 293.
8. Rogers, "A Comparison of the Structural Features of the Disturbed Districts in Europe and America," announced in the society's proceedings in the *Quarterly Journal of the Geological Society of London* 5 (1849): 130. According to Murchison, Rogers had with him a map comparing the Alps and Appalachians, and I assume he used it here. Murchison, "On the Geological Structure of the Alps, Apennines, and Carpathians, More Especially to Prove a Transition from Secondary to Tertiary Rocks, and the Development of Eocene Deposits in Southern Europe," *Quarterly Journal of the Geological Society of London* 5 (1849): 249.
9. HDR to Sir Henry de la Beche, June 18, 1849, National Museum of Wales; *Proceedings of the American Association for the Advancement of Science, 2nd Meeting, August 1849* (1850; reprint, Boston: Henry Flanders, 1885), pp. 113–118; Rogers, *The Geology of Pennsylvania: A Government Survey*, 2:901–902.
10. Henry de la Beche, "Anniversary Address of the President," *Quarterly Journal of the Geological Society of London* 5 (1849): xix–cxvi, especially lxx–lxxi.
11. Roderick I. Murchison, *The Geology of Russia in Europe and the Ural Mountains* (London: John Murray; Paris: P. Bertrand, 1845), p. 462.
12. Murchison, "On the Geological Structure of the Alps," p. 250.
13. HDR to RER, December 1, 1848, *LL1*, p. 293.
14. HDR to Murchison, December 1, 1848, Murchison Papers, London Geological Society.
15. Murchison, "On the Geological Structure of the Alps," pp. 249–252.
16. Lesley to Susan Lyman, January 22, 1849, Ames, *Life and Letters of Peter and Susan Lesley*, p. 210; *Boston Daily Advertiser*, February 17, 1849.
17. HDR to de la Beche, June 18, 1849, National Museum of Wales.
18. A bitter priority argument between Agassiz and Forbes followed their initial observations of the ribbons and is summarized by Jules Marcou, *Life and Letters of Louis Agassiz* (New York: Macmillan, 1896), pp. 195–201. The controversy is also mentioned in Lurie, *Louis Agassiz*, pp. 104–105.
19. For a study of Forbes's theory and other opposing arguments derived, as was his, from physics, see J. S. Rowlinson, "The Theory of Glaciers," *Notes and Records of the Royal Society of London* 26 (1971): 189–204.
20. HDR to WBR, September 26, 1848, Rogers Papers (MC2), MIT.
21. Rowlinson, "The Theory of Glaciers," pp. 192–193. For more on Tyndall's ideas, see his *The Glaciers of the Alps, Being a Narrative of Excursions and Ascents, an Account of the Origin and Phenomena of Glaciers, and an Exposition of the Physical Principles to Which They are Related* (Boston: Ticknor & Fields, 1861; London: John Murray, 1860).
22. JBR to WBR, September 18, 1849, Rogers Papers (MC1), MIT.
23. For information on Agassiz and the appointment, see Lurie, *Louis Agassiz*, pp. 136–141.

24. For details of the Desor-Agassiz conflict, see Lurie, *Louis Agassiz,* especially pp. 153–161. By the early 1850s it had been proved to the satisfaction of nearly all that Desor had maliciously lied to hurt and discredit Agassiz.

25. HDR to WBR, November 14, 1849, *LL1,* p. 309.

26. WBR to his brothers, September 21, 1849, *LL1,* pp. 304–308. There is at MIT part of a diary kept by William on this trip.

27. HDR to WBR, November 14, 1849, *LL1,* p. 310; editorial comment, *LL1,* p. 310; HDR to WBR, November 15, 1850, Rogers Papers (MC2), MIT.

28. HDR to WBR, March 1850, *LL1,* p. 313.

29. HDR to WBR, December 22, 1849, *LL1,* p. 311.

30. HDR to Dr. James Rush, March 8, 1850, Rush correspondence, HSP. The paper is not extant.

31. Joseph Henry to WBR, May 31, 1849, *LL1,* p. 296.

32. HDR to Benjamin Silliman, December 13 and 31, 1850, Silliman Family Papers, Yale University Library.

33. For information on the Helfensteins and the development of the coal area, see Herbert C. Bell, ed., *History of Northumberland County, Pa.* (Chicago: Brown, Runk, 1891), pp. 368–371; George Korson, *Black Rock: Mining Folklore of the Pennsylvania Dutch* (Baltimore: Johns Hopkins University Press, 1960), p. 80; *Genealogical and Biographical Annals of Northumberland County, Pa.* (Chicago: J. L. Floyd, 1911), pp. 18–20. Rogers was also hired by Edward Gratz to do an analysis of coal in the Rennie Coal Mines near Lykens, Pennsylvania. Gratz appears to have been an independent operator in the area, and there was some kind of trouble between him and the Helfensteins, to which HDR referred in a letter to Gratz. Gratz and Helfenstein reached financial accord on their dispute, and Rogers asked for his rightful share of the settlement. HDR to Gratz, June 3, 1850, and January 6, 1851, Simon Gratz Coll, HSP. An example of Rogers's work on iron is his *Report on the Iron Ores of the Pottsville Coal Basin* (Philadelphia: L. R. Bailey, Printer, 1850), which I have not seen.

34. On Sheafer's work for companies in the area, see Bell, *History of Northumberland County,* p. 370; on Lesley, see HDR to Lesley, August 1, 1850, Lesley Papers, APS, and Lesley, unpublished Autobiographical Sketch, Society Collections, HSP.

35. Rogers, *Report on the Coal Lands of the Zerbe's Run and Shamokin Improvement Company* (Boston: Thurston, Torry, & Company, printers, 1850), p. 3.

36. Lesley, Autobiographical Sketch, Society Collections, HSP.

37. Rogers, *Report on the Coal Lands* (Boston: Thurston, Torry, Printers, 1850); *Report on the Combustible Qualities* (Boston: Thurston, Torry, & Emerson, Printers, 1851); *Report on the Coal Lands of Mt. Carmel* (New York: Snowden, Printer, 1851).

38. HDR to WBR, November 15, 1850, and December 21 [1850], Rogers Papers (MC2), MIT.

39. Rogers, "On the Position and Character of the Reptilian Foot-Prints in the Carboniferous Red Shale Formations of Eastern Pennsylvania," *Proceedings of the American Association for the Advancement of Science, Fourth Meeting,* pp. 250–251; Lea, "On Reptilian Footmarks in the Gorge of the Sharpe Mountain Near Pottsville, Pennsylvania," *Proceedings of the American Philosophical Society* 5 (1849): 91–94. The footprints proved to be those of a dinosaur, much younger than the Carboniferous.

40. The paper on salt was abstracted in the *Proceedings of the American Association for the Advancement of Science, Fourth Meeting,* p. 126. An abstract of the paper on coal is on pp. 65–70. Rogers's comments to the Boston Society were summarized in *Proceedings of the Boston Society of Natural History* 3 (1848–1851): 259–260, March 6 and March 20, 1850. Several years later Rogers published "On the Relation of Deposits of Common Salt to Climate" in the *Proceedings of the Philosophical Society of Glasgow* 5 (1860–1864), pp. 7–9, November 21, 1860.

41. Agassiz's comments are in the *Proceedings of the American Association for the Advancement of Science, Fourth Meeting,* pp. 126–127. Baird's request is in a letter, WBR to JBR, Oct. 24, 1850, Rogers Papers (MC1), MIT; the abstract appears in *Scientific American* 5 (1850): 402.

42. John Cadwalader to HDR, July 15, 1848, and WBR to Cadwalader, July 28, 1848, John Cadwalader Collection, HSP.

43. For details of Hall's involvement, see Clarke, *James Hall of Albany,* pp. 204–213, and Schneer, "The Great Taconic," p. 186.

44. Horsford to Hall, November 22, 1849, and HDR to Hall, November 23, 1849, James Hall Papers, NYSL.

45. Mather to Hall, March 19, 1850, and April 17, 1850, James Hall Papers, NYSL.

46. HDR to Hall, November 23, 1850, George P. Merrill Collection, Library of Congress; HDR to Hall, November 27, 1850, James Hall Papers, NYSL.

47. HDR to Hall, November 23, 1850, George P. Merrill Collection, Library of Congress; HDR to WBR, September 26 and October 13, 1850, Rogers Papers (MC2), MIT. The New Orleans trip is discussed in letters HDR to WBR, September 26, November 15, December 2 and 20, 1850, Rogers Papers (MC2), MIT; HDR to Hall, November 27, 1850, James Hall Papers, NYSL; HDR to Lesley, December 10, 1850, Lesley Papers, APS.

48. HDR to Desor, July 4, 1851, Desor Papers, AN.

49. HDR to Lesley, December 10, 1850, Lesley Papers, APS.

50. As reported in HDR to Silliman, December 31, 1850, Silliman Manuscripts, HSP. Rogers's notification of Joseph Henry is mentioned in HDR to WBR, December 2, 1850, Rogers Papers (MC2), MIT.

51. HDR to WBR, November 15, 1850, Rogers Papers (MC2), MIT.

52. Silliman to HDR, January 3, [1851], Silliman Manuscripts, HSP. The date on this letter is 1850, but it should clearly be 1851.

53. HDR to Lesley, December 10, 1850, Lesley Papers, APS; HDR to WBR, December 20, 1850, Rogers Papers (MC2), MIT.

13. In Pursuit of a Great Objective, 1845–1852

1. *Journal of the Fifty-fifth House of Representatives of the Commonwealth of Pennsylvania* (Harrisburg, 1845), 1:169.

2. "Communication from the Secretary of the Commonwealth, Relative to the Final Report of the State Geologist," *Journal of the Fifty-fifth House of Representatives,* pp. 580–582.

3. The state later set the exact expenditure at $76,657.87.

4. *Journal of the Fifty-sixth House of Representatives of the Commonwealth of Pennsylvania* (Harrisburg, 1846), January 16, p. 42, and March 7, p. 372.

5. Document No. 82, "Report of the Select Committee Appointed to Examine into the Condition of the Geological Specimens," *Journal of the Fifty-sixth House of Representatives*, 2:498–499. Mrs. George Hillard assisted him with the catalog for one of the cabinets. HDR to WBR, January 16, [1845], Rogers Papers (MC2), MIT. The year is not given on this letter, but the letter mentions a pending European trip. Rogers did go to Europe in 1848, but since, according to his report, he had completed the three cabinets and a catalog by February 1845, the letter must refer to 1845, at which time Rogers planned to visit Europe but did not. Rogers's plans in 1845 are discussed in Lesley's journal of his own travels in Europe that year in Ames, *Life and Letters of Peter and Susan Lesley* (New York: G. P. Putnam's Sons, 1909), 1:125. The entry is dated March 21, 1845.

6. Lesley to Mrs. S. E. Lesley, November 23, 1846, in Ames, *Life and Letters of Peter and Susan Lesley*, 1:141. Lesley still preached occasionally in Boston while working for Rogers.

7. "The Governor's Annual Message," *Journal of the Fifty-seventh House of Representatives of the Commonwealth of Pennsylvania* (Harrisburg, 1847), 2:10.

8. *Journal of the Fifty-seventh House of Representatives*, 2:393–395 (read March 16, 1847).

9. "The Governor's Annual Message," *Journal of the Fifty-eighth House of Representatives of the Commonwealth of Pennsylvania* (Harrisburg, 1848), 2:12.

10. HDR to WBR, March 3, 1848, Rogers Papers (MC2), MIT.

11. HDR to Haldeman, February 21, 1848, Haldeman Papers, ANSP.

12. *Journal of the Senate of the Commonwealth of Pennsylvania [1848]* (Harrisburg, 1848), 1:479–80 (March 13) and p. 691 (April 5).

13. *Laws of the General Assembly of the Commonwealth of Pennsylvania, Passed at the Session of 1848 in the Seventy-second Year of Independence* (Harrisburg, 1848), pp. 559–560.

14. "Resolutions Relative to the Publication of the Final Report of the Geological Survey of the State," House of Representatives Document 388, *House of Representatives File* (Harrisburg: J. M. G. Lescure, Printer, 1848), p. 2, February 12, 1848.

15. JBR to HDR, July 31, 1848, Rogers Papers (MC1), MIT.

16. Lesley to his father, September 11, 1848, Ames, *Life and Letters of Susan and Peter Lesley*, 1:159.

17. HDR to RER, December 1, 1848, *LL1*, p. 293.

18. He and Johnston corresponded about corrections for a chart two years later. HDR to WBR, November 7, 1850, Rogers Papers (MC2), MIT.

19. HDR to JBR, February 1, [1850], Rogers Papers (MC2), MIT.

20. Report presented to the Franklin Institute by its Committee on Minerals, June 20, 1850, signed by Charles M. Wetherill as chairman, Franklin Institute archives. This catalog has not survived.

21. *Journal of the Sixtieth House of Representatives of the Commonwealth of Pennsylvania* (Harrisburg, 1850), 1:460.

22. *Report to the Directors of the Pequa Railroad and Improvement Company*, signed by Isaac Lea (Philadelphia: T. K. & P. G. Collins, 1849), p. 4.

23. Rogers's work is discussed in several places in Taylor's *Statistics of Coal*, for example, on pp. 348, 352, 353, 372, 385. The first edition of this work was published in 1848, and Taylor was preparing a second edition in 1851 when he died. This edition was completed by Haldeman and was published in

1855. Citations are to the 1855 edition, but favorable references to Rogers appear in both editions. For a general review of Foulke's activities, see J. P. Lesley, *Memoir of William Parker Foulke* (Philadelphia, 1869). Foulke's attachment to the academy led him to write a brief history of it: *Discourse in Commemoration of the Founding* (Philadelphia, 1854).

24. Memo on Rogers, Foulke Papers, APS.
25. Rogers to Foulke, January 17 and January 28, 1851, Foulke Papers, APS.
26. Sheafer to Foulke, February 12, 1851, Foulke papers, APS.
27. Sheafer's political connections are mentioned in Wallace, *St. Clair*, p. 204.
28. Foulke to Sheafer, February 5, and Sheafer to Foulke, February 5 and 12, 1851, Foulke Papers, APS.
29. Sheafer to Foulke, February 6, Foulke Papers, APS.
30. Foulke to HDR, January 22, 1851, Foulke Papers, APS; Rogers to Foulke, January 28, 1851, Foulke Papers, APS.
31. Sheafer to Foulke, February 12, 1851, and Foulke to Sheafer, February 13, 1851, Foulke Papers, APS.
32. Foulke to Morton, February 5, 1851, Morton Manuscripts, HSP; Morton to Foulke, February 7, 1851, and Booth to Foulke, February 12, 1851, Foulke Papers, APS. The latter has a postscript from Frazer. A copy of the "Memorial of the American Philosophical Society," is in the Foulke papers.
33. Sheafer to Foulke, February 14, 1851, Foulke Papers, APS.
34. Telegram, Foulke to HDR, February 17, 1851, Foulke Papers, APS.
35. Foulke to HDR, February 18, 1851, Foulke Papers, APS; Telegram, Foulke to HDR, February 18, Foulke Papers, APS; HDR to WBR, February 19, 1851, Rogers Papers (MC2), MIT.
36. HDR to WBR, February 19, 1851, Rogers Papers (MC2), MIT.
37. Sheafer to Foulke, February 19, 1851, Foulke Papers, APS.
38. Foulke, "Memoranda of What Was Done," February 26 and March 2, 1851, Foulke Papers, APS.
39. On Virginia's consideration of William's report, see WBR to HDR, February 24, 1851, *LL1*, p. 316; HDR to Lesley, March 5, 1851, Lesley Papers, APS; Foulke, "Memoranda of What Was Done," March 2, Foulke Papers, APS; Foulke to HDR, March 10, Foulke Papers, APS.
40. Document 9, "Report of the Joint Committee on the Publication of the Geological Survey," *Journal of the Senate of the Commonwealth of Pennsylvania* (Harrisburg, 1851), 2:131–44.
41. Sheafer had encouraged Rogers to ask the legislature for at least $32,000, which may account for the increase over Hart's proposed $30,000. See Foulke's memo on the bottom of a copy of a letter he sent to Rogers on March 5, Foulke Papers, APS.
42. *Journal of the Sixty-first House of Representatives of the Commonwealth of Pennsylvania* (Harrisburg, 1851), 1:500–501, March 10, 1851; "Memoranda of What Was Done," February 25 and 26, 1851, Foulke Papers, APS. A draft of this memorial in Foulke's hand is in the Foulke Papers, APS.
43. Foulke to Sheafer, March 14, 1851, Foulke Papers, APS.
44. *Journal of the Senate of the Commonwealth of Pennsylvania* (1851), 1:491, March 17, 1851.
45. Foulke, "Memoranda of What Was Done," March 20, 1851, Foulke Papers, APS.
46. Ibid., March 21. Foulke conveyed all of this to Rogers on March 21 in a

letter. Foulke Papers, APS. The secretive nature of this twist was mentioned later by Lesley in connection with the house vote in his *Memoir of William Parker Foulke,* p. 46.

47. *Journal of the Senate* (1851), 1:615.

48. HDR to Lesley, April 3, 1851, Lesley Papers, APS; HDR to WBR, April 5, Rogers Papers (MC2), MIT.

49. Act 341, *Laws of the General Assembly of the Commonwealth of Pennsylvania, Passed at the Session of 1851* (Harrisburg, 1851), p. 636; *Journal of the Sixty-first House of Representatives* (1851), 1:894–895. A distribution book prepared at the time, giving a list of all the people who ordered the report and whether the person ordered maps, volume 1, and/or volume 2, is in the Department of State, Secretary of the Commonwealth, General Records, 1789–1915, Pennsylvania State Archives, Harrisburg, Pennsylvania. It is arranged by counties and individuals within the counties and is followed by a section listing officials who were to receive copies. Although it was 1858 before the report was published, this book was marked to show when the volumes were sent and how. There is also a list of state governments that received it in accordance with an entitlement act of 1855.

50. Peter Sheafer's presence is mentioned frequently, as is Desor's. John Sheafer is mentioned in HDR to Edward Desor, July 9, 1851, Desor Papers, AN. Lesley mentions them all in his *Historical Sketch,* p. 125.

51. Foulke, "Memoranda of What Was Done," July 14, 1851, Foulke Papers, APS; HDR to WBR, May 10 and 26, July 18, 1851, Rogers Papers (MC2), MIT.

52. HDR to Desor, June 27, July 4 and July 9, Desor Papers, AN.

53. Foulke, "Memoranda of What Was Done," July 14, 1851, Foulke Papers, APS.

54. Ibid.

55. HDR to WBR, July 18, 1851, Rogers Papers (MC2), MIT.

56. Kohlstedt, *The Formation of the American Scientific Community,* p. 157. Kohlstedt presents an excellent discussion of the changes taking place in the AAAS, especially in chapter 7.

57. HDR to Desor, August 28, 1851, Desor Papers, AN.

58. John D. Holmfeld, "From Amateurs to Professionals in American Science: The Controversy over the Proceedings of an 1853 Scientific Meeting," *Proceedings of the American Philosophical Society* 114 (1970); Kohlstedt, *The Formation of the American Scientific Community,* pp. 141–142.

59. Kohlstedt, *The Formation of the American Scientific Community,* pp. 174–176.

60. *Proceedings of the American Association for the Advancement of Science, Sixth Meeting, Held at Albany (N.Y.), August 1851* (Washington D.C.: S. F. Baird; New York: G. P. Putnam, 1852), pp. 297, 304, 306.

61. Lesley to Susan Lesley, June 10, 1851, in Ames, *Life and Letters of Peter and Susan Lesley,* pp. 246–248.

62. HDR to Desor, August 28, 1851, Desor Correspondence, AN. Mrs. Lesley mentioned Colter as being part of the corps. Susan Lesley to Catherine Robbins, September 11, 1851, in Ames, *Life and Letters of Peter and Susan Lesley,* p. 254.

63. Joseph Ewen, "Leo Lesquereux," *Dictionary of Scientific Biography,* vol. 8 (New York: Scribners, 1973). Lesquereux published a three-volume work on the coal plants of Pennsylvania between 1879 and 1884 for the Second Penn[

sylvania State Geological Survey: *Description of the Coal Flora of Pennsylvania, and of the Carboniferous Formation Throughout the United States* (Harrisburg: Board of Commissioners for the Second Geological Survey, 1879–1884).

64. HDR to Benjamin Silliman, Jr., December 24, 1851, Charles Roberts Autograph Letters Collection, Haverford College, Haverford, Pennsylvania.
65. Agreement with J. Peter Lesley, June 27, 1851, Lesley Papers, APS; HDR to Lesley, July 20, 1851, Lesley Papers, APS.
66. Lesley to HDR, July 25, 1851, Lesley Papers, APS.
67. HDR to Lesley, July 29, 1851, Lesley Papers, APS.
68. Lesley to Susan Lesley, August 17, 1851, in Ames, *Life and Letters of Peter and Susan Lesley*, p. 253.
69. Susan Lesley to her Aunt Catherine Robbins, September 28, 1851, in Ames, *Life and Letters of Peter and Susan Lesley*, 1:253 and 257.
70. Lesley, *Memoir of William Parker Foulke*, p. 47.
71. Lesley to HDR, November 3, December 3 and 11, 1851; telegram Rogers to Lesley, November 4, 1851; HDR to Lesley, November 5, Lesley Papers, APS.
72. Lesley to HDR, March 25 and March 28, 1852, HDR to Lesley, March 26 and March 30, 1852, Lesley Papers, APS.
73. HDR to Desor, April 22, 1852, Desor Papers, AN.
74. Ibid.
75. HDR to Desor, March 22, 1852, Desor Papers, AN.
76. Desor to Lesley, May 16, 1852, Lesley Papers, APS.
77. Lesquereux to Lesley, January 8, 1852, Lesley Papers, APS.
78. For Lesquereux's comments and feelings about Rogers, see, for example, Lesquereux to Lesley, August 28 and December 18, 1852, and April 16 and May 18, 1853, Lesley Papers, APS.
79. Lesquereux to Lesley, November 7, 1852, April 16 and May 18, 1853, Lesley Papers, APS.
80. "Introduction." Lesquereux, "New Species of Fossil Plants from the Anthracite and Bituminous Coal-fields of Pennsylvania," which appeared in the society's *Boston Journal of Natural History* 6 (1854): 409–413. Rogers had presented Lesquereux's observations on the coal measures of Ohio to the Boston Society of Natural History several months earlier. *Proceedings of the Boston Society of Natural History* 4 (1851–1854): 175–179, January 7, 1852. Among the things Rogers stressed in his introduction were the similarities in the fossils of the European and American Carboniferous and the similar age of the deposits. Rogers made this point again in *Proceedings of the Boston Society of Natural History* 4 (1851–1854): 189–191, February 18, 1852.
81. Lyman, "Biographical Notice of J. Peter Lesley," p. 462.
82. In a later account of the legislature on the copies of the final report to be distributed, H. B. Hall, W. H. Boyd, P. Daly, and James Clarkson are mentioned as assistants. When they served is not stated, but it may have been in this period. *Laws of the General Assembly of the State of Pennsylvania, Passed at the Session of 1858* (Harrisburg, 1858), p. 397.

14. Few to Take an Interest in My Volumes, 1852–1855

1. HDR to WBR, May 2, 1851, Rogers Papers (MC2), MIT.
2. Taylor, *Statistics of Coal . . . Haldeman*, p. 385.

3. Henry D. Rogers, *Reports of Professor Henry D. Rogers on Wheatley, Brookdale, and Charleston Mines, Phoenixville, Chester County, Pennsylvania* (Philadelphia: T. K. & P. G. Collins, 1853). Two copies of a manuscript on the Wheatley and Charleston lodes are in the Wheatley Papers at the APS. One is in Rogers's hand, the other in the hand of another person, probably an amanuensis.

4. R. C. Goodrich and B. Silliman, eds., *The World of Science, Art, and Industry Illustrated from Examples in the New York Exhibition, 1853–54* (New York: G. P. Putnam, 1854), pp. 57–58.

5. HDR to Desor, August 16, 1853, Desor Papers, AN.

6. HDR to Desor, February 21, 1853, Desor Papers, AN.

7. Ibid.

8. Goodrich and Silliman, eds., *The World of Science*, p. 12.

9. C. R. Goodrich, ed., *Science and Mechanism: Illustrated by Examples from the New York Exhibition, 1853–4, Including Extended Descriptions of the Most Important Contributions in Various Departments, with Annotations and Notes Relative to the Progress and Present State of Applied Science and the Useful Arts* (New York: G. P. Putnam, 1854), Hall, pp. 1–9, Wetherill, pp. 44–58.

10. The manuscript for the textbook is in the Rogers Papers at MIT and is discussed in note 21 of Chapter 9. Henry told William about his changes in a letter dated December 16, 1852, *LL1*, p. 329. Henry and William frequently talked about writing a textbook of geology. The manuscript was no doubt an effort toward that goal.

11. For the New York view, see Hall, *Geology of New York*, pt. 4, p. 24.

12. Schneer, "The Great Taconic," p. 178.

13. Sedgwick, "On the Classification and Nomenclature of the Lower Paleozoic Rocks of England and Wales," *Quarterly Journal of the Geological Society of London* 8 (1852): 165. Sedgwick did not mention Rogers by name, but the context of his remarks makes the reference clear.

14. HDR to WBR, December 16, 1852, *LL1*, pp. 328–329. Sedgwick's three-volume work is *A Synopsis of the Classification of the British Palaeozoic Rocks, with a Systematic Description of the British Palaeozoic Fossils in the Geological Museum of the University of Cambridge*, 3 vols. (Cambridge: Cambridge University Press, 1851–1854).

15. HDR to Sedgwick, February 15, 1853, and Sedgwick to Phillips, Monday, [1853], Sedgwick Papers (ADD 7763), CUL. The date 1852 has been added to the last letter, but I am certain that it should be 1853.

16. HDR to Charles Sumner, January 2, 1853, Sumner Papers, Houghton Library, Harvard.

17. HDR to Baird, February 24, 1854, Spencer F. Baird Papers, 1833–1889 (RU 7002), Smithsonian Archives.

18. For information on Marcou, see Edward Lurie, "Jules Marcou," *Dictionary of Scientific Biography*, vol. 9 (New York: Scribners, 1974).

19. HDR to Desor, February 21, 1853, Desor Papers, AN. For an interesting sense of the American reaction to Marcou's map, see James Dwight Dana's "Review of Murchison's *Siluria*" in the *American Journal of Science* 19 (1855): 379n.

20. Rogers showed his map for Johnston at a meeting of the Boston Society on May 16, 1855, at which time it was complete. *Proceedings of the Boston Society of Natural History* 5 (1854–1856): 207, May 16, 1855. There are several refer-

ences in Rogers's correspondence to the larger map. See, for example, HDR to Sedgwick, December 29, 1855, Sedgwick Papers (ADD 7682 IIA 16a), CUL.

21. Alexander Keith Johnston, *The Physical Atlas of Natural Phenomena, A New and Enlarged Edition* (Edinburgh: William Blackwood & Sons, 1856). A year after the *Atlas* was published, Rogers assisted Johnston with several maps for Johnston's *Atlas of the United States of North America, Canada, New Brunswick, Nova Scotia, Newfoundland, Mexico, Central America, Cuba, and Jamaica. On a Uniform Scale, from the Most Recent State Documents, Marine Surveys, and Unpublished Materials, with Plans of the Principal Cities and Sea-Ports, and an Introductory Essay on the Physical Geography, Products, and Resources of North America* (London: John Murray; Edinburgh: W. & A. K. Johnston, 1857). This atlas, or some part of it, was also published in 1857 in London by Edward Stanford.

22. See editorial comment, *LL1*, pp. 333–334, on William's decision to leave.

23. HDR to Desor, August 16, 1853, Desor Papers, AN.

24. The record of their marriage is on file in the Massachusetts Archives at Columbia Point, Boston, Massachusetts.

25. The history of the problem is reconstructed from accounts in the Report of the Select Committee of the Senate of Pennsylvania in Relation to the Progress and Present Condition of the State Geological Survey, *Journal of the Senate of the Commonwealth of Pennsylvania of the Session Begun at Harrisburg on the Second Day of January, A.D. 1855* (Harrisburg, 1855), reprinted in Lesley's *Historical Sketch*, pp. 128–131; Act 249 "Relative to the Courts of Common Pleas of Berks and Tioga Counties; to Hawking and Peddling in Berks County; to the Conestoga Turnpike Company; to the Collection of Taxes in Spring, Windsor, and Brecknock Townships, Berks County; to the Publication of the Geological Survey, and Providing for the Presentation of the Laws to the Smithsonian Institute" and Act 335 "To Provide for the Ordinary Expenses of Government, the Repair of the Public Canals and Railroads, and Other General and Special Appropriations," *Laws of the General Assembly of the Commonwealth of Pennsylvania, Passed at the Session of 1852, in the Seventy-Sixth Year of Independence* (Harrisburg, 1852), pp. 388 and 554 (also in the *Journal of the Senate . . . 1852*, 1:845); Act 345 "To Provide for the Ordinary Expenses of Government, the Repair of the Public Canals and Railroads, and Other General and Special Appropriations," *Laws of the General Assembly of the Commonwealth of Pennsylvania, Passed at the Session of 1853, in the Seventy-Seventh Year of Independence* (Harrisburg, 1853), p. 598. The name Hogan and Company comes from the microfilmed "Records of the Department of the Auditor General. Geological Survey Administrative Records, 1851–1857," in the Pennsylvania State Archives.

26. The firm of Parrish, Dunning, and Mears is identified by name in a deposition from N. B. Brown taken in a lawsuit brought by Rogers in 1857 and described in detail in Chapter 15. Rogers mentioned the sudden call to Harrisburg in a letter to Charles Wheatley, March 24, 1854, Wheatley Papers, APS. The committee's demands are clear in *Journal of the House of Representatives of the Commonwealth of Pennsylvania of the Session Begun at Harrisburg, on the Third Day of January A.D. 1854* (Harrisburg, 1854), p. 764, April 15, and p. 956, April 28.

27. HDR to Desor, August 5, 1854, AN.

28. "To the Committee in Charge of the Geological Survey" from HDR, *Journal of the Senate . . . 1855*, p. 399.

29. Report of the Select Committee, p. 398.

30. "To the Committee in Charge of the Geological Survey," p. 399.

31. *Journal of the Senate . . . 1855*, p. 398. On the legislative dickering, see *Journal of the Senate . . . 1855*, pp. 396–401; *Journal of the House of Representatives of the Commonwealth of Pennsylvania for the Session Begun on the Second Day of January, A.D. 1855* (Harrisburg, 1855), pp. 639, 829, 834, 849, 937, 995, April 9, 23, 24, 25, May 1 and 4.

32. Act No. 440, "Supplementary to an Act, Entitled 'An Act to Incorporate the Byberry and Poqueson Turnpike Road Company, and Relative to the Publication of the Final Report of the Geological Survey of the State,' Approved April Fourteenth, One Thousand Eight Hundred and Fifty-One," *Laws of the General Assembly of the State of Pennsylvania, Passed at the Session of 1855, in the Seventy-Ninth Year of Independence* (Harrisburg, 1855), pp. 412–413.

33. *Journal of the Senate . . . 1855*, pp. 399–400.

34. Letters to Desor, May 23, 1855, Desor Papers, AN, to Charles Wheatley, May 24, 1855, Wheatley Papers, APS, and to William, June 14, 1855, Rogers Papers (MC3), MIT. The latter stresses the need for secrecy.

35. HDR to A. K. Johnston, July 3, 1855, Blackwood Correspondence (MS 4112), National Library of Scotland, Edinburgh, Scotland.

36. The Carbondale work, apparently undertaken in 1856, was for the Delaware and Hudson Canal and Coal Company, and the Scranton project may have been for the same company. The company reneged on payment for the Carbondale job, and Rogers blamed Willie for not getting an advance. Letters to Willie Rogers, October 30 and November 16, 1855, February, March 7, May 8, June 13 and 15, 1856, Rogers Papers (MC3), MIT.

37. HDR to Willie Rogers, August 20, 1855, Rogers Papers (MC3), MIT.

38. In a letter addressed to Lesley by Willie Rogers, September 11, 1877, and printed in J. P. Lesley, "Preface," *Historical Sketch of Geological Explorations in Pennsylvania and Other States*, new issue with new preface (Harrisburg, 1878), pp. vi–ix.

39. Sedgwick to John Phillips, October 26, 1855, Sedgwick Papers (ADD 7763), CUL. On Rogers papers, see Proceedings, *Report of the Twenty-fifth Meeting of the British Association for the Advancement of Science, Held at Glasgow in September 1855* (London: John Murray, 1856), p. 95.

40. HDR to Sedgwick, December 29, 1855, Sedgwick Papers (ADD 7682 IIA 16a), CUL.

41. WBR to HDR, August 12, November 20, and December 25, 1855, *LL1*, pp. 341–342, 359, and 362.

42. WBR to HDR, Christmas morning, 1855, *LL1*, p. 362.

15. To Leave a Land Sterile of Friendship, 1855–1857

1. Editorial comment, *LL1*, p. 358.

2. James Coutts, *A History of the University of Glasgow from Its Foundation in 1451 to 1909* (Glasgow: James Maclehose & Sons, 1909), pp. 340–343.

3. HDR to RER, April 12, 1856, Rogers Papers (MC3), MIT. This was the eighth duke of Argyll, and according to the National Register of Archives in Scotland, almost none of his papers survive.

4. HDR to Desor, December 22, 1853, Desor Papers, AN.

5. Rogers, "On the Geology and Physical Geography of North America," *Notices of the Proceedings at the Meetings of the Members of the Royal Institution of Great Britain with the Abstracts of the Discourses Delivered at the Meetings* 2 (1854–1858): 167–187; "On the Laws of Structure of the More Disturbed Zones of the Earth's Crust," *Transactions of the Royal Society of Edinburgh* 2 (1857): 431–471; "On the Correlation of the North American and British Paleozoic Strata," *Report of the Twenty-sixth Meeting of the British Association for the Advancement of Science Held at Cheltenham in August 1856* (London: John Murray, 1857), pp. 175–186. Rogers told Desor that the last paper was ninety pages in length. HDR to Desor, August 23, 1856, Desor Papers, AN.

6. Charles Darwin to Daniel Sharpe, November 1, 1846, October 16, 1851, in Frederick Burkhardt and Sydney Smith, eds., *The Correspondence of Charles Darwin*, Vol. 3: *1844–1846* (Cambridge: Cambridge University Press, 1987), p. 361.

7. HDR to Desor, August 12, 1855, Desor Papers, AN.

8. Rogers, "On the Laws of Structure," p. 471.

9. HDR to Desor, August 23, 1856, Desor Papers, AN.

10. Rogers, "On the Correlation," pp. 185–186.

11. Rogers's appointment met with some criticism and ridicule in the United States. Joseph Henry to Frazer, August 13, 1856, Frazer Papers, APS.

12. Rogers anticipated its appearance in the *Journal* in January 1857 and again in April. He also thought the BAAS would print it in full. HDR to Sedgwick, September 24, 1856, Sedgwick Papers (ADD 7652 CC 13), CUL.

13. "On the Discovery of Paradoxides in the Altered Rocks of Eastern Massachusetts," *Edinburgh New Philosophical Journal* 4 (July–October 1856): 301–304.

14. "Geological Map of the United States and British North America Constructed from the Most Recent Documents and Unpublished Material for Keith Johnston's Physical Atlas by Prof. H. D. Rogers, Boston, U.S., 1855." The essay on work by the Rogers brothers and Desor is on pp. 29–32 as numbered, but two pages are numbered "32" and two as "33." The essay ends on the second page "32," so it is actually two pages longer than is indicated by the sequential numbering. Other material by Rogers includes an original geological section to accompany a map of the physical features of North and South America by Johnston, an essay on the physical features of North America, a map entitled "The Arctic Basin: Its Limits, Features, Drainage, Currents, Winds, and Climates," and a very short article called "Salt Lakes of Continental Basins and Their Origins."

15. William J. Hamilton, "Anniversary Address of the President," *Quarterly Journal of the Geological Society of London* 12 (1856): cvi; Schneer, "The Great Taconic," p. 178.

16. Hall to Lesley, May 20, 1856, Lesley Papers, APS. The Canadian map is in Logan, Geological Survey of Canada, *Report of Progress from Commencement to 1863: Atlas of Maps and Sections with an Introduction and Appendix* (Montreal:

Dawson Brothers; London, Paris & New York: Balliere, 1863). Hunt recited some of the differences that struck him in his letter to Hall, which Hall quoted in another letter to Lesley, May 20, 1856, Lesley Papers, APS.

17. Donald Hoskins, current state geologist of Pennsylvania, located these records in Philadelphia at the city hall archives. He found copies of the map at the American Philosophical Society and the Library of Congress. This map bears the 1844 nomenclature Rogers was still using in 1850.

18. Lesley to Hall, August 11, 1856, James Hall Papers, NYSL.

19. Lesley to Hall, July 24, 1856, James Hall Papers, NYSL.

20. J. P. Lesley, *Manual of Coal and Its Topography, Illustrated by Original Drawings, Chiefly of Facts in the Geology of the Appalachian Region of the United States of North America* (Philadelphia: J. B. Lippincott, 1856), p. 179.

21. Ibid., p. vi.

22. Ibid., p. 200.

23. "Misc. Intelligence," *American Journal of Science*, 2d ser., 22 (1856), p. 302.

24. Whelpley to Lesley, March 27, 1852, Lesley Papers, APS.

25. WBR to HDR, March 21, 1865, *LL2*, p. 229.

26. Lesley to Mrs. S. E. Lesley, June 28, 1848, and to his brother Allen, May 24, 1848, Ames, *Life and Letters of Peter and Susan Lesley*, pp. 150 and 154.

27. HDR to Lesley, February 25, 1852, Lesley Papers, APS.

28. HDR to Sedgwick, August 18, 1857, Sedgwick Papers (ADD 7652 II JKC), CUL.

29. HDR to Sedgwick, August 12 and 18, 1857, Sedgwick Papers (ADD 7652 IIJ 16b and II JKC), CUL. Quoted by permission of the Syndics of Cambridge University Library.

30. An undated draft of one of the letters Sedgwick wrote is in the Sedgwick Papers (ADD 7652 II J16a), CUL.

31. WBR to his wife, September 4, 1857, *LL1*, p. 371; Senate Minutes, November 9, 1857 (GUA 26705, Clerk's Press 90), University of Glasgow Archives, Glasgow, Scotland.

32. WBR to his wife, September 4, 1857, *LL1*, p. 371. Sir Archibald Geike told J. W. Gregory that Rogers's appointment was very unpopular at Glasgow, although he singled out no particular reasons. Gregory, *Henry Darwin Rogers*, p. 14.

33. HDR to Sedgwick, June 25, 1857, Sedgwick Papers (ADD 7652 II I24), CUL.

34. Rogers's only comments at the meeting were on the Pennsylvania Survey. *Report of the Twenty-Seventh Meeting of the British Association for the Advancement of Science, Held at Dublin in August and September 1857* (London: John Murray, 1853), p. 89. Several members of the BAAS, including Henry Rogers, received honorary degrees from the University of Dublin (or Trinity College, as it is also called and as Rogers's diploma reads) during the meeting. His diploma is at the Archives of the University of Pennsylvania.

35. Senate Minutes, November 9 and December 7, 1857 (GUA 26705 Clerk's Press 90), University of Glasgow Archives.

16. The Facts Are Better than the Theory, 1857–1858

1. See, for example, HDR to Willie Rogers, August 6, October 2, and October 17, 1857, Rogers Papers (MC3), MIT.

2. Eliza Rogers to Desor, March 25, 1858, Desor Papers, AN.

3. HDR to Willie Rogers, November 2, [1857], Rogers Papers (MC3), MIT.

4. "Preface," *The Geology of Pennsylvania: A Government Survey*, p. iii. An attempt to secure reimbursement from the legislature is suggested in a letter to Sedgwick, February 19, 1859, Sedgwick Papers (ADD 7652 II P7), CUL.

5. "Act No. 426 Relative to the Distribution of the Final Report of the Geological Survey of the State," *Laws of the General Assembly of the State of Pennsylvania, Passed at the Session of 1858* (Harrisburg, 1858), pp. 396–397.

6. *Journal of the Senate of the Commonwealth of Pennsylvania for the Session Begun at Harrisburg on the Sixth Day of October, A.D. 1857* (Harrisburg, 1857), pp. 895, 935, 1132, 1135, 1136, and 1182.

7. HDR to Charles Sumner, May 19, 1858, Sumner Papers (bMS AM1), Houghton Library, Harvard University. He mentions additional work that he did after returning to America in volume 2 of the final report, p. 413.

8. Charles Darwin to J. D. Hooker, May 19, 1846, and to Robert Mallet, August 26, 1846, in *The Correspondence of Charles Darwin*, 3:319–320, 335.

9. See certificate of a Candidate for Election, Royal Society of London Archives, London, England; Gassiot to WBR, April 16, 1858, *LL1*, p. 387.

10. His formal admission was announced in the *Proceedings of the Royal Society of London 9* (1857–1858): 498, as having occurred between November 1857 and November 1858.

11. *Proceedings of the Boston Society of Natural History 5* (1854–1856): 190–191, April 18, 1855.

12. Markes E. Johnson, "The Second Geological Career of Ebenezer Emmons: Success and Failure in the Southern States, 1851–1860," in *The Geological Sciences in the Antebellum South*, ed. James X. Corgan (University: University of Alabama Press, 1982), pp. 158–166.

13. WBR to HDR, March 9, 1838, *LL1*, p. 385.

14. This discovery and controversy is summarized in Merrill, *The First One Hundred Years of American Geology*, pp. 368–370, and discussed recently in more detail in Mike Foster, "The Permian Controversy of 1858: An Affair of the Heart," *Proceedings of the American Philosophical Society 133* (1989): 370–390.

15. HDR to Sedgwick, August 7, 1858, Sedgwick Papers (ADD 7652 H7), CUL.

16. HDR to WBR, October 1, 1858, *LL1*, p. 391; HDR to Meek, October 8, 1858, George P. Merrill Collection, Smithsonian Institution; Rogers, *The Geology of Pennsylvania: A Government Survey*, 2:759.

17. The mid-December completion date is given in WBR to HDR, December 10, 1858, *LL1*, p. 395.

18. Rogers, *The Geology of Pennsylvania: A Government Survey*, 1:v. Lesley was one of the first persons to introduce contour lines to geological mapping. These lines allow a more accurate portrayal of vertical scale. Lesley first used them on a map that he did for the Pennsylvania Rail Road Company. For Lesley on mapping, see his *Manual on Coal*, especially pp. 207–209, and his unpublished "Autobiographical Sketch," Society Collection, HSP.

19. The map is between pp. 1018 and 1019; Rogers's comments are in the preface, 1:viii.

20. "Scientific Intelligence," *American Journal of Science 28* (1859): 149–151.

21. *Saturday Review of Politics, Literature, Science, and Art,* April 30, 1859, pp. 530–531. In a review of opinions on fossil footprints, the *Edinburgh Review* (110 [1859]: 60–61) called Rogers's report "one of the most valuable recent contributions to geological science." Lesquereux to Lesley, April 4, 1860, Lesley Papers, APS.

22. Eli Bowen, *The Physical History of the Creation of the Earth and Its Inhabitants; or, A Vindication of the Cosmogony of the Bible from the Assaults of Modern Science* (Philadelphia: W. S. Laird, 1861), p. 51. This volume was republished with some additions as *Coal and Coal Oil; or, The Geology of the Earth, Being a Popular Description of Minerals and Mineral Combustibles* (Philadelphia: T. B. Peterson & Brothers, 1865).

23. For Rogers's classification, see "The Classification of the Metamorphic Strata of the Atlantic Slope of the Middle and Southern States," *Proceedings of the Boston Society of Natural History* 6 (1856–1859): 140–145, February 18, 1857; *The Geology of Pennsylvania, A Government Survey,* 2:777. Hunt, "On Some Points on American Geology," *American Journal of Science* 31 (1861): 394. The only thing the Canadian geologists and Rogers agreed on with regard to the pre-Paleozoic rocks was that evidence of life might be found in them. Rogers suggested that such evidence might be found in a letter to Sedgwick, February 15, 1853, Sedgwick Papers (ADD 7652 III B 73), CUL, and in his essay accompanying his map in Johnston's atlas. For the position of Hunt and Logan on the issue, see Morris Zaslow, *Reading the Rocks: The Story of the Geological Survey of Canada, 1842–1972* (Toronto: Macmillan of Canada, 1975), pp. 86–87.

24. "Contributions to the Geological History of the North American Continent" (presidential address to the American Association for the Advancement of Science, 1857), *Proceedings of the American Association for the Advancement of Science* 31 (1882): 29–71, and *Paleontology,* Vol. 3: *Containing Descriptions and Figures of the Organic Remains of the Lower Helderberg Group and the Oriskany Sandstone, 1855–1859,* Pt. 1: *Text* (Albany: Printed by C. Van Benthuysen, 1859), especially pp. 86–96.

25. For studies of Hall in the context of elevation theorists, see Mott Green, *Geology in the Nineteenth Century,* especially chap. 5. See also Rodger T. Faill, "Evolving Tectonic Concepts," pp. 27–28.

26. For a summary of the geosyncline and isostasy, see Dwight E. Mayo, "Mountain-Building Theory: The Nineteenth Century Origins of Isostasy and the Geosyncline," in *Geologists and Ideas: A History of North American Geology,* ed. Ellen T. Drake and William M. Jordan (Boulder, Colo.: Geological Society of America, 1985), pp. 1–18. For a history of the geosyncline, see R. H. Dott, Jr., "The Geosyncline: First Major Geological Concept 'Made in America,'" in *Two Hundred Years of Geology in America: Proceedings of the New Hampshire Bicentennial Conference on the History of Geology* (Hanover, N.H.: University Press of New England for the University of New Hampshire, 1979), pp. 237–264.

27. Gregory, *Henry Darwin Rogers,* p. 23.

28. Draft in the collection of the Bureau of Topographic and Geologic Survey of Pennsylvania, Harrisburg, Pennsylvania.

29. Lesley, *The Iron Manufacturers' Guide to the Furnaces, Forges, and Rolling Mills of the United States with Discussions of Iron As a Chemical Element, an American*

Ore, and a Manufactured Article, in Commerce and in History (New York: John Wiley, 1859), p. ix.

30. Ibid.
31. Lesley, *Memoir of William Parker Foulke*, p. 48.
32. [Willie Rogers], *A Few Facts Regarding the Geological Survey of Pennsylvania, Exposing the Erroneous Statements and Claims of J. P. Lesley, Secretary of the American Iron Association* (Philadelphia: Collins, Printer, 1859).
33. Ibid., p. 3.
34. Ibid., p. 8.
35. Ibid., pp. 16–17. The similarity between the published map and the 1850 map is discussed in William M. Jordan and Norman A. Pierce, "J. Peter Lesley and the Second Geological Survey of Pennsylvania," *Northeast Geology* 3 (1981): 78.
36. [Willie Rogers], *A Few Facts*, p. 22, for the association's resolution.
37. Lesley, *Historical Sketch*, pp. 81–84; Rogers, *Second Annual Report*, p. 28.
38. Lesley, *Historical Sketch*, p. 85.
39. J. P. Lesley, "Note on the Geological Age of the New Jersey Highlands as Held by Prof. H. D. Rogers," *American Journal of Science*, n.s. 39 (1865): 221–223. A summary of the paper by Logan and Hall, read by Logan at a meeting of the Natural History Society of Montreal on October 24, 1864, appeared in the *American Journal of Science*, n.s. 39 (1865): 96–97.
40. Lesley, *Historical Sketch*, pp. 127, 171–174, 179.
41. J. P. Lesley, "Memoir of Samuel Stedman Haldeman," p. 145.
42. Henry D. Rogers, *The Geology of Pennsylvania: A Government Survey* (New York: D. Van Nostrand, 1868). For information on the introduction of the Bessemer process, see Temin, *Iron and Steel in Nineteenth-Century America*, chapter 6. Ten years later, Lesley urged Van Nostrand to issue their remaining copies so that copies would be available to mining engineers. Lesley, *Historical Sketch*, p. 155.
43. "Report of the Luzerne County Medical Society," in *Transactions of the Medical Society of the State of Pennsylvania*, 5th ser., pt. 1 (Philadelphia, 1868), p. 144.
44. Edward Suess, *The Face of the Earth*, translated by H. B. C. Sollas (Oxford: Clarendon Press, 1904), 1:553–554. This was originally published in 1888 as *Das Antlitz der Erde*.
45. Bailey Willis, "The Mechanics of Appalachian Structure," in J. W. Powell, *Thirteenth Annual Report of the United States Geological Survey to the Secretary of the Interior, 1891–92*, Pt. 2: *Geology* (Washington, D.C.: U.S. Government Printing Office, 1893), p. 274.

17. A Greatly Respected Man, 1859–1866

1. HDR to Desor, August 8, 1858, Desor Correspondence, AN.
2. Eliza Rogers to Desor, March 25, 1858, and HDR to Desor, August 8, 1858, Desor Correspondence, AN.
3. For a study of Thomson's concerns with geology, see Joe D. Burchfield, *Lord Kelvin and the Age of the Earth* (New York: Science History Publications, 1975).
4. HDR to WBR, July 15 and July 22, 1859, *LL2*, pp. 10–11.
5. "College Life at Glasgow," pp. 505–506.

6. Minutes of the Hunterian Museum, February 21, 1859, University of Glasgow Library Special Collection AC2.

7. Minutes of the Hunterian Museum, April 26, 1859, University of Glasgow Library Special Collection AC2.

8. Senate Minutes, March 29, 1861 (GUA 26706 Clerk's Press 91), University of Glasgow Archives. Receipts continued to increase through the years. Rogers noted in 1864 that, since he and John Young had been at the museum, the annual ticket sales had increased from 55 pounds to 145 pounds. Rogers felt the museum was constantly improving through the addition of new specimens, some of which he was able to add as a result of his trips to other places like London. (See the report on the Hunterian Museum made by Rogers and read to the senate on February 15, 1864, University of Glasgow, Hunterian Museum.)

9. Certificate of Candidate for Election, on file at the Royal Geographical Society, London, England; Gregory, *Henry Darwin Rogers*, pp. 12–15.

10. HDR to WBR, February 24, 1860, *LL2*, p. 24.

11. HDR to WBR, December 23, 1859, *LL2*, p. 18.

12. Charles Darwin to Charles Lyell, February 25, 1860, *Life and Letters of Charles Darwin, Including an Autobiographical Chapter*, ed. Francis Darwin (New York: D. Appleton, 1887), 2:85.

13. For a contemporary summary of this issue, see Charles Lyell, *The Geological Evidences of the Antiquity of Man, with Remarks on Theories of the Origin of Species by Variation*, 2d ed. (Philadelphia: George W. Childs, 1863), especially chapter 6.

14. Henry Rogers, "The Reputed Traces of Primeval Man," *Edinburgh Magazine* 88 (1860): 422–439.

15. HDR to WBR, October 6, 1860, and WBR to HDR, October 30, 1860, *LL2*, p. 43.

16. HDR to WBR, November 22, 1860, *LL2*, p. 52.

17. Rogers, "On the Origin of the Parallel Roads of Lochaber," pp. 341–345.

18. "On the Distribution and Probable Origin of Petroleum, or Rock Oil, of Western Pennsylvania, New York and Ohio," *Proceedings of the Philosophical Society of Glasgow* 4 (1855–1860): 355–359, May 2, 1860; "On the Relations of Deposits of Common Salt to Climate," *Proceedings of the Philosophical Society of Glasgow* 5 (1860–1864): 7–9, November 21, 1860. "On Some Phenomena of Metamorphism in Coal in the United States," *Report of the Thirtieth Meeting of the British Association for the Advancement of Science, Held at Oxford in June and July 1860* (London: John Murray, 1861), p. 101. Rogers also gave a paper on cyclones in response to another society member, John Taylor, who thought that cyclones were caused by heat. Rogers argued that they resulted from the tangential action of trade winds on the relatively still air of the equatorial zone. "On the Origin of Cyclones," *Proceedings of the Philosophical Society of Glasgow* 5 (1860–1864): 57–60, March 13, 1861.

19. Gregory, *Henry Darwin Rogers*, p. 11; Burchfield, *Lord Kelvin*, p. 27; David Murray, *Memoirs of the Old College of Glasgow: Some Chapters in the History of the University* (Glasgow: Jackson, Wylie, 1927), p. 272.

20. For a study of Sumner, see David Donald, *Charles Sumner and the Coming of the Civil War* (New York: Alfred A. Knopf, 1961).

21. Clark to Combe, October 9, 1857, and HDR to Combe, October 12 and 16,

1857, George Combe Correspondence (MSS 7360 ff 111–116 and MS 7367 ff 84–85), National Library of Scotland; HDR to Sumner, October 12, 1857, Sumner Papers (bMS AM 1), Houghton Library, Harvard University. A study of Combe is in Roger Cooter, *The Cultural Meaning of Popular Science: Phrenology and the Organization of Consent in Nineteenth-Century Britain* (Cambridge: Cambridge University Press, 1984), chapter 4.

22. HDR to Desor, January 20, 1865, Desor Correspondence, AN.

23. WBR to HDR, January 30 and February 10, 1863, *LL2*, pp. 148–149.

24. Information on class structure is based on the Glasgow University Calendar for the session 1863–1864 (Glasgow: Printed at Glasgow University Press, 1863, 1864), pp. 43–44.

25. HDR to WBR, April 4, 1862, *LL2*, p. 109.

26. For a sense of the antislavery movement in Great Britain as related to that in the United States, see C. Duncan Rice, *The Scots Abolitionists, 1833–1861* (Baton Rouge: Louisiana State University Press, 1981), and Betty Fladeland, *Men and Brothers: Anglo-American Antislavery Cooperation* (Urbana: University of Illinois Press, 1972).

27. Clark, *Life and Letters of Sedgwick,* pp. 392–394.

28. Gregory, *Henry Darwin Rogers,* pp. 13–15. HDR to WBR, September 13, 1862, *LL2*, p. 132.

29. HDR to WBR, November 13, 1862, *LL2*, p. 135.

30. HDR to WBR, October 11 and November 13, 1862, April 24 and October 31, 1863, *LL2*, pp. 134–135, 160, 180. For these lectures, he expected to use diagrams on the progress of free and slave states copied by Johnston from the *National Almanac of the United States* (Philadelphia: George W. Childs, 1863).

31. HDR to WBR, April 24, 1863, *LL2*, p. 160.

32. His papers are "Coal and Petroleum" and "Coal: Its Nature, Origin, Distribution, and Mechanical Efficacy," *Good Words* 4 (1863): 247–250 and 374–379.

33. HDR to WBR, October 3, 1863, *LL2*, p. 179.

34. HDR to WBR, June 23, 1865, *LL2*, p. 244.

35. HDR to WBR, April 10, 1863, *LL2*, p. 158.

36. HDR to WBR, August 7, 1863, *LL2*, p. 171.

37. Ibid., p. 170; John Hanning Speke, *Journal of the Discovery of the Source of the Nile* (Edinburgh: W. Blackwood & Sons, 1863).

38. HDR to WBR, August 7, 1863, *LL2*, p. 170.

39. HDR to Blackwoods, November 13, 1863, Blackwood Correspondence (MSS 4185), National Library of Scotland.

40. Senate Minutes, February 12, 1864 (GUA 26706 Clerk's Press #91), University of Glasgow Archives; "Report on Hunterian Museum," dated April 25, 1864, University of Glasgow, Hunterian Museum.

41. *Report of the Thirty-fourth Meeting of the British Association for the Advancement of Science, Held at Bath in September 1864* (London: John Murray, 1865), p. 66.

42. HDR to WBR, March 2, 1865, *LL2*, p. 227.

43. Ibid.; James Henry Bennet, *Mentone, the Riviera, Corsica, and Biarritz As Winter Climates* (London: J. Churchill & Sons, 1862); James Henry Bennet, *Winter in the South of Europe; or, Mentone, the Riviera, Corsica, Sicily, and Biarritz, As Winter Climates,* 3d ed. (London: John Churchill & Sons, 1865). When Bennet wrote his first book on the area, Mentone was part of Italy. By

the time the more detailed book was written, Mentone was part of France and was called Menton. Many people continued, nevertheless, to refer to it as "Mentone."

44. Bennet, *Winter in the South of Europe*, pp. 45–47; Lesley, *Historical Sketch*, pp. 171–174.

45. HDR to WBR, March 15, 1865, LL2, p. 228.

46. HDR to WBR, April 29, 1865, LL2, pp. 235–236.

47. HDR to WBR, June 30, 1865, LL2, pp. 245–246.

48. David Livingstone to Harriette L. Livingstone, June 6, 1865, Livingstone Family Papers (on microfilm), Library of Congress.

49. HDR to WBR, June 23, 1865, LL2, p. 244.

50. Rogers, "On Petroleum," *Proceedings of the Philosophical Society of Glasgow* 6 (1865–1868): 48–60, December 13, 1865. It summarizes his ideas on oil that appeared in his article in *Good Words*. He also wrote at least one article for the Boston Society of Natural History. This paper is mentioned in WBR to HDR, April 24, 1866, LL2, p. 258. For correspondence with friends while he was in Boston, see, for example, letters to Spencer Baird, March 14 and March 27, 1866, Spencer F. Baird Incoming Correspondence, 1850–1877 (RU52), Smithsonian Institution Archives.

51. Editorial comment, LL2, pp. 259–260.

52. Senate Minutes, June 7, 1866 (GUA 26707 Clerk's Press 92), University of Glasgow Archives. The obituaries, in the order cited: *Glasgow Herald*, May 30, 1866; *American and Gazette*, June 15, 1866; *Proceedings of the Royal Society of Edinburgh* 6 (1866–1869): 22–23; *Proceedings of the Royal Society of London* 16 (1868): xxxv–xxxvi; *Quarterly Journal of the Geological Society of London* 23 (1867): xxxvii–xxxviii; *American Journal of Science and the Arts* 42 (1866): 136–138. Part of the notice in the *Glasgow Herald* is reprinted in LL2, pp. 260–261, as is part of the notice in the *American and Gazette* that is identified in the *Life and Letters* as having come from the *Philadelphia Journal* (p. 262). According to LL2, the notice in the *Glasgow Herald* also appeared in the *London Times*. The American Philosophical Society announced Rogers's death on June 15, 1866, and appointed Robert Rogers to prepare an obituary notice, but Robert did not prepare it. The Franklin Institute maintained a notable silence.

53. Warrington Smith, "Anniversary Address of the President," *Quarterly Journal of the Geological Society of London* 23 (1867): xxxviii.

54. For information on Rogers's collection, see WBR address to the Society of Arts, May 21, 1868, Society of Arts Records, MIT; Professor Alpheus Hyatt's Report on the society's collections, *Proceedings of the Boston Society of Natural History* 18 (1875–1876): 2–3, May 5, 1875; letter to the author from William Kochanczyk of the Boston Science Museum, April 6, 1990. The Society of Arts was one of three proposed schools that were to make up MIT. Its identity within the institute was never clear, and it was dissolved in 1962 (from a brief history of the society provided by the MIT Archives). A list of the specimens donated to the Hunterian Museum is at the museum.

55. The *Philadelphia Evening Bulletin* for May 11, 1936, erroneously reported an early theft of the Harrisburg cabinet. Lesley, in his *Historical Sketch* (p. 179), implied that the collections were never deposited in Philadelphia, but this suggestion is incorrect. On the official disposition of the Harrisburg cabinet,

see *Laws of the General Assembly of the State of Pennsylvania Passed at the Session of 1857* (Harrisburg, 1857), pp. 609–610. There is a "Catalog, First Geological Survey of Pennsylvania, 1836–1858," in Collection 235 (Geology and Paleontology Department) at the Academy of Natural Sciences. It is a 1959 photostat of an original identified as being in Harrisburg.

56. Huxley to WBR, June 8, 1887, *LL2,* p. 338.

Bibliography

Manuscript Sources

Academy of Natural Sciences of Philadelphia, Philadelphia, Pennsylvania
  Academy records
  Association of American Geologists and Naturalists Papers
  Samuel Haldeman Papers
Alexander Turnbull Library, Wellington, New Zealand
  Gideon Mantell Collection
American Association for the Advancement of Science, Washington, D.C.
American Philosophical Society, Philadelphia, Pennsylvania
  Correspondence, misc.
  George W. Featherstonhaugh Papers on microfilm (originals at the Minnesota
    Historical Society)
  William Parker Foulke Papers
  J. P. Lesley Papers
  Charles Wheatley Papers
Archive de l'Etat, Neuchâtel, Switzerland
  Edward Desor Papers
Cambridge University Libraries, Cambridge, England
  Adam Sedgwick Papers
Dickinson College Archives, Carlisle, Pennsylvania
  Minutes of the Board of Trustees
  College catalogs
Franklin Institute, Philadelphia, Pennsylvania
  Board of Managers Reports
  Minutes of the Committee on Minerals
Geological Society of London, London, England
  Minute Books
  Roderick I. Murchison Papers

Harvard University Archives, Cambridge, Massachusetts
  College Papers
  Corporation Papers
  Incoming Letters of the University Treasurer
Harvard Medical Library in the Francis A. Countway Library of Medicine
  Wyman Family Papers
Harvard University, Gray Herbarium
  Asa Gray Papers
Harvard University, Houghton Library
  Benjamin Peirce Correspondence
  Charles Sumner Papers
Haverford College Library, Haverford, Pennsylvania
  Charles Roberts Autograph Letters Collection
Historical Society of Pennsylvania, Philadelphia, Pennsylvania
  Cadwalader Collection
  Roswell Colt Collection
  Simon Gratz Collection
  James Hamilton, Jr., Papers
  Library Company of Philadelphia Collection
  Samuel George Morton Papers
  Physicians of Pennsylvania Collection
  Benjamin Rush Correspondence
  Benjamin Silliman Manuscripts
  Society Collections
Library of Congress, Washington, D.C.
  Nicholas Biddle Correspondence
  Livingstone Family Papers
  George P. Merrill Collection
Massachusetts Archives at Columbia Point, Boston, Massachusetts
  Marriage records
Massachusetts Historical Society, Boston, Massachusetts
  Nathan Appleton Papers
  John C. Warren Papers
Massachusetts Institute of Technology Libraries, Institute Archives, and Special
    Collections, Cambridge, Massachusetts
  William Barton Rogers Papers, 1804–1911 (MC1)
  Rogers Family Papers, 1811–1904 (MC2)
  William Barton Rogers Papers, 1817–1886 (MC3)
  Society of Arts Records
National Library of Scotland, Edinburgh, Scotland
  Blackwood Correspondence
  George Combe Correspondence
New York Botanical Garden, Bronx, New York
  John Torrey Correspondence
New York Historical Society, New York, New York
  Miscellaneous Manuscripts
New York State Library Manuscripts and Special Collections, Albany, New York
  James Hall Papers

Pennsylvania Bureau of Topographic and Geological Survey, Harrisburg, Pennsylvania
First Survey records
Pennsylvania State Archives, Harrisburg, Pennsylvania
Department of State, Secretary of the Commonwealth, General Records, 1789–1915
Records of the Department of the Auditor General, Geological Survey, Financial Records, 1837–1875
Pennsylvania State University Archives, State College, Pennsylvania
R. M. S. Jackson Papers
Philadelphia City Hall Archives, Philadelphia, Pennsylvania
Rensselaer Polytechnic University, Troy, New York
Eben Horsford Papers
Royal Geographical Society, London, England
Royal Society of London, London, England
Smithsonian Institution, Washington, D.C.
Alexander Dallas Bache Papers
Spencer F. Baird Papers
Joseph Henry Papers
St. Andrews University, St. Andrews, Scotland
Edward Forbes Papers
University of Delaware, Newark, Delaware
James C. Booth Papers
University of Edinburgh, Edinburgh, Scotland
Charles Lyell Letters
University of Glasgow Archives, Glasgow, Scotland
Minutes of the Hunterian Museum
Senate Minutes
University of Michigan, William L. Clements Library, Ann Arbor, Michigan
Lucius Lyon Papers
University of Pennsylvania Archives, Philadelphia, Pennsylvania
Minutes of the University Trustees
Yale University Library, Manuscripts and Archives, New Haven, Connecticut
Silliman Family Papers
Southard Hay Collection

Selected Bibliography of Published Material and Dissertations

Aldrich, Michele A. L. "American State Geological Surveys." In *Two Hundred Years of Geology in America,* ed. Cecil J. Schneer. Hanover, N.H.: University Press of New England for the University of New Hampshire, 1979.
———. "New York Natural History Survey, 1836–1845." Ph.D. diss., University of Texas at Austin, 1974.
Ames, Mary Ellen, ed. *Life and Letters of Peter and Susan Lesley.* 2 vols. New York: G. P. Putnam's Sons, 1909.
Anon. "Report upon the Geological Survey of the State." *Hazard's Register of Pennsylvania* 40, no. 15 (April 13, 1833).
Association of American Geologists and Naturalists. "Abstract of the Proceed-

ings of the Fourth Session of the Association of American Geologists and Naturalists." *American Journal of Science* 45 (1843): 135–165, 310–352.

———. *Abstract of the Proceedings of the Sixth Annual Meeting of the Association of American Geologists and Naturalists Held in New Haven, Conn., April 1845.* New Haven: Printed by B. L. Hamlen, 1845.

———. *Reports of the First, Second, and Third Meetings of the Association of Geologists and Naturalists at Philadelphia in 1840 and 1841 and at Boston in 1842, Embracing Its Proceedings and Transactions.* Boston: Gould, Kendall, & Lincoln, 1843.

Bache, Alexander D., and Henry Rogers. "Analysis of Some Coals of Pennsylvania." *Journal of the Academy of Natural Sciences of Philadelphia* 7, pt. 1 (1834): 158–177.

Bell, Herbert C., ed. *History of Northumberland County, Pa.* Chicago: Brown, Runk, 1891.

Bennet, James Henry. *Winter in the South of Europe; or, Mentone, the Riviera, Corsica, Sicily, and Biarritz as Winter Climates.* 3d ed. London: John Churchhill & Sons, 1865.

Berkeley, Edmund, and Dorothy Smith Berkeley. *George William Featherstonhaugh: The First U.S. Government Geologist.* Tuscaloosa: University of Alabama Press, 1988.

Bowen, Eli. *Coal and Coal Oil; or, The Geology of the Earth, Being a Popular Description of Minerals and Mineral Combustibles.* Philadelphia: T. B. Peterson & Brothers, 1865.

Bowler, Peter J. *Evolution: The History of an Idea.* Rev. ed. Berkeley: University of California Press, 1989.

Brown, Thomas, M.D. *Lectures on the Philosophy of the Human Mind by the Late Thomas Brown, M.D.* 2 vols. Hallowell: Glazier, 1828.

Browne, Peter. *An Address Intended to Promote a Geological and Mineralogical Survey of Pennsylvania.* Philadelphia: P. M. Lafourcade, 1826.

Bruce, Robert V. *The Launching of Modern American Science, 1846–1876.* New York: Alfred A. Knopf, 1987.

Burchfield, Joe D. *Lord Kelvin and the Age of the Earth.* New York: Science History Publications, 1975.

Burkhardt, Frederick, and Sydney Smith, eds. *The Correspondence of Charles Darwin.* Vols. 3 and 5. Cambridge: Cambridge University Press, 1987, 1989.

Carson, Joseph. *A History of the Medical Department of the University of Pennsylvania.* Philadelphia: Lindsay & Blakiston, 1869.

———. *A Memoir of the Life and Character of James B. Rogers, M.D.* Philadelphia: T. K. & P. G. Collins, 1851.

Chandler, Alfred D., Jr. "Anthracite Coal and the Beginnings of the Industrial Revolution in the United States." *Business History Review* 46 (1972): 141–181.

Cheyney, Edward Potts. *History of the University of Pennsylvania.* Philadelphia: University of Pennsylvania Press, 1940.

Chorley, Richard J., Antony J. Dunn, and Robert P. Beckinsale. *The History of the Study of Landforms; or, The Development of Geomorphology,* Vol. 1: *Geomorphology Before Davis.* London: Methuen; New York: John Wiley, 1964.

Clarke, John M. *James Hall of Albany: Geologist and Paleontologist, 1811–1898.* Albany, 1923.

Conrad, Timothy. "Observations on Characteristic Fossils and upon a Fall of Temperature in Different Geological Epochs." *American Journal of Science* 35 (1839): 237–251.

———. "Observations on the Silurian and Devonian Systems of the United States, with Descriptions of New Organic Remains." *Journal of the Academy of Natural Sciences of Philadelphia* 8, pt. 2 (1842): 228–280.

———. *Second Annual Report of T. A. Conrad on the Paleontological Department of the Survey.* [New York] Assembly Document 275, February 27, 1839.

Cooter, Roger. *The Cultural Meaning of Popular Science: Phrenology and the Organization of Consent in Nineteenth-Century Britain.* Cambridge: Cambridge University Press, 1984.

Cordell, Eugene Fauntleroy. *The Medical Annals of Maryland, 1799–1899.* Baltimore, 1903.

Couthouy, J. P. "Reply of J. P. Couthouy, to the Accusation of J. D. Dana." In "Proceedings of the Association of American Geologists and Naturalists." *American Journal of Science* 45 (1843): 378–389.

Coutts, James. *A History of the University of Glasgow from Its Foundation in 1451 to 1909.* Glasgow: James Maclehose & Sons, 1909.

Dana, James Dwight. "On the Temperature Limiting the Distribution of Corals." *American Journal of Science* 45 (1843): 130–131.

Daniels, George H. *American Science in the Age of Jackson.* New York: Columbia University Press, 1968.

———. "The Process of Professionalization in American Science: The Emergent Period, 1820–1860," *Isis* 58 (1967): 151–166.

Darwin, Charles. "On the Connexion of Certain Volcanic Phenomena in South America and on the Formation of Mountain Chains and Volcanos, as the Effect of the Same Power by Which Continents are Elevated." *Transactions of the Geological Society of London* 5 (1840): 601–631.

Darwin, Francis, ed. *The Life and Letters of Charles Darwin.* Vol. 2. New York: D. Appleton, 1887.

DeJong, John A. "American Attitudes Toward Evolution Before Darwin." Ph.D. diss., Iowa State University, 1962.

de la Beche, Henry. "Anniversary Address of the President." *Quarterly Journal of the Geological Society of London* 5 (1849): xix–cxvi.

Donald, David. *Charles Sumner and the Coming of the Civil War.* New York: Alfred A. Knopf, 1961.

Dott, R. H., Jr. "The Geosyncline: First Major Geological Concept 'Made in America.'" In *Two Hundred Years of Geology in America,* ed. Cecil J. Schneer. Hanover, N.H.: University Press of New England for the University of New Hampshire, 1979.

———. "James Hall's Discovery of the Craton." In *Geologists and Ideas: A History of North American Geology,* ed. Ellen T. Drake and William M. Jordan. Boulder, Colo.: Geological Society of America, 1985.

Dupree, A. Hunter. *Asa Gray, 1810–1888.* Cambridge, Mass.: Belknap Press, 1959.

Emmons, Ebenezer. *Geology of New York, Pt. 2: Comprising the Survey of the Second Geological District.* Albany: W. & A. White & J. Visscher, Printers, 1842.

Faill, Rodger T. "Evolving Tectonic Concepts of the Central and Southern Appalachians." In *Geologists and Ideas: A History of North American Geology,* ed. Ellen T. Drake and William M. Jordan. Boulder, Colo.: Geological Society of America, 1985.

Fairholme, George. "On the Falls of the Niagara." *Philosophical Magazine,* 3d ser. 5 (1834): 11–25.

Fladeland, Betty. *Men and Brothers: Anglo-American Antislavery Cooperation.* Urbana: University of Illinois Press, 1972.

Foster, Mike. "The Permian Controversy of 1858: An Affair of the Heart." *Proceedings of the American Philosophical Society* 133 (1989): 370–390.

Franklin Institute. "Annual Report of the Board of Managers." *Journal of the Franklin Institute* 13 (1834): 226–231.

———. *For the Establishment of a School of Arts: Memorial of the Franklin Institute, of the State of Pennsylvania, for the Promotion of the Mechanic Arts, to the Legislature of Pennsylvania.* Philadelphia: Printed by J. Crissy, 1837.

———. "Report on the Stone of Leiper's Quarries." *Journal of the Franklin Institute* 19 (1837): 367–369.

———. "Review of a 'Geological Report of an Examination, Made in 1834, of the Elevated Country Between the Missouri and Red Rivers. By G. W. Featherstonhaugh, U.S. Geologist, Published by Order of Both Houses of Congress, Washington: Printed by Gales & Seaton, 1835.'" *Journal of the Franklin Institute* 17 (1836): 109–117, 184–190.

Geological Society of Pennsylvania. "Mr. Featherstonhaugh's Geological Report." *Transactions of the Geological Society of Pennsylvania* 1, pt. 2 (1835): 413.

Gerstner, Patsy A. "A Dynamic Theory of Mountain Building: Henry Darwin Rogers, 1842." *Isis* 66 (1975): 26–37.

———. "Henry Darwin Rogers and William Barton Rogers on the Nomenclature of the American Paleozoic Rocks." In *Two Hundred Years of Geology in America,* ed. Cecil J. Schneer. Hanover, N.H.: University Press of New England for the University of New Hampshire, 1979.

———. "Vertebrate Paleontology, an Early Nineteenth-Century Transatlantic Science." *Journal of the History of Biology* 3 (1970): 137–148.

Goodrich, C. R., ed. *Science and Mechanism: Illustrated by Examples from the New York Exhibition, 1853–4, Including Extended Descriptions of the Most Important Contributions in Various Departments with Annotations and Notes Relative to the Progress and Present State of Applied Science, and the Useful Arts.* New York: G. P. Putnam, 1854.

Goodrich, C. R., and B. Silliman, eds. *The World of Science, Art, and Industry Illustrated from Examples in the New York Exhibition, 1853–54.* New York: G. P. Putnam, 1854.

Gray, Jane Loring. *Letters of Asa Gray.* 2 vols. London: Macmillan, 1893.

Green, Mott T. *Geology in the Nineteenth Century: Changing Views of a Changing World.* Ithaca: Cornell University Press, 1982.

Gregory, J. W. *Henry Darwin Rogers, An Address to the Glasgow University Geological Society, 20th January, 1916.* Glasgow: James MacLehose & Son, 1916.

Guralnick, Stanley M. *Science and the Ante-Bellum American College.* Philadelphia: American Philosophical Society, 1975.

Hall, James. *Geology of New-York, Pt. 4: Comprising the Survey of the Fourth Geological District.* Albany: Carroll & Cook, 1843.

———. *Natural History of New York*, Pt. 6: *Paleontology*. Vol. 3. Albany: C. Van Benthuysen, 1859.

———. "Notes Explanatory of a Section from Cleveland, Ohio, to the Mississippi River, in a Southwest Direction, with Remarks upon the Identity of the Western Formations with Those of New York." *Reports of the First, Second, and Third Meetings of the Association of Geologists and Naturalists at Philadelphia, in 1840 and 1841, and at Boston in 1842, Embracing Its Proceedings and Transactions*. Boston: Gould, Kendall, & Lincoln, 1843.

———. "Notes upon the Geology of the Western States." *American Journal of Science* 42 (1841): 51–62.

Hamilton, William J. "The Anniversary Address of the President." *Quarterly Journal of the Geological Society of London* 12 (1856): xxvi–cxix.

Harlan, Richard. *Remarks on Prof. Rogers' Geological Report to the British Association for the Advancement of Science During Their Recent Meeting Held in Edinburgh.* Philadelphia: Printed for the Publisher, 1835.

Hendrickson, Walter B. *David Dale Owen, Pioneer Geologist of the Middle West.* Indiana Historical Collections, vol. 27. Indianapolis: Indiana Historical Bureau, 1943.

Hitchcock, Edward. *First Anniversary Address Before the Association of American Geologists at Their Second Annual Meeting in Philadelphia, Ap. 5, 1841.* New Haven: Printed by B. L. Hamlen, 1841.

Holmfeld, John D. "From Amateurs to Professionals in American Science: The Controversy over the Proceedings of an 1853 Scientific Meeting." *Proceedings of the American Philosophical Society* 114 (1970): 22–36.

Hoskins, Donald. "The First Geological Survey of Pennsylvania: The Discovery Years." *Pennsylvania Geology* 18 (February 1987): 1–2. (Condensed from "Celebrating a Century and a Half: The Geologic Survey," *Pennsylvania Heritage* 12 (Summer 1986): 26–31.)

———. "Henry Rogers and James Hall of the Pennsylvania and New York Geological Surveys." *Earth Sciences History* 6 (1987): 14–23.

Hovenkamp, Herbert. *Science and Religion in America, 1800–1860.* Philadelphia: University of Pennsylvania Press, 1978.

Hunt, T. Sterry. "On Some Points in American Geology." *American Journal of Science*, n.s. 31 (1861): 392–414.

———. "On Some Points in Chemical Geology." *American Journal of Science*, n.s. 30 (1860): 133–137.

Johnson, Markes E. "The Second Geological Career of Ebenezer Emmons: Success and Failure in the Southern States, 1851–1860." In *The Geological Sciences in the Antebellum South*, ed. James X. Corgan. University: University of Alabama Press, 1982.

Johnston, Alexander Keith. *Atlas of the United States of North America, Canada, New Brunswick, Nova Scotia, Newfoundland, Mexico, Central America, Cuba, and Jamaica, on a Uniform Scale, from the Most Recent State Documents, Marine Surveys, and Unpublished Materials, with Plans of the Principal Cities and Sea-Ports, and an Introductory Essay on the Physical Geography, Products, and Resources of North America.* London: John Murray; Edinburgh: W. & A. K. Johnston, 1857.

———. *The Physical Atlas of Natural Phenomena: A New and Enlarged Edition.* Edinburgh: William Blackwood & Sons, 1856.

Jordan, William M. "Geology and the Industrial Transportation Revolution in Early to Mid Nineteenth-Century Pennsylvania." In *Two Hundred Years of Geology in America*, ed. Cecil J. Schneer. Hanover, N.H.: University Press of New England for the University of New Hampshire, 1979.

————. "J. Peter Lesley and the Second Geological Survey of Pennsylvania." *Northeast Geology* 3 (1981): 75–85.

Judd, J. W. [Presidential Address (Geology).] In *Report of the Fifty-fifth Meeting of the British Association for the Advancement of Science*. London: J. Murray, 1885.

Klein, Philip, and Ari Hoogenboom. *A History of Pennsylvania*. New York: McGraw-Hill, 1973.

Kohlstedt, Sally Gregory. *The Formation of the American Scientific Community: The American Association for the Advancement of Science, 1848–1860*. Urbana: University of Illinois Press, 1976.

————. "A Step Toward Scientific Self-Identity in the United States: The Failure of the National Institute, 1844." *Isis* 62 (1971): 339–362.

Korson, George. *Black Rock: Mining Folklore of the Pennsylvania Dutch*. Baltimore: Johns Hopkins University Press, 1960.

Lea, Isaac. [On Reptilian Footmarks in the Gorge of the Sharpe Mountain Near Pottsville, Pennsylvania.] *Proceedings of the American Philosophical Society* 5 (1849): 91–94.

————. *Report to the Directors of the Pequa Railroad and Improvement Company*. Philadelphia: T. K. & P. G. Collins, 1849.

Leopold, Richard William. *Robert Dale Owen: A Biography*. Cambridge, Mass.: Harvard University Press, 1940.

Lesley, J. Peter. *Historical Sketch of Geological Exploration in Pennsylvania and Other States*. 1876. Reprint. Harrisburg: Published by the Board of Commissioners for the Second Geological Survey, 1878.

————. *The Iron Manufacturers' Guide to the Furnaces, Forges, and Rolling Mills of the United States, with Discussions of Iron as a Chemical Element, an American Ore, and a Manufactured Article, in Commerce and in History*. New York: John Wiley, 1859.

————. *Manual of Coal and Its Topography: Illustrated by Original Drawings, Chiefly of Facts in the Geology of the Appalachian Region of the United States of North America*. Philadelphia: J. B. Lippincott, 1856.

————. "Memoir of Samuel Stedman Haldeman, 1812–1880." *National Academy of Sciences Biographical Memoirs*. Vol. 2. Washington, D.C.: Judd & Detweiler, Printers, 1886.

————. *Memoir of William Parker Foulke*. Philadelphia, 1869.

————. "Note on the Geological Age of the New Jersey Highlands as Held by Prof. H. D. Rogers." *American Journal of Science*, n.s. 39 (1865): 221–223.

————. *A Summary Description of the Geology of Pennsylvania in Three Volumes*. Harrisburg, 1892.

Lewis, Virginia E. *Russell Smith, Romantic Realist*. Pittsburgh: University of Pittsburgh Press, 1956.

Logan, William E. *Geological Survey of Canada: Report of Progress from Its Commencement to 1863* and *Atlas of Maps and Sections with an Introduction and Appendix*. Montreal: Dawson Brothers; London: Balliere, 1863.

————. *Louis Agassiz: A Life in Science*. Chicago: University of Chicago Press, 1960.

———. "On the Coal-Fields of Pennsylvania and Nova Scotia." *Proceedings of the Geological Society of London* 3 (1838–1842): 707–712. (A report on this paper appeared in the *Athenaeum: London Literary and Critical Journal*, April 16, 1842, pp. 345–346.)

Lyell, Charles. "A Letter Addressed to Dr. Fitton, by Mr. Lyell, and Dated Boston the 15th of October, 1841." *Proceedings of the Geological Society of London* 3 (1838–1842): 554–558.

———. *The Geological Evidences of the Antiquity of Man, with Remarks on Theories of the Origin of Species by Variation.* 2d Amer. ed. Philadelphia: George W. Childs, 1863.

———. "On the Probable Age and Origin of a Bed of Plumbago and Anthracite Occurring in the Mica Schist Near Worcester, Massachusetts." *Proceedings of the Geological Society of London* 4 (1842–1845): 416–420.

———. *Principles of Geology, Being an Attempt to Explain the Former Changes of the Earth's Surface, by Reference to Causes Now in Operation.* Vol. 3. London: John Murray, 1833.

———. *Travels in North America, in the Years 1841–2; with Geological Observations on the United States, Canada, and Nova Scotia.* New York: Wiley & Putnam, 1845.

Lyman, Benjamin Smith. "Biographical Notice of J. Peter Lesley." In Mary Ellen Ames, *Life and Letters of Peter and Susan Lesley.* Vol. 2. New York: G. P. Putnam's Sons, 1909.

McDowell, R. B. *Ireland in the Age of Imperialism and Revolution, 1760–1801.* Oxford: Clarendon Press, 1979.

Mammat, Edward. *A Collection of Geological Facts and Practical Observations, Intended to Elucidate the Formation of the Ashby Coal Fields.* Ashby-de-la-Zouch: George Lawford, London, 1834.

Mather, W. W. *Geology of New York,* Pt. 1: *Comprising the Geology of the First Geological District.* Albany: Printed by Carroll & Cook, 1843.

———. "On the Origin of the Sedimentary Rocks of the United States and on the Causes That Have Led to Their Elevation Above the Level of the Sea." In Henry Rogers, *Address Delivered at the Meeting of the Association of American Geologists and Naturalists, Held in Washington, May 1844, with an Abstract of the Proceedings at Their Meeting.* New Haven: B. L. Hamlen, 1844.

Mayo, Dwight E. "Mountain-Building Theory: The Nineteenth-Century Origins of Isostasy and the Geosyncline." In *Geologists and Ideas: A History of North American Geology,* ed. Ellen T. Drake and William M. Jordan. Boulder, Colo.: Geological Society of America, 1985.

Merrill, George P. *Contributions to a History of American State Geological and Natural History Surveys.* U.S. National Museum Bulletin 109. Washington, D.C.: U.S. Government Printing Office, 1920.

———. *One Hundred Years of American Geology.* New York: Hafner, 1964.

Milici, Robert C., and C. R. Bruce Hobbs, Jr. "William Barton Rogers and the First Geological Survey of Virginia, 1835, 1841." *Earth Sciences History* 6 (1987): 3–13.

Millbrooke, Anne. "The Geological Society of Pennsylvania, 1832–1836, Pt. 1: Founding the Society." *Pennsylvania Geology* 7/6 (1976): 7–11.

———. "The Geological Society of Pennsylvania, 1832–1836, Pt. 2: Promoting a State Survey." *Pennsylvania Geology* 8 (1977): 12–16.

————. "State Geological Surveys of the Nineteenth Century." Ph.D. diss., University of Pennsylvania, 1981.

Morton, Samuel George. *Synopsis of the Organic Remains of the Cretaceous Group of the United States.* Philadelphia: Key & Biddle, 1834.

Murchison, Roderick I. "Anniversary Address of the President." *Proceedings of the Geological Society of London* 4 (1842–1845): 65–151.

————. *The Geology of Russia in Europe and the Ural Mountains.* London: John Murray; Paris: P. Bertrand, 1845.

————. "On the Geological Structure of the Alps, Apennines, and Carpathians, More Especially to Prove a Transition from Secondary to Tertiary Rocks, and the Development of Eocene Deposits in Southern Europe." *Quarterly Journal of the Geological Society of London* 5 (1849): 157–311.

Murray, David. *Memories of the Old College of Glasgow: Some Chapters in the History of the University.* Glasgow: Jackson, Wylie, 1927.

New Jersey, State of. *Journal of the Proceedings of the Legislative Council of the State of New-Jersey.* Burlington, N.J., 1837.

————. *Votes and Proceedings of the Fifty-ninth General Assembly of the State of New Jersey.* Freehold, N.J., 1835.

————. *Votes and Proceedings of the Sixtieth General Assembly of the State of New Jersey.* Freehold, N.J., 1836.

Owen, Richard. "Notes on Remains of Fossil Reptiles Discovered by Professor Henry Rogers of Pennsylvania, U.S., in Greensand Formations of New Jersey." *Quarterly Journal of the Geological Society of London* 5 (1849): 380–383.

Pennsylvania, State of. Journals of the House of Representatives of the Commonwealth of Pennsylvania. Harrisburg, 1830–1855.

————. Journals of the Senate of the Commonwealth of Pennsylvania. Harrisburg, 1835–1857.

————. Laws of the General Assembly of the Commonwealth of Pennsylvania. Harrisburg, 1836–1858.

————. "Resolutions Relative to the Publication of the Final Report of the Geological Survey of the State." Document 388. *House of Representatives File, Rd. Feb. 12, 1848.* Harrisburg: J.M.G. Lescure, Printer, 1848.

Phillips, John. *A Treatise on Geology.* 2 vols. London: Longman, Orme, Brown, Green & Longmans, 1839.

Pickett, Thomas E. "James C. Booth and the First Delaware Geological Survey, 1837–1841." In *Two Hundred Years of Geology in America,* ed. Cecil J. Schneer. Hanover, N.H.: University Press of New England for the University of New Hampshire, 1979.

Quinan, John R., M.D. *Medical Annals of Baltimore.* Baltimore, 1884.

Reingold, Nathan, ed. *The Papers of Joseph Henry.* Vol. 2 Washington, D.C.: Smithsonian Institution Press, 1975.

————. *Science in Nineteenth Century America: A Documentary History.* New York: Hill & Wang, 1964.

Rice, C. Duncan. *The Scots Abolitionists, 1833–1861.* Baton Rouge: Louisiana State University Press, 1981.

Roberts, S. W. "An Obituary Notice of Charles B. Trego." *Proceedings of the American Philosophical Society* 14 (1875): 356–358.

Rogers, Emma, ed. *Life and Letters of William Barton Rogers.* 2 vols. Boston: Houghton, Mifflin, 1896.

Rogers, Henry Darwin. *Address Delivered at the Meeting of the Association of American Geologists and Naturalists: Held in Washington, May 1844.* New York: Wiley & Putnam, 1844.

———. *Address Delivered at the Meeting of the Association of American Geologists and Naturalists, Held in Washington, May, 1844, with an Abstract of the Proceedings at Their Meeting.* New Haven: Printed by B. L. Hamlen, 1844. (Also in *American Journal of Science* 47 (1844): 137–160, 247–278.)

———. *Address on the Recent Progress of Geological Research in the United States Delivered at the Annual Meeting of the Association of American Geologists and Naturalists Held at Washington City, May 1844.* Philadelphia: C. Sherman, 1844.

———. "An Inquiry into the Origin of the Appalachian Coal Strata, Bituminous and Anthracitic." *Reports of the First, Second, and Third Meetings of the Association of American Geologists and Naturalists at Philadelphia, in 1840 and 1841, and at Boston in 1842, Embracing Its Proceedings and Transactions.* Boston: Gould, Kendall, & Lincoln, 1843.

———. "Anthracite Coal Region of Pennsylvania." *Messenger of Useful Knowledge* 1, no. 8 (March 1831): 113–115.

———. "The Classification of the Metamorphic Strata of the Atlantic Slope of the Middle and Southern States." *Proceedings of the Boston Society of Natural History* 6 (1856–1859): 140–145.

———. "Coal and Petroleum." *Good Words* 4 (1863): 374–379.

———. "Coal: Its Nature, Origin, Distribution, and Mechanical Efficiency." *Good Words* 4 (1863): 247–250.

———. *Description of the Geology of the State of New Jersey, Being a Final Report.* Philadelphia: Sherman & Company, Printers, 1840.

———. "The Duration of Our Coal Fields." *Good Words* 5 (1864): 334–339.

———. "Education." *Messenger of Useful Knowledge* 1 (December 1830): 65–70.

———. "Education-Essay." *Free Enquirer,* April 28; May 5, 12; June 2, 30, 1832.

———. *Fifth Annual Report of the Geological Survey of Pennsylvania.* Harrisburg: James S. Wallace, Printer, 1841.

———. *First Annual Report of the State Geologist.* Harrisburg: Printed by Emanuel Guyer, 1836.

———. *Fourth Annual Report of the Geological Survey of the State of Pennsylvania.* Harrisburg: Holbrook, Henlock, & Bratton, printers, 1840.

———. "Geology [U.S.]." In Hugh Murray, *The Encyclopedia of Geography, Comprising a Complete Description of the Earth, Physical, Statistical, Civil, and Political, Exhibiting Its Relation to the Heavenly Bodies, Its Physical Structure, the Natural History of Each Country, and the Industry, Commerce, Political Institutions, and Civil and Social State.* Vol. 3. Philadelphia: Carey, Lea, & Blanchard, 1837.

———. *The Geology of Pennsylvania: A Government Survey.* 2 vols. Philadelphia: J. B. Lippincott; Edinburgh and London: W. Blackwood and Sons, 1858. A later issue was published by Van Nostrand in New York in 1868.

———. *A Guide to a Course of Lectures.* Philadelphia: Printed by W. P. Gibbons, 1835.

———. [Introduction to "New Species of Fossil Plants from the Anthracite and Bituminous Coal-Fields of Pennsylvania, Collected and Described by Leo

Lesquereux, with Introductory Observations by Henry Darwin Rogers"]. *Boston Journal of Natural History* 6 (1854): 409–413.

————. [Observations on Ohio coal measures, presented in the name of Lesquereux]. *Proceedings of the Boston Society of Natural History* 4 (1851–1854): 175–179.

————. [On a collection of specimens from the Smithsonian]. *Proceedings of the Boston Society of Natural History* 5 (1854–1856): 56.

————. [On earthquakes]. In "Celebration of the Hundredth Anniversary." *Proceedings of the American Philosophical Society* 3 (1843): 64–67.

————. [On fossil impressions in the red shale]. *Proceedings of the Boston Society of Natural History* 5 (1854–1856): 182–186.

————. [On his new geological map of the United States]. *Proceedings of the Boston Society of Natural History* 5 (1854–1856): 207.

————. [On lignite deposits]. *Proceedings of the Boston Society of Natural History* 5 (1854–1856): 190–191.

————. [On M. Barrande]. *Proceedings of the Boston Society of Natural History* 7 (1859–1861): 419–422.

————. "On Petroleum." *Proceedings of the Philosophical Society of Glasgow* 6 (1865–1868): 48–60.

————. "On Some Phenomena of Metamorphism in Coal in the United States." *Report of the Thirtieth Meeting of the British Association for the Advancement of Science, Held at Oxford in June and July 1860.* London: John Murray, 1861.

————. [On the age of coal in the United States and France]. *Proceedings of the Boston Society of Natural History* 4 (1851–1854): 189–191.

————. [On the Berkshires]. *Proceedings of the American Philosophical Society* 2 (January 1841–June 1843): 3–4.

————. "On the Connection of the Deposits of Common Salt with Climate." In *Proceedings of the American Association for the Advancement of Science, Fourth Meeting, Held at New Haven, Connecticut, Aug. 1850.* Washington, D.C.: Published by S. F. Baird; New York: G. R. Putnam, 1851.

————. "On the Correlation of the North American and British Paleozoic Strata." *Report of the Twenty-sixth Meeting of the British Association for the Advancement of Science Held at Cheltenham in August 1856.* London: John Murray, 1857.

————. [Abstract of] "On the Direction of Slaty Cleavage in the Strata of the Southeastern Belts of the Appalachian Chain, and the Parallelism of the Cleavage Dip with the Planes of Maximum Temperature." *Abstract of the Proceedings of the Sixth Annual Meeting of the Association of American Geologists and Naturalists Held in New Haven, Conn., April 1845.* New Haven: Printed by B. L. Hamlen, 1845.

————. "On the Distribution and Probable Origin of Petroleum, or Rock Oil, of Western Pennsylvania, New York and Ohio." *Proceedings of the Philosophical Society of Glasgow* 4 (1855–1860): 355–359.

————. [On the epoch of the mammoth, or *Elephas primigenius*]. *Proceedings of the Boston Society of Natural History* 5 (1854–1856): 22–23.

————. "On the Falls of the Niagara and the Reasonings of Some Authors Respecting Them." *American Journal of Science* 27 (1835): 326–335.

————. "On the Geology and Physical Geography of North America." *Notices of the Proceedings at the Meetings of the Members of the Royal Institution of Great*

Britain with the Abstracts of the Discourses Delivered at the Meetings 2 (1854–1858): 167–187.

———. "On the Geology of Pennsylvania." *Report of the Eighteenth Meeting of the British Association for the Advancement of Science, Held at Swansea in August, 1848.* London: John Murray, 1849.

———. [On the geology of the eastern base of the Rocky Mountains]. *Proceedings of the Boston Society of Natural History* 5 (1854–1856): 190–191.

———. "On the Laws of Structure of the More Disturbed Zones of the Earth's Crust." *Transactions of the Royal Society of Edinburgh* 2 (1857): 431–471.

———. "On the Origin of Cyclones." *Proceedings of the Philosophical Society of Glasgow* 5 (1860–1864): 57–60.

———. [On the origin of salt lakes]. *Proceedings of the Boston Society of Natural History* 3 (1848–1851): 259–260.

———. "On the Origin of the Parallel Roads of Lochaber (Glen Roy) Scotland." *Proceedings of the Royal Institution* 3 (1858–1862): 341–345.

———. "On the Position and Character of the Reptilian Foot-Prints in the Carboniferous Red Shale Formation of Eastern Pennsylvania." In *Proceedings of the American Association for the Advancement of Science, Fourth Meeting, Held at New Haven, Connecticut, Aug. 1850.* Washington, D.C.: Published by S. F. Baird; New York: G. R. Putnam, 1851.

———. [On the position of the *Mastodon giganteum*]. *Proceedings of the Boston Society of Natural History* 5 (1854–1856): 111–117.

———. "On the Proposed Method of Analysing Mineral Waters by Alcohol." *Journal of the Philadelphia College of Pharmacy* 5 (1833): 279–284.

———. "On the Relations of Deposits of Common Salt to Climate." *Proceedings of the Philosophical Society of Glasgow* 5 (1860–1864): 7–9.

———. [On the state of geology], *Proceedings of the Boston Society of Natural History* 6 (1856–1859): 183–184.

———. [On vertebrates in the Silurian]. *Proceedings of the Boston Society of Natural History* 4 (1851–1854): 169.

———. [Abstract of] "Structural Features of the Appalachians and Alps." *Proceedings of the American Association for the Advancement of Science, 2nd Meeting, August 1849.* 1850. Reprint. Boston: Henry Flanders, 1885. (A similar paper was given to the Boston Society two years later: *Proceedings of the Boston Society of Natural History* 4 (1851–1854): 3.)

———. "Report Exhibiting the Geological Features of the Country between Massachusetts and the Narragansett Waters." In *Report of the Board of Directors to the Stockholders of the Boston and Providence Rail-Road Company, Submitting the Report of Their Engineer, with Plans and Profiles, Illustrative of the Surveys, and Estimates of the Cost of a Rail-Road from Boston to Providence to Which Are Annexed the Acts of Incorporation.* Boston: J. E. Hinckley, 1832.

———. *Report on the Coal Lands of the Mt. Carmel and Shamokin Rail Road Co.* New York: Snowden, Printer, 1851.

———. *Report on the Coal Lands of the Zerbe's Run and Shamokin Improvement Company.* Boston: Thurston, Torry, Printers, 1850.

———. *Report on the Combustible Qualities of the Semi-Anthracites of the Zerbes's Run Coal Fields.* Boston: Thurston, Torry, & Emerson, Printers, 1851.

———. *Report on the Geological Survey of the State of New Jersey.* Philadelphia: Desilver, Thomas, 1836.

———. "Report on the Geology of North America." *Report of the Fourth Meeting of the British Association for the Advancement of Science Held at Edinburgh in 1834.* London: John Murray, 1835.

———. *Reports of Professor Henry D. Rogers on Wheatley, Brookdale, and Charleston Mines, Phoenixville, Chester County, Pennsylvania.* Philadelphia: T. K. & P. G. Collins, 1853.

———. "The Reputed Traces of Primeval Man." *Edinburgh Magazine* 88 (1860): 422–439.

———. *Second Annual Report of the Geological Exploration of the State of Pennsylvania.* Harrisburg: Printed by Packer, Barrett & Parke, 1838.

———. *Sixth Annual Report on the Geological Survey of Pennsylvania.* Harrisburg: Henlock & Bratton, Printers, 1842.

———. [Abstract of] "Some Facts in the Geology of the Central and Western Portions of North America, Collected Principally from Statements and Unpublished Notices of Recent Travellers." *Proceedings of the Geological Society of London* 2 (1833–1838): 103–106.

———. *Third Annual Report of the Geological Survey of the State of Pennsylvania.* Harrisburg: Boas & Coplan, Printers, 1839.

Rogers, Henry Darwin, and Martin Boye. "Upon a New Compound of Deuto-Chloride of Platinum Nitric Oxide, and Chlorohydric Acid." *Transactions of the American Philosophical Society,* n.s. 7 (1841): 59–66.

Rogers, Henry Darwin, and William Barton Rogers. "An Account of Two Remarkable Trains of Angular Erratic Blocks in Berkshire, Massachusetts, with an Attempt at an Explanation of the Phenomena." *Boston Journal of Natural History* 5 (1845–1847): 310–330.

———. "Contributions to the Geology of the Tertiary Formations of Virginia." *Transactions of the American Philosophical Society,* n.s. 6 (1839): 329–341, 347–370.

———. "Experimental Enquiry into Some of the Laws of the Elementary Voltaic Battery." *American Journal of Science* 27 (1835): 39–61.

———. "Observations on the Geology of the Western Peninsula of Upper Canada and the Western Part of Ohio." *Transactions of the American Philosophical Society,* n.s. 8 (1843): 273–284.

———. "On the Geological Age of the White Mountains." *American Journal of Science,* n.s. 1 (1846): 1–11.

———. "On the Physical Structure of the Appalachian Chain, As Exemplifying the Laws Which Have Regulated the Elevation of Great Mountain Chains, Generally." *Reports of the First, Second, and Third Meetings of the Association of American Geologists and Naturalists at Philadelphia in 1840 and 1841 and at Boston in 1842.* Boston: Gould, Kendall, & Lincoln, 1843. (A report on this paper is in "Abstract of the Proceedings of the British Association for the Advancement of Science," *Athenaeum: London Literary and Critical Journal,* July 2, 1842, pp. 591–592, and in "Abstract of the Proceedings of the Twelfth Meeting of the British Association for the Advancement of Science Condensed from the Report in the London Athenaeum," *American Journal of Science* 44 (1843): 359–365.)

Rogers, Patrick. [Autobiographical fragment]. In *Life and Letters of William Barton Rogers,* ed. Emma Rogers. Boston: Houghton, Mifflin, 1896.

[Rogers, Willie]. *A Few Facts Regarding the Geological Survey of Pennsylvania, Ex-*

posing the Erroneous Statements and Claims of J. P. Lesley, Secretary of the American Iron Association. Philadelphia: Collins, Printer, 1859.

Rossiter, Margaret W. *The Emergence of Agricultural Science: Justus Liebig and the Americans, 1840–1880.* New Haven: Yale University Press, 1975.

Rowlinson, J. S. "The Theory of Glaciers." *Notes and Records of the Royal Society of London* 26 (1971): 189–204.

Rudwick, Martin J. S. *The Great Devonian Controversy: The Shaping of Scientific Knowledge Among Gentlemanly Specialists.* Chicago: University of Chicago Press, 1985.

Ruffner, W. H. "The Brothers Rogers." In *The Scotch-Irish in America: Proceedings and Addresses of the Seventh Congress of the Scotch Irish at Lexington, Virginia, June 20–23, 1895.* Nashville, Tenn.: Barber & Smith, Agents, 1895. (Also published in the *Alumni Bulletin [University of Virginia]* (1898): 1–13.)

Ruschenberger, W. S. W. "A Sketch of the Life of Robert E. Rogers, M.D., LL.D., with Biographical Notices of His Father and Brothers." *Proceedings of the American Philosophical Society* 23 (1885): 104–146.

Scharf, J. Thomas, and Thompson Westcott. *History of Philadelphia 1609–1884.* Philadelphia: L. H. Everts, 1884.

Schneer, Cecil. "Ebenezer Emmons and the Foundations of American Geology." *Isis* 60 (1969): 439–450.

———. "The Great Taconic Controversy." *Isis* 69 (1978): 173–191.

Secord, James A. *Controversy in Victorian Geology: The Cambrian–Silurian Dispute.* Princeton, N.J.: Princeton University Press, 1986.

Sedgwick, Adam. "On the Classification and Nomenclature of the Lower Paleozoic Rocks of England and Wales." *Quarterly Journal of the Geological Society of London* 8 (1852): 136–168.

———. "Remarks on the Structure of Large Mineral Masses, and Especially on the Chemical Changes Produced in the Aggregation of Stratified Rocks During Different Periods after Their Deposition." *Transactions of the Geological Society of London* 3 (1835): 461–486.

Sellers, Charles Coleman. *Dickinson College: A History.* Middletown, Conn.: Wesleyan University Press, 1973.

Sidar, Jean Wilson. *George Hammell Cook: A Life in Agriculture and Geology.* New Brunswick, N.J.: Rutgers University Press, 1976.

———. "New Jersey Geological Surveys in the Nineteenth Century." *Northeast Geology* 3 (1981): 52–55.

Sinclair, Bruce. *Philadelphia's Philosopher Mechanics: A History of the Franklin Institute, 1824–1865.* Baltimore: Johns Hopkins University Press, 1974.

Smith, Crosbie. "William Hopkins and the Shaping of Dynamical Geology, 1830–1860." *British Journal of the History of Science* 22 (1989): 27–52.

Smith, Edgar Fahs. *Biographical Memoir of Robert Empie Rogers, 1813–1884.* Washington, D.C., 1904.

Smith, Robert Walter. *History of Armstrong County, Pennsylvania.* 2 vols. Chicago: Waterman, Walker, etc., 1914.

Snyder, Frank E., and Brian H. Guss. *The District: A History of the Philadelphia District U.S. Army Corps of Engineers, 1866–1971.* Philadelphia: U.S. Army Engineers, District Philadelphia, January 1974.

Stapleton, Darwin. "The Origin of American Railroad Technology, 1825–1840." *Railroad History* 139 (Autumn 1978): 65–77.

Suess, Edward. *The Face of the Earth,* translated by H. B. C. Sollas. 5 vols. Oxford: Clarendon Press, 1904.

Sullivan, Robert. *The Disappearance of Dr. Parkman.* Boston: Little, Brown, 1971.

Swank, James M. *History of the Manufacture of Iron in All Ages and Particularly in the United States from Colonial Times to 1891.* Philadelphia: American Iron and Steel Association, 1892.

Taylor, Richard. "On the Mineral Basin or Coal Field of Blossburg, on the Tioga River, Tioga County, Pennsylvania." *Transactions of the Geological Society of Pennsylvania* 1, pt. 2 (1835): 204–219.

———. *Statistics of Coal: Including Mineral Bituminous Studies Employed in Arts and Manufacturers, with Their Geological, Geographical, and Commercial Distribution and Art of Production and Consumption on the American Continent, Second Edition Revised and Brought Down to 1854 by S. S. Haldeman.* Philadelphia: J. W. Moore, 1854.

Trego, Charles B. *A Geography of Pennsylvania.* Philadelphia: Edward C. Biddle, 1843.

———. *Report of the Committee Appointed on So Much of the Governor's Message As Related to a Geological and Mineralogical Survey of the State of Pennsylvania, Read in the House of Representatives, Feb. 3, 1836.* Harrisburg, 1836.

Tyler, Alice Felt. *Freedom's Ferment: Phases of American Social History from the Colonial Period to the Outbreak of the Civil War.* New York: Harper Torchbooks, 1962.

Vanuxem, Lardner. *Geology of New York,* Pt. 3: *Comprising the Survey of the Third Geological District.* Albany: W. & A. White & J. Visscher, Printers, 1842.

Vose, George L. *Orographic Geology; or, The Origin and Structure of Mountains: A Review.* Boston: Lee & Shepard, 1866.

Wallace, Anthony F C. *St. Clair: A Nineteenth-Century Coal Town's Experience with a Disaster Prone Industry.* New York: Alfred A. Knopf, 1987.

Whewell, William. *The History of the Inductive Sciences.* 3 vols. London: John W. Parker, 1837.

Willis, Bailey. "The Mechanics of Appalachian Structure." In J. W. Powell, *Thirteenth Annual Report of the United States Geological Survey to the Secretary of the Interior, 1891–1892,* Pt. 2: *Geology.* Washington, D.C.: U.S. Government Printing Office, 1893.

Wilson, Leonard G. "The Emergence of Geology As a Science in the United States." *Journal of World History* 10 (1967): 416–437.

Yearley, C. K., Jr. *Enterprise and Anthracite: Economics and Democracy in Schuylkill County, 1820–1875.* Baltimore: Johns Hopkins University Press, 1961.

Zaslow, Morris. *Reading the Rocks: The Story of the Geological Survey of Canada, 1842–1972.* Toronto: Macmillan of Canada, 1975.

Index